Texts in Computational Science and Engineering

23

Editors

More information about this series at http://www.springer.com/series/5151

Tom Lyche • Georg Muntingh • Øyvind Ryan

Exercises in Numerical Linear Algebra and Matrix Factorizations

Springer

Tom Lyche
Blindern
University of Oslo
Oslo, Norway

Georg Muntingh
SINTEF ICT
Oslo, Norway

Øyvind Ryan
Blindern
University of Oslo
Oslo, Norway

ISSN 1611-0994 ISSN 2197-179X (electronic)
Texts in Computational Science and Engineering
ISBN 978-3-030-59791-7 ISBN 978-3-030-59789-4 (eBook)
https://doi.org/10.1007/978-3-030-59789-4

Mathematics Subject Classification (2010): 15-XX, 65-XX

Cover illustration: A steepest descent iteration (left) and a conjugate gradient iteration (right) to find the minimum of a quadratic surface shown by some level curves.

This Springer imprint is published by the registered company Springer Nature Switzerland AG
The registered company address is: Gewerbestrasse 11, 6330 Cham, Switzerland

To our families:

Annett, Max, Ella, and Mina;

Natalia, Katrina, and Alina.

Preface

The starting point for these solutions is a course in numerical linear algebra at the University of Oslo. Tom Lyche wrote and gradually expanded a text for this course over a period of more than 30 years. He also expanded the text with many new exercises, and many exam questions were made over the years as well. This work ultimately ended in the book "Numerical Linear Algebra and Matrix Factorizations" (Springer, 2020). Several assistants worked on the exercise sessions for the course over the years, and a number of professors taught the course. Their combined effort formed the foundation for this solution manual, which underwent a one and a half year preparation process by the authors prior to publication.

The solutions repeat the exercise statements as they are in Tom Lyche's book. Since these solutions repeatedly refer to this book, it will simply be mentioned as "the book" in the following. Several exercises refer to labeled equations in the book. To distuinguish between labels in this solution manual, roman numbering is used for equations in the solutions (the book uses non-roman numbering). Otherwise there is little room for confusion, since objects like Theorems, lemmas, and definitions appear only in the book (algorithms and figures specify whether they appear in the book or not, as confusion is possible here).

There are some small differences from the book. The exercise text is here shown within a grey box, so that one easily can distinguish between exercise text appearing in the book, and their solutions. Also, hints are shown in their own type of grey box, rather than as footnotes as in the book. The grey hint- and exercise boxes also have their own icons.

Many solutions contain code listings. The corresponding files can be found online together with the code from the book at `http://folk.uio.no/tom/numlinalg/code`. The listings also include the corresponding file names. All code listed in the solutions is MATLAB code, but the code directory also contains a Python module `numlinalg.py`, which contains the main functions from the first five chapters translated to Python. The code is very similar, but the Python versions naturally take advantage of several things in the Python language. As an example, parameters in Python are passed by reference, not by value.

The reader may notice that some exercises, in particular older exam exercises, are more difficult than others. A partial explanation to this may be that all aids were permitted in older exams. The reader may also notice that some of those older exams use pseudocode, and are formulated in a way that encourages students to write loops, where one nowadays would prefer vectorized code. Time may again be a partial explanation: The oldest exercises originated way back in the infancy of languages like MATLAB, when modern programming standards were still not established.

Acknowledgments

Many should be mentioned when it comes to contributions to this material. Are Magnus Bruaset and Njål Foldnes contributed with solutions to some of the exercises. Several others contributed with exam exercises, including Nils Henrik Risebro, Ragnar Winther, Aslak Tveito, and Michael Floater. A special thanks goes to Christian Schulz, who guided the exercise sessions in the course for several years and provided many solutions, which formed the basis for many solutions in this solution manual.

Oslo, Norway
February 2020

Tom Lyche, Georg Muntingh, and Øyvind Ryan.

Contents

Listings

Chapter 1
A Short Review of Linear Algebra

Exercises section 1.1

Exercise 1.1: Strassen multiplication (Exam exercise 2017-1)

(By arithmetic operations we mean additions, subtractions, multiplications and divisions.)

Let $\boldsymbol{A}$ and $\boldsymbol{B}$ be $n \times n$ real matrices.

a) With $\boldsymbol{A}, \boldsymbol{B} \in \mathbb{R}^{n\times n}$, how many arithmetic operations are required to form the product $\boldsymbol{AB}$?

Solution. Each entry in $\boldsymbol{AB}$ requires n multiplications and $n-1$ additions, i.e., a total of $2n-1$ operations. Since there are n^2 entries to compute, the total number of operations is $n^2(2n-1) = 2n^3 - n^2$.

b) Consider the $2n \times 2n$ block matrix

$$\begin{bmatrix} \boldsymbol{W} & \boldsymbol{X} \\ \boldsymbol{Y} & \boldsymbol{Z} \end{bmatrix} = \begin{bmatrix} \boldsymbol{A} & \boldsymbol{B} \\ \boldsymbol{C} & \boldsymbol{D} \end{bmatrix} \begin{bmatrix} \boldsymbol{E} & \boldsymbol{F} \\ \boldsymbol{G} & \boldsymbol{H} \end{bmatrix},$$

where all matrices $\boldsymbol{A}, \ldots, \boldsymbol{Z}$ are in $\mathbb{R}^{n\times n}$. How many operations does it take to compute $\boldsymbol{W}$, $\boldsymbol{X}$, $\boldsymbol{Y}$ and $\boldsymbol{Z}$ by the obvious algorithm?

Solution. We have that

$$\boldsymbol{W} = \boldsymbol{AE} + \boldsymbol{BG}.$$

So it takes $2(2n^3 - n^2) + n^2 = 4n^3 - n^2$ operations to compute this. We must compute 4 such matrices, hence the total operation cost is $16n^3 - 4n^2$. note that this is the same $2(2n)^3 - (2n)^2$, so the number of operations in block matrix multiplication is the same as that of ordinary matrix multiplication. This should not be surprising, since block matrix multiplication computes the same multiplications as the latter. The difference is simply that the summations are done in a different order.

T. Lyche et al., *Exercises in Numerical Linear Algebra and Matrix Factorizations*, Texts in Computational Science and Engineering 23, https://doi.org/10.1007/978-3-030-59789-4_1

c) An alternative method to compute $\boldsymbol{W}$, $\boldsymbol{X}$, $\boldsymbol{Y}$ and $\boldsymbol{Z}$ is to use Strassen's formulas:

$$
\begin{aligned}
&\mathbf{P}_1 = (\boldsymbol{A}+\boldsymbol{D})\,(\boldsymbol{E}+\boldsymbol{H})\,, && \\
&\mathbf{P}_2 = (\boldsymbol{C}+\boldsymbol{D})\,\boldsymbol{E}, && \mathbf{P}_5 = (\boldsymbol{A}+\boldsymbol{B})\,\boldsymbol{H}, \\
&\mathbf{P}_3 = \boldsymbol{A}\,(\boldsymbol{F}-\boldsymbol{H})\,, && \mathbf{P}_6 = (\boldsymbol{C}-\boldsymbol{A})\,(\boldsymbol{E}+\boldsymbol{F})\,, \\
&\mathbf{P}_4 = \boldsymbol{D}\,(\boldsymbol{G}-\boldsymbol{E})\,, && \mathbf{P}_7 = (\boldsymbol{B}-\boldsymbol{D})\,(\boldsymbol{G}+\boldsymbol{H})\,, \\
&\boldsymbol{W} = \mathbf{P}_1+\mathbf{P}_4-\mathbf{P}_5+\mathbf{P}_7, && \boldsymbol{X} = \mathbf{P}_3+\mathbf{P}_5, \\
&\boldsymbol{Y} = \mathbf{P}_2+\mathbf{P}_4, && \boldsymbol{Z} = \mathbf{P}_1+\mathbf{P}_3-\mathbf{P}_2+\mathbf{P}_6.
\end{aligned}
$$

You do not have to verify these formulas. What is the operation count for this method?

Solution. We assume here that the $n \times n$ matrix multiplications are computed directly, i.e., that Strassen multiplication is not used further recursively. $\mathbf{P}_1$, $\mathbf{P}_6$ and $\mathbf{P}_7$ each need $2n^3 + n^2$ operations. $\mathbf{P}_2$, $\mathbf{P}_3$, $\mathbf{P}_4$ and $\mathbf{P}_5$ each need $2n^3$ operations. Hence forming the $\mathbf{P}$'s needs $3(2n^3 + n^2) + 4 \cdot 2n^3 = 14n^3 + 3n^2$ operations. To find the final result demands $8n^2$ operations. Thus the total cost is $14n^3 + 11n^2$. For large n this is clearly less than what we obtained in **b)**.

d) Describe a recursive algorithm, based on Strassen's formulas, which given two matrices $\boldsymbol{A}$ and $\boldsymbol{B}$ of size $m \times m$, with $m = 2^k$ for some $k \geq 0$, calculates the product $\boldsymbol{AB}$.

Solution.

code/strassen.m

```
function Z=strassen(A,B)
[m,~]=size(A);
if m==1
  Z=A*B; return;
end
one=1:m/2; two=m/2+1:m;
P1=strassen(A(one,one)+A(two,two),B(one,one)+B(two,two));
P2=strassen(A(two,one)+A(two,two),B(one,one));
P3=strassen(A(one,one),B(one,two)-B(two,two));
P4=strassen(A(two,two),B(two,one)-B(one,one));
P5=strassen(A(one,one)+A(one,two),B(two,two));
P6=strassen(A(two,one)-A(one,one),B(one,one)+B(one,two));
P7=strassen(A(one,two)-A(two,two),B(two,one)+B(two,two));
Z=[P1+P4-P5+P7,P3+P5; P2+P4,P1+P3-P2+P6];
```

Listing 1.1: Recursive algorithm based on Strassen's formulas for multiplication of two $m \times m$ matrices, with $m = 2^k$ for some $k \geq 0$.

e) Show that the operation count of the recursive algorithm is $\mathcal{O}\left(m^{\log_2(7)}\right)$. Note that $\log_2(7) \approx 2.8 < 3$, so this is less costly than straightforward matrix multiplication.

Solution. Let s_k be the number of operations in recursive Strassen multiplication of two matrices of size 2^k. Then

$$s_{k+1} = 7s_k + 18 \cdot 2^{2k},$$

since one Strassen multiplication is split into 7 Strassen multiplications of size $2^k \times 2^k$, and 18 additions/substractions of matrices of size $2^k \times 2^k$. A particular solution to this difference equation is easily found as $s_k^{(p)} = -6 \cdot 4^k$, and the general solution to the homogeneous equation is clearly $s_k^{(h)} = \gamma \cdot 7^k$, where γ is a constant. We thus obtain the general solution

$$s_k = \gamma \cdot 7^k - 6 \cdot 4^k.$$

Here $\gamma \cdot 7^k$ is the dominating term (note that $\gamma > 0$ is clear, since we otherwise would have that $s_k < 0$). Since $m = 2^k$,

$$7^k = 2^{k \log_2 7} = 2^{\log_2 m \log_2 7} = m^{\log_2(7)}.$$

The result follows.

Exercises section 1.3

Exercise 1.2: The inverse of a general 2×2 matrix

Show that

$$\begin{bmatrix} a & b \\ c & d \end{bmatrix}^{-1} = \alpha \begin{bmatrix} d & -b \\ -c & a \end{bmatrix}, \qquad \alpha = \frac{1}{ad - bc},$$

for any a, b, c, d such that $ad - bc \neq 0$.

Solution. A straightforward computation yields

$$\frac{1}{ad - bc} \begin{bmatrix} d & -b \\ -c & a \end{bmatrix} \begin{bmatrix} a & b \\ c & d \end{bmatrix} = \frac{1}{ad - bc} \begin{bmatrix} ad - bc & 0 \\ 0 & ad - bc \end{bmatrix} = \begin{bmatrix} 1 & 0 \\ 0 & 1 \end{bmatrix},$$

showing that the two matrices are inverse to each other.

Exercise 1.3: The inverse of a special 2×2 matrix

Find the inverse of

$$A = \begin{bmatrix} \cos\theta & -\sin\theta \\ \sin\theta & \cos\theta \end{bmatrix}.$$

Solution. By Exercise 1.2, and using that $\cos^2\theta + \sin^2\theta = 1$, the inverse is given by

$$\begin{bmatrix} \cos\theta & \sin\theta \\ -\sin\theta & \cos\theta \end{bmatrix}.$$

Exercise 1.4: Sherman-Morrison formula

Suppose $\boldsymbol{A} \in \mathbb{C}^{n\times n}$, and $\boldsymbol{B}, \boldsymbol{C} \in \mathbb{R}^{n\times m}$ for some $n, m \in \mathbb{N}$. If $(\boldsymbol{I} + \boldsymbol{C}^{\mathrm{T}}\boldsymbol{A}^{-1}\boldsymbol{B})^{-1}$ exists then

$$(\boldsymbol{A} + \boldsymbol{B}\boldsymbol{C}^{\mathrm{T}})^{-1} = \boldsymbol{A}^{-1} - \boldsymbol{A}^{-1}\boldsymbol{B}(\boldsymbol{I} + \boldsymbol{C}^{\mathrm{T}}\boldsymbol{A}^{-1}\boldsymbol{B})^{-1}\boldsymbol{C}^{\mathrm{T}}\boldsymbol{A}^{-1}.$$

Solution. A direct computation yields

$$\begin{aligned}
&(\boldsymbol{A} + \boldsymbol{B}\boldsymbol{C}^{\mathrm{T}})(\boldsymbol{A}^{-1} - \boldsymbol{A}^{-1}\boldsymbol{B}(\boldsymbol{I} + \boldsymbol{C}^{\mathrm{T}}\boldsymbol{A}^{-1}\boldsymbol{B})^{-1}\boldsymbol{C}^{\mathrm{T}}\boldsymbol{A}^{-1})\\
=&\,\boldsymbol{I} - \boldsymbol{B}(\boldsymbol{I} + \boldsymbol{C}^{\mathrm{T}}\boldsymbol{A}^{-1}\boldsymbol{B})^{-1}\boldsymbol{C}^{\mathrm{T}}\boldsymbol{A}^{-1} + \boldsymbol{B}\boldsymbol{C}^{\mathrm{T}}\boldsymbol{A}^{-1}\\
&\quad - \boldsymbol{B}\boldsymbol{C}^{\mathrm{T}}\boldsymbol{A}^{-1}\boldsymbol{B}(\boldsymbol{I} + \boldsymbol{C}^{\mathrm{T}}\boldsymbol{A}^{-1}\boldsymbol{B})^{-1}\boldsymbol{C}^{\mathrm{T}}\boldsymbol{A}^{-1}\\
=&\,\boldsymbol{I} + \boldsymbol{B}\boldsymbol{C}^{\mathrm{T}}\boldsymbol{A}^{-1} - \boldsymbol{B}(\boldsymbol{I} + \boldsymbol{C}^{\mathrm{T}}\boldsymbol{A}^{-1}\boldsymbol{B})(\boldsymbol{I} + \boldsymbol{C}^{\mathrm{T}}\boldsymbol{A}^{-1}\boldsymbol{B})^{-1}\boldsymbol{C}^{\mathrm{T}}\boldsymbol{A}^{-1}\\
=&\,\boldsymbol{I} + \boldsymbol{B}\boldsymbol{C}^{\mathrm{T}}\boldsymbol{A}^{-1} - \boldsymbol{B}\boldsymbol{C}^{\mathrm{T}}\boldsymbol{A}^{-1}\\
=&\,\boldsymbol{I},
\end{aligned}$$

showing that the two matrices are inverse to each other.

Exercise 1.5: Inverse update (Exam exercise 1977-1)

a) Let $\boldsymbol{u}, \boldsymbol{v} \in \mathbb{R}^n$ and suppose $\boldsymbol{v}^{\mathrm{T}}\boldsymbol{u} \neq 1$. Show that $\boldsymbol{I} - \boldsymbol{u}\boldsymbol{v}^{\mathrm{T}}$ has an inverse given by $\boldsymbol{I} - \tau\boldsymbol{u}\boldsymbol{v}^{\mathrm{T}}$, where $\tau := 1/(\boldsymbol{v}^{\mathrm{T}}\boldsymbol{u} - 1)$.

Solution. We have $(\boldsymbol{I} - \boldsymbol{u}\boldsymbol{v}^{\mathrm{T}})^{-1} = \boldsymbol{I} - \tau\boldsymbol{u}\boldsymbol{v}^{\mathrm{T}}$ since

$$\begin{aligned}
(\boldsymbol{I} - \boldsymbol{u}\boldsymbol{v}^{\mathrm{T}})(\boldsymbol{I} - \tau\boldsymbol{u}\boldsymbol{v}^{\mathrm{T}}) &= \boldsymbol{I} - \boldsymbol{u}\boldsymbol{v}^{\mathrm{T}} - \tau\boldsymbol{u}\boldsymbol{v}^{\mathrm{T}} + \tau\boldsymbol{u}\boldsymbol{v}^{\mathrm{T}}\boldsymbol{u}\boldsymbol{v}^{\mathrm{T}}\\
&= \boldsymbol{I} - (1 + \tau - \tau\boldsymbol{v}^{\mathrm{T}}\boldsymbol{u})\boldsymbol{u}\boldsymbol{v}^{\mathrm{T}} = \boldsymbol{I}.
\end{aligned}$$

b) Let $\boldsymbol{A} \in \mathbb{R}^{n\times n}$ be nonsingular with inverse $\boldsymbol{C} := \boldsymbol{A}^{-1}$, and let $\boldsymbol{a} \in \mathbb{R}^n$. Let $\overline{\boldsymbol{A}}$ be the matrix which differs from $\boldsymbol{A}$ by replacing the ith row of $\boldsymbol{A}$ with $\boldsymbol{a}^{\mathrm{T}}$, i.e., $\overline{\boldsymbol{A}} = \boldsymbol{A} - \boldsymbol{e}_i(\boldsymbol{e}_i^{\mathrm{T}}\boldsymbol{A} - \boldsymbol{a}^{\mathrm{T}})$, where $\boldsymbol{e}_i$ is the ith column in the identity matrix $\boldsymbol{I}$. Show that if

$$\lambda := \boldsymbol{a}^{\mathrm{T}}\boldsymbol{C}\boldsymbol{e}_i \neq 0 \tag{1.20}$$

then $\overline{\boldsymbol{A}}$ has an inverse $\overline{\boldsymbol{C}} = \overline{\boldsymbol{A}}^{-1}$ given by

$$\overline{\boldsymbol{C}} = \boldsymbol{C}(\boldsymbol{I} + \frac{1}{\lambda}\boldsymbol{e}_i(\boldsymbol{e}_i^{\mathrm{T}} - \boldsymbol{a}^{\mathrm{T}}\boldsymbol{C})) \tag{1.21}$$

Solution. One has

$$\begin{aligned}\overline{\boldsymbol{A}} &= \boldsymbol{A} - \boldsymbol{e}_i(\boldsymbol{e}_i^{\mathrm{T}}\boldsymbol{A} - \boldsymbol{a}^{\mathrm{T}}) \\ &= \big(\boldsymbol{I} - \boldsymbol{e}_i(\boldsymbol{e}_i^{\mathrm{T}} - \boldsymbol{a}^{\mathrm{T}}\boldsymbol{C})\big)\boldsymbol{A} = (\boldsymbol{I} - \boldsymbol{u}\boldsymbol{v}^{\mathrm{T}})\boldsymbol{A},\end{aligned}$$

where $\boldsymbol{u} := \boldsymbol{e}_i$ and $\boldsymbol{v} := \boldsymbol{e}_i - \boldsymbol{C}^{\mathrm{T}}\boldsymbol{a}$. We find

$$\begin{aligned}\boldsymbol{v}^{\mathrm{T}}\boldsymbol{u} &= 1 - \boldsymbol{a}^{\mathrm{T}}\boldsymbol{C}\boldsymbol{e}_i = 1 - \lambda \neq 1 \qquad \Longleftrightarrow \qquad \lambda \neq 0 \\ \tau &= 1/(\boldsymbol{v}^{\mathrm{T}}\boldsymbol{u} - 1) = 1/(1 - \lambda - 1) = -1/\lambda \\ \overline{\boldsymbol{C}} &= \boldsymbol{A}^{-1}(\boldsymbol{I} - \boldsymbol{u}\boldsymbol{v}^{\mathrm{T}})^{-1} = \boldsymbol{C}(\boldsymbol{I} + \frac{1}{\lambda}\boldsymbol{u}\boldsymbol{v}^{\mathrm{T}}),\end{aligned}$$

and (1.21) follows.

c) Write an algorithm which to given $\boldsymbol{C}$ and $\boldsymbol{a}$ checks if (1.20) holds and computes $\overline{\boldsymbol{C}}$ provided $\lambda \neq 0$.

Hint.
Use (1.21) to find formulas for computing each column in $\overline{\boldsymbol{C}}$.

Solution. Let $\boldsymbol{c}_j$ and $\overline{\boldsymbol{c}}_j$, $j = 1, \ldots, n$ be the columns of $\boldsymbol{C}$ and $\overline{\boldsymbol{C}}$, respectively. We find

$$\overline{\boldsymbol{c}}_i = \overline{\boldsymbol{C}}\boldsymbol{e}_i = \boldsymbol{C}\big(\boldsymbol{I} + \frac{1}{\lambda}\boldsymbol{u}\boldsymbol{v}^{\mathrm{T}}\big)\boldsymbol{e}_i = \boldsymbol{C}\big(\boldsymbol{e}_i + \frac{1}{\lambda}\boldsymbol{e}_i(1-\lambda)\big) = \frac{1}{\lambda}\boldsymbol{c}_i.$$

If $j \neq i$ then $\boldsymbol{e}_i^{\mathrm{T}}\boldsymbol{e}_j = 0$ and so

$$\overline{\boldsymbol{c}}_j = \overline{\boldsymbol{C}}\boldsymbol{e}_j = \boldsymbol{C}\big(\boldsymbol{I} + \frac{1}{\lambda}\boldsymbol{e}_i(\boldsymbol{e}_i^{\mathrm{T}} - \boldsymbol{a}^{\mathrm{T}}\boldsymbol{C})\big)\boldsymbol{e}_j = \boldsymbol{C}\big(\boldsymbol{e}_j - \frac{1}{\lambda}\boldsymbol{e}_i\boldsymbol{a}^{\mathrm{T}}\boldsymbol{C}\boldsymbol{e}_j\big).$$

In other words, all columns in $\overline{\boldsymbol{C}}$ except the ith column coincide with those of

$$\boldsymbol{C}\big(\boldsymbol{I} - \frac{1}{\lambda}\boldsymbol{e}_i\boldsymbol{a}^{\mathrm{T}}\boldsymbol{C}\big) = \boldsymbol{C} - \frac{1}{\lambda}\boldsymbol{c}_i\boldsymbol{a}^{\mathrm{T}}\boldsymbol{C} = \boldsymbol{C} - \overline{\boldsymbol{c}}_i\boldsymbol{a}^{\mathrm{T}}\boldsymbol{C}.$$

When computing the product $\overline{\boldsymbol{c}}_i\boldsymbol{a}^{\mathrm{T}}\boldsymbol{C}$, it is more efficient to compute $\boldsymbol{a}^{\mathrm{T}}\boldsymbol{C}$ first, and afterwards multiplying with $\overline{\boldsymbol{c}}_i$ from the left. It is easily checked that the other multiplication order yields an operation count of order n^3. This gives no benefit for a rank one update of the inverse over a direct computation of the inverse of $\overline{\boldsymbol{A}}$: Such a direct implementation is typically based on Gaussian elimination, which we will see in Chapter 3 also has an operation count of order n^3.

Computing $\boldsymbol{a}^{\mathrm{T}}\boldsymbol{C}$ first actually gives an operation count of order n^2, which we will now show. Also, an implementation needs to compute and store this, and this is the only vector requiring extra storage. The rest of the algorithm stores the result directly into the input memory (i.e., $\boldsymbol{C}$). This is possible since the computation of $\overline{\boldsymbol{c}}_j$ with $j \neq i$ does not affect $\overline{\boldsymbol{c}}_i$. Note that the ith entry in $\boldsymbol{a}^{\mathrm{T}}\boldsymbol{C}$ is λ. As shown above, the ith column in $\boldsymbol{C}$ simply needs to be scaled by this number. In fact, $\overline{\boldsymbol{c}}_i$ can overwrite $\boldsymbol{c}_i$ at the start of the algorithm.

The operations we need are as follows.

1. Computing $\boldsymbol{a}^{\mathrm{T}}\boldsymbol{C}$ needs $n(2n-1) = 2n^2 - n$ operations.
2. Computing $\frac{1}{\lambda}\boldsymbol{c}_i$ needs n divisions.
3. Computing $n-1$ columns in $\overline{\boldsymbol{c}}_i\boldsymbol{a}^{\mathrm{T}}\boldsymbol{C}$ needs $n(n-1) = n^2 - n$ multiplications.
4. Computing $n-1$ columns in $\boldsymbol{C} - \overline{\boldsymbol{c}}_i\boldsymbol{a}^{\mathrm{T}}\boldsymbol{C}$ needs $n(n-1) = n^2 - n$ subtractions.

Adding these, the total number of operations is $4n^2 - 2n$. An implementation can be found in Listing 1.2.

code/inverseupdate.m

```
function C=inverseupdate(C,a,i)
    res = a'*C;
    if res(i)~=0
        C(:,i) = C(:,i)/res(i);
        inds = 1:size(C,1); inds(i)=[];
        C(:,inds) = C(:,inds) - C(:,i)*res(inds);
    end
end
```

Listing 1.2: Compute the inverse $\overline{\boldsymbol{C}}$ of the matrix $\overline{\boldsymbol{A}}$ obtained from the matrix $\boldsymbol{A}$ by replacing the ith column by $\boldsymbol{a}$.

Listing 1.3 tests this implementation on a random 4×4-matrix (a random matrix is non-singular with high probability. Why?). The code compares the result with a direct computation of the inverse:

code/test_inverseupdate.m

```
A=rand(4);
C=inv(A);
i=2;
a=rand(4,1);

```

```
Abar=A;
Abar(i,:)=a';

max(max(abs(inverseupdate(C,a,i)-inv(Abar))))
```

Listing 1.3: Test Listing 1.2 by computing the maximum error for a random 4×4 matrix.

Exercise 1.6: Matrix products (Exam exercise 2009-1)

Let $\boldsymbol{A}, \boldsymbol{B}, \boldsymbol{C}, \boldsymbol{E} \in \mathbb{R}^{n \times n}$ be matrices where $\boldsymbol{A}^{\mathrm{T}} = \boldsymbol{A}$. In this exercise an (arithmetic) operation is an addition or a multiplication. We ask about exact numbers of operations.

a) How many operations are required to compute the matrix product $\boldsymbol{BC}$? How many operations are required if $\boldsymbol{B}$ is lower triangular?

Solution. For each of the n^2 elements in $\boldsymbol{B}$ we have to compute an inner product of length n. This requires n multiplications and $n-1$ additions. Therefore to compute $\boldsymbol{BC}$ requires $n^2(2n-1) = 2n^3 - n^2$ operations.

If $\boldsymbol{B}$ is lower triangular then row k of $\boldsymbol{B}$ contains k non-zero elements, $k = 1, \ldots, n$. Therefore, to compute an element in the k-th row of $\boldsymbol{BC}$ requires k multiplications and $k-1$ additions. Hence in total we need $n\sum_{k=1}^{n}(2k-1) = n^3$ operations.

b) Show that there exists a lower triangular matrix $\boldsymbol{L} \in \mathbb{R}^{n \times n}$ such that $\boldsymbol{A} = \boldsymbol{L} + \boldsymbol{L}^{\mathrm{T}}$.

Solution. We have $\boldsymbol{A} = \boldsymbol{A_L} + \boldsymbol{A_D} + \boldsymbol{A_R}$, where $\boldsymbol{A_L}$ is lower triangular with 0 on the diagonal, $\boldsymbol{A_D} = \operatorname{diag}(a_{11}, \ldots, a_{nn})$, and $\boldsymbol{A_R}$ is upper triangular with 0 on the diagonal. Since $\boldsymbol{A}^{\mathrm{T}} = \boldsymbol{A}$, we have $\boldsymbol{A_R} = \boldsymbol{A_L}^{\mathrm{T}}$. If we let $\boldsymbol{L} := \boldsymbol{A_L} + \frac{1}{2}\boldsymbol{A_D}$ we obtain $\boldsymbol{A} = \boldsymbol{L} + \boldsymbol{L}^{\mathrm{T}}$.

c) We have $\boldsymbol{E}^{\mathrm{T}}\boldsymbol{AE} = \boldsymbol{S} + \boldsymbol{S}^{\mathrm{T}}$ where $\boldsymbol{S} = \boldsymbol{E}^{\mathrm{T}}\boldsymbol{LE}$. How many operations are required to compute $\boldsymbol{E}^{\mathrm{T}}\boldsymbol{AE}$ in this way?

Solution. We need n operations to compute the diagonal in $\boldsymbol{L}$. We need n^3 operations to compute $\boldsymbol{LE}$ and after that $2n^3 - n^2$ operations to compute $\boldsymbol{E}^{\mathrm{T}}(\boldsymbol{LE})$. Finally we need n^2 operations to compute the sum $\boldsymbol{S} + \boldsymbol{S}^{\mathrm{T}}$. This totals $3n^3$ operations. Direct computation of $\boldsymbol{E}^{\mathrm{T}}\boldsymbol{AE}$ requires $4n^3 - 2n^2$ operations.

Exercises section 1.4

Exercise 1.7: Cramer's rule; special case

Solve the following system by Cramer's rule:

$$\begin{bmatrix} 1 & 2 \\ 2 & 1 \end{bmatrix} \begin{bmatrix} x_1 \\ x_2 \end{bmatrix} = \begin{bmatrix} 3 \\ 6 \end{bmatrix}.$$

Solution. Cramer's rule yields

$$x_1 = \begin{vmatrix} 3 & 2 \\ 6 & 1 \end{vmatrix} / \begin{vmatrix} 1 & 2 \\ 2 & 1 \end{vmatrix} = 3, \qquad x_2 = \begin{vmatrix} 1 & 3 \\ 2 & 6 \end{vmatrix} / \begin{vmatrix} 1 & 2 \\ 2 & 1 \end{vmatrix} = 0.$$

Exercise 1.8: Adjoint matrix; special case

Show that if

$$\boldsymbol{A} = \begin{bmatrix} 2 & -6 & 3 \\ 3 & -2 & -6 \\ 6 & 3 & 2 \end{bmatrix},$$

then

$$\operatorname{adj}(\boldsymbol{A}) = \begin{bmatrix} 14 & 21 & 42 \\ -42 & -14 & 21 \\ 21 & -42 & 14 \end{bmatrix}.$$

Moreover,

$$\operatorname{adj}(\boldsymbol{A})\boldsymbol{A} = \begin{bmatrix} 343 & 0 & 0 \\ 0 & 343 & 0 \\ 0 & 0 & 343 \end{bmatrix} = \det(\boldsymbol{A})\boldsymbol{I}.$$

Solution. Computing the cofactors of $\boldsymbol{A}$ gives

$$\operatorname{adj}_{\boldsymbol{A}}^{\mathrm{T}} = \begin{bmatrix} (-1)^{1+1}\begin{vmatrix} -2 & -6 \\ 3 & 2 \end{vmatrix} & (-1)^{1+2}\begin{vmatrix} 3 & -6 \\ 6 & 2 \end{vmatrix} & (-1)^{1+3}\begin{vmatrix} 3 & -2 \\ 6 & 3 \end{vmatrix} \\ (-1)^{2+1}\begin{vmatrix} -6 & 3 \\ 3 & 2 \end{vmatrix} & (-1)^{2+2}\begin{vmatrix} 2 & 3 \\ 6 & 2 \end{vmatrix} & (-1)^{2+3}\begin{vmatrix} 2 & -6 \\ 6 & 3 \end{vmatrix} \\ (-1)^{3+1}\begin{vmatrix} -6 & 3 \\ -2 & -6 \end{vmatrix} & (-1)^{3+2}\begin{vmatrix} 2 & 3 \\ 3 & -6 \end{vmatrix} & (-1)^{3+3}\begin{vmatrix} 2 & -6 \\ 3 & -2 \end{vmatrix} \end{bmatrix}$$

$$= \begin{bmatrix} 14 & 21 & 42 \\ -42 & -14 & 21 \\ 21 & -42 & 14 \end{bmatrix}^{\mathrm{T}}.$$

One checks directly that $\operatorname{adj}_{\boldsymbol{A}}\boldsymbol{A} = \det(\boldsymbol{A})\boldsymbol{I}$, with $\det(\boldsymbol{A}) = 343$.

Exercise 1.9: Determinant equation for a plane

Show that

$$\begin{vmatrix} x & y & z & 1 \\ x_1 & y_1 & z_1 & 1 \\ x_2 & y_2 & z_2 & 1 \\ x_3 & y_3 & z_3 & 1 \end{vmatrix} = 0$$

is the equation for a plane through three points (x_1, y_1, z_1), (x_2, y_2, z_2) and (x_3, y_3, z_3) in space.

Solution. Let $ax + by + cz + d = 0$ be an equation for a plane through the points (x_i, y_i, z_i), with $i = 1, 2, 3$. There is precisely one such plane if and only if the points are not collinear. Then $ax_i + by_i + cz_i + d = 0$ for $i = 1, 2, 3$, so that

$$\begin{bmatrix} x & y & z & 1 \\ x_1 & y_1 & z_1 & 1 \\ x_2 & y_2 & z_2 & 1 \\ x_3 & y_3 & z_3 & 1 \end{bmatrix} \begin{bmatrix} a \\ b \\ c \\ d \end{bmatrix} = \begin{bmatrix} 0 \\ 0 \\ 0 \\ 0 \end{bmatrix}.$$

Since the coordinates a, b, c, d of the plane are not all zero, the above matrix is singular, implying that its determinant is zero. Computing this determinant by cofactor expansion of the first row gives the equation

$$+\begin{vmatrix} y_1 & z_1 & 1 \\ y_2 & z_2 & 1 \\ y_3 & z_3 & 1 \end{vmatrix} x - \begin{vmatrix} x_1 & z_1 & 1 \\ x_2 & z_2 & 1 \\ x_3 & z_3 & 1 \end{vmatrix} y + \begin{vmatrix} x_1 & y_1 & 1 \\ x_2 & y_2 & 1 \\ x_3 & y_3 & 1 \end{vmatrix} z - \begin{vmatrix} x_1 & y_1 & z_1 \\ x_2 & y_2 & z_2 \\ x_3 & y_3 & z_3 \end{vmatrix} = 0$$

of the plane.

Exercise 1.10: Signed area of a triangle

Let $P_i = (x_i, y_i)$, $i = 1, 2, 3$, be three points in the plane defining a triangle T. Show that the area of T is

$$A(T) = \frac{1}{2} \begin{vmatrix} 1 & 1 & 1 \\ x_1 & x_2 & x_3 \\ y_1 & y_2 & y_3 \end{vmatrix}.$$

The area is positive if we traverse the vertices in counterclockwise order.

Hint.

One has $A(T) = A(ABP_3P_1) + A(P_3BCP_2) - A(P_1ACP_2)$; c.f. Figure 1.1.

Solution. Let T denote the triangle with vertices P_1, P_2, P_3. Since the area of a triangle is invariant under translation, we can assume $P_1 = A = (0, 0)$, $P_2 = (x_2, y_2)$, $P_3 = (x_3, y_3)$, $B = (x_3, 0)$, and $C = (x_2, 0)$. As is clear from Figure 1.1, the area $A(T)$ can be expressed as

$$\begin{aligned} A(T) &= A(ABP_3) + A(P_3BCP_2) - A(ACP_2) \\ &= \frac{1}{2}x_3y_3 + (x_2 - x_3)y_2 + \frac{1}{2}(x_2 - x_3)(y_3 - y_2) - \frac{1}{2}x_2y_2 \\ &= \frac{1}{2}\begin{vmatrix} 1 & 1 & 1 \\ 0 & x_2 & x_3 \\ 0 & y_2 & y_3 \end{vmatrix}, \end{aligned}$$

which is what needed to be shown.

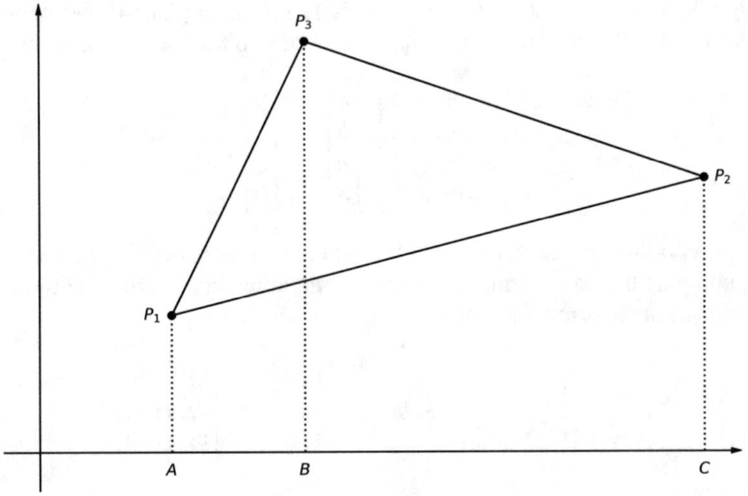

Figure 1.1: The triangle T defined by the three points P_1, P_2 and P_3.

Exercise 1.11: Vandermonde matrix

Show that

$$\begin{vmatrix} 1 & x_1 & x_1^2 & \cdots & x_1^{n-1} \\ 1 & x_2 & x_2^2 & \cdots & x_2^{n-1} \\ \vdots & \vdots & \vdots & & \vdots \\ 1 & x_n & x_n^2 & \cdots & x_n^{n-1} \end{vmatrix} = \prod_{i>j}(x_i - x_j),$$

where $\prod_{i>j}(x_i - x_j) = \prod_{i=2}^{n}(x_i - x_1)(x_i - x_2)\cdots(x_i - x_{i-1})$. This determinant is called the *Vandermonde determinant.*

Hint.

Subtract x_n^k times column k from column $k+1$ for $k = n-1, n-2, \ldots, 1$.

Solution. For any $n = 1, 2, \ldots$, let

$$D_n := \begin{vmatrix} 1 & x_1 & x_1^2 & \cdots & x_1^{n-1} \\ 1 & x_2 & x_2^2 & \cdots & x_2^{n-1} \\ 1 & x_3 & x_3^2 & \cdots & x_3^{n-1} \\ \vdots & \vdots & \vdots & \ddots & \vdots \\ 1 & x_n & x_n^2 & \cdots & x_n^{n-1} \end{vmatrix}$$

be the determinant of the Vandermonde matrix in the exercise. Clearly the formula

$$D_N = \prod_{1 \le j < i \le N} (x_i - x_j) \tag{1.i}$$

holds for $N = 1$ (in which case the product is empty and defined to be 1) and $N = 2$.

Let us assume (1.i) holds for some $N = n - 1 \le 2$. Since the determinant is an alternating multilinear form, adding a scalar multiple of one column to another does not change the value of the determinant. Subtracting x_n times column k from column $k + 1$ for $k = n - 1, n - 2, \ldots, 1$, we find

$$D_n = \begin{vmatrix} 1 & x_1 - x_n & x_1^2 - x_1 x_n & \cdots & x_1^{n-1} - x_1^{n-2} x_n \\ 1 & x_2 - x_n & x_2^2 - x_2 x_n & \cdots & x_2^{n-1} - x_2^{n-2} x_n \\ 1 & x_3 - x_n & x_3^2 - x_3 x_n & \cdots & x_3^{n-1} - x_3^{n-2} x_n \\ \vdots & \vdots & \vdots & \ddots & \vdots \\ 1 & x_n - x_n & x_n^2 - x_n x_n & \cdots & x_n^{n-1} - x_n^{n-2} x_n \end{vmatrix}.$$

Next, by cofactor expansion along the last row and by the multilinearity in the rows,

$$\begin{aligned} D_n &= (-1)^{n-1} \cdot 1 \cdot \begin{vmatrix} x_1 - x_n & x_1^2 - x_1 x_n & \cdots & x_1^{n-1} - x_1^{n-2} x_n \\ x_2 - x_n & x_2^2 - x_2 x_n & \cdots & x_2^{n-1} - x_2^{n-2} x_n \\ \vdots & \vdots & \ddots & \vdots \\ x_{n-1} - x_n & x_{n-1}^2 - x_{n-1} x_n & \cdots & x_{n-1}^{n-1} - x_{n-1}^{n-2} x_n \end{vmatrix} \\ &= (-1)^{n-1}(x_1 - x_n)(x_2 - x_n) \cdots (x_{n-1} - x_n) D_{n-1} \\ &= (x_n - x_1)(x_n - x_2) \cdots (x_n - x_{n-1}) \prod_{1 \le j < i \le n-1} (x_i - x_j) \\ &= \prod_{1 \le j < i \le n} (x_i - x_j). \end{aligned}$$

By induction, we conclude that (1.i) holds for any $N = 1, 2, \ldots$

Exercise 1.12: Cauchy determinant (1842)

Let $\boldsymbol{\alpha} = [\alpha_1, \ldots, \alpha_n]^{\mathrm{T}}$, $\boldsymbol{\beta} = [\beta_1, \ldots, \beta_n]^{\mathrm{T}}$ be in $\mathbb{R}^n$.

a) Consider the matrix $\boldsymbol{A} \in \mathbb{R}^{n\times n}$ with elements $a_{i,j} = 1/(\alpha_i + \beta_j)$, $i, j = 1, 2, \ldots, n$. Show that

$$\det(\boldsymbol{A}) = Pg(\boldsymbol{\alpha})g(\boldsymbol{\beta})$$

where $P = \prod_{i=1}^n \prod_{j=1}^n a_{ij}$, and for $\boldsymbol{\gamma} = [\gamma_1, \ldots, \gamma_n]^{\mathrm{T}}$

$$g(\boldsymbol{\gamma}) = \prod_{i=2}^{n} (\gamma_i - \gamma_1)(\gamma_i - \gamma_2)\cdots(\gamma_i - \gamma_{i-1})$$

Hint.
Multiply the ith row of $\boldsymbol{A}$ by $\prod_{j=1}^n (\alpha_i + \beta_j)$ for $i = 1, 2, \ldots, n$. Call the resulting matrix $\boldsymbol{C}$. Each element of $\boldsymbol{C}$ is a product of $n-1$ factors $\alpha_r + \beta_s$. Hence $\det(\boldsymbol{C})$ is a sum of terms where each term contain precisely $n(n-1)$ factors $\alpha_r + \beta_s$. Thus $\det(\boldsymbol{C}) = q(\alpha, \beta)$ where q is a polynomial of degree at most $n(n-1)$ in α_i and β_j. Since $\det(\boldsymbol{A})$ and therefore $\det(\boldsymbol{C})$ vanishes if $\alpha_i = \alpha_j$ for some $i \neq j$ or $\beta_r = \beta_s$ for some $r \neq s$, we have that $q(\boldsymbol{\alpha}, \boldsymbol{\beta})$ must be divisible by each factor in $g(\boldsymbol{\alpha})$ and $g(\boldsymbol{\beta})$. Since $g(\boldsymbol{\alpha})$ and $g(\boldsymbol{\beta})$ is a polynomial of degree $n(n-1)$, we have $q(\boldsymbol{\alpha}, \boldsymbol{\beta}) = kg(\boldsymbol{\alpha})g(\boldsymbol{\beta})$ for some constant k independent of $\boldsymbol{\alpha}$ and $\boldsymbol{\beta}$. Show that $k = 1$ by choosing $\beta_i + \alpha_i = 0$, $i = 1, 2, \ldots, n$.

Solution. Let $[\alpha_1, \ldots, \alpha_n]^{\mathrm{T}}, [\beta_1, \ldots, \beta_n]^{\mathrm{T}} \in \mathbb{R}^n$ and let

$$\boldsymbol{A} = (a_{i,j})_{i,j} = \left(\frac{1}{\alpha_i + \beta_j}\right)_{i,j} = \begin{bmatrix} \frac{1}{\alpha_1+\beta_1} & \frac{1}{\alpha_1+\beta_2} & \cdots & \frac{1}{\alpha_1+\beta_n} \\ \frac{1}{\alpha_2+\beta_1} & \frac{1}{\alpha_2+\beta_2} & \cdots & \frac{1}{\alpha_2+\beta_n} \\ \vdots & \vdots & \ddots & \vdots \\ \frac{1}{\alpha_n+\beta_1} & \frac{1}{\alpha_n+\beta_2} & \cdots & \frac{1}{\alpha_n+\beta_n} \end{bmatrix}.$$

Multiplying the ith row of $\boldsymbol{A}$ by $\prod_{k=1}^n (\alpha_i + \beta_k)$ for $i = 1, 2, \ldots, n$ gives a matrix

$$\boldsymbol{C} = (c_{i,j})_{i,j}, \qquad c_{i,j} = \prod_{\substack{k=1 \\ k\neq j}}^{n} (\alpha_i + \beta_k).$$

The determinant of an $n \times n$ matrix is a homogeneous polynomial of degree n in the entries of the matrix. Since each entry of $\boldsymbol{C}$ is a homogeneous polynomial of degree $n - 1$ in the variables α_i, β_j, the determinant of $\boldsymbol{C}$ must be a homogeneous polynomial of degree $n(n - 1)$ in α_i, β_j.

By the multilinearity of the determinant, $\det \boldsymbol{C} = \prod_{i,j=1}^n (\alpha_i + \beta_j) \det \boldsymbol{A}$. Since $\boldsymbol{A}$ vanishes whenever $\alpha_i = \alpha_j$ or $\beta_i = \beta_j$ for $i \neq j$, the homogeneous polynomial $\det \boldsymbol{C}$ contains factors $(\alpha_i - \alpha_j)$ and $(\beta_i - \beta_j)$ for $1 \leq i < j \leq n$. As there are precisely $2 \cdot \binom{n}{2} = (n-1)n$ such factors, necessarily

$$\det \boldsymbol{C} = k \prod_{1 \le i < j \le n} (\alpha_i - \alpha_j) \prod_{1 \le i < j \le n} (\beta_i - \beta_j) \tag{1.ii}$$

for some constant k. To determine k, we can evaluate $\det \boldsymbol{C}$ at a particular value, for instance any $\{\alpha_i, \beta_j\}_{i,j}$ satisfying $\alpha_1 + \beta_1 = \cdots = \alpha_n + \beta_n = 0$. In that case $\boldsymbol{C}$ becomes a diagonal matrix with determinant

$$\begin{aligned}\det \boldsymbol{C} &= \prod_{i=1}^{n} \prod_{\substack{k=1 \\ k \ne i}}^{n} (\alpha_i + \beta_k) = \prod_{i=1}^{n} \prod_{\substack{k=1 \\ k \ne i}}^{n} (\alpha_i - \alpha_k) \\ &= \prod_{1 \le i < k \le n} (\alpha_i - \alpha_k) \prod_{1 \le i < k \le n} (\alpha_k - \alpha_i).\end{aligned}$$

Comparing with (1.ii) shows that $k = 1$. We conclude that

$$\det \boldsymbol{A} = \frac{\prod_{1 \le i < j \le n} (\alpha_i - \alpha_j) \prod_{1 \le i < j \le n} (\beta_i - \beta_j)}{\prod_{i,j=1}^{n} (\alpha_i + \beta_j)}. \tag{1.iii}$$

b) Notice that the cofactor of any element in the above matrix $\boldsymbol{A}$ is the determinant of a matrix of similar form. Use the cofactor and determinant of $\boldsymbol{A}$ to represent the elements of $\boldsymbol{A}^{-1} = (b_{j,k})$.

Hint.
Answer:

$$b_{j,k} = (\alpha_k + \beta_j) A_k(-\beta_j) B_j(-\alpha_k),$$

where

$$A_k(x) = \prod_{s \ne k} \left(\frac{\alpha_s - x}{\alpha_s - \alpha_k} \right), \qquad B_k(x) = \prod_{s \ne k} \left(\frac{\beta_s - x}{\beta_s - \beta_k} \right).$$

Solution. Deleting row l and column k from $\boldsymbol{A}$, results in the matrix $\boldsymbol{A}_{l,k}$ associated to the vectors $[\alpha_1, \ldots, \alpha_{l-1}, \alpha_{l+1}, \ldots, \alpha_n]$ and $[\beta_1, \ldots, \beta_{k-1}, \beta_{k+1}, \ldots, \beta_n]$. By the adjoint formula for the inverse $\boldsymbol{A}^{-1} = (b_{k,l})$ and by (1.iii),

$$b_{k,l} := (-1)^{k+l} \frac{\det \boldsymbol{A}_{l,k}}{\det \boldsymbol{A}}$$

$$= (-1)^{k+l} \frac{\displaystyle\prod_{i,j=1}^{n} (\alpha_i + \beta_j) \prod_{\substack{1 \le i < j \le n \\ i,j \ne l}} (\alpha_i - \alpha_j) \prod_{\substack{1 \le i < j \le n \\ i,j \ne k}} (\beta_i - \beta_j)}{\displaystyle\prod_{\substack{i,j=1 \\ i \ne l \\ j \ne k}}^{n} (\alpha_i + \beta_j) \prod_{1 \le i < j \le n} (\alpha_i - \alpha_j) \prod_{1 \le i < j \le n} (\beta_i - \beta_j)}$$

$$= (\alpha_l + \beta_k) \frac{\displaystyle\prod_{\substack{s=1 \\ s \ne l}}^{n} (\alpha_s + \beta_k)}{\displaystyle\prod_{\substack{s=1 \\ s \ne l}}^{n} (\alpha_s - \alpha_l)} \frac{\displaystyle\prod_{\substack{s=1 \\ s \ne k}}^{n} (\beta_s + \alpha_l)}{\displaystyle\prod_{\substack{s=1 \\ s \ne k}}^{n} (\beta_s - \beta_k)}$$

$$= (\alpha_l + \beta_k) \prod_{\substack{s=1 \\ s \ne l}}^{n} \frac{\alpha_s + \beta_k}{\alpha_s - \alpha_l} \prod_{\substack{s=1 \\ s \ne k}}^{n} \frac{\beta_s + \alpha_l}{\beta_s - \beta_k},$$

which is what needed to be shown.

Exercise 1.13: Inverse of the Hilbert matrix

Let $\boldsymbol{H}_n = (h_{i,j})$ be the $n \times n$ matrix with elements $h_{i,j} = 1/(i+j-1)$. Use Exercise 1.12 to show that the elements $t_{i,j}^n$ in $\boldsymbol{T}_n = \boldsymbol{H}_n^{-1}$ are given by

$$t_{i,j}^n = \frac{f(i)f(j)}{i+j-1},$$

where

$$f(i+1) = \left(\frac{i^2 - n^2}{i^2}\right) f(i), \quad i = 1, 2, \ldots, \quad f(1) = -n.$$

Solution. If we write

$$\alpha = [\alpha_1, \ldots, \alpha_n] = [1, 2, \ldots, n], \qquad \beta = [\beta_1, \ldots, \beta_n] = [0, 1, \ldots, n-1],$$

then the Hilbert matrix is of the form $\boldsymbol{H}_n = (h_{i,j}) = \big(1/(\alpha_i + \beta_j)\big)$. By Exercise 1.12.b), its inverse $\boldsymbol{T}_n = (t_{i,j}^n) := \boldsymbol{H}_n^{-1}$ is given by

$$t_{i,j}^n = (i+j-1) \prod_{\substack{s=1 \\ s \ne j}}^{n} \frac{s+i-1}{s-j} \prod_{\substack{s=1 \\ s \ne i}}^{n} \frac{s+j-1}{s-i}, \qquad 1 \le i, j \le n.$$

We wish to show that

$$t^n_{i,j} = \frac{f(i)f(j)}{i+j-1}, \qquad 1 \le i, j \le n, \tag{1.iv}$$

where $f : \mathbb{N} \longrightarrow \mathbb{Q}$ is the sequence defined by

$$f(1) = -n, \qquad f(i+1) = \left(\frac{i^2 - n^2}{i^2}\right) f(i), \qquad \text{for } i = 1, 2, \dots.$$

Clearly (1.iv) holds when $i = j = 1$. Suppose that (1.iv) holds for some (i, j). Then

$$\begin{aligned}
t^n_{i+1,j} &= (i+j) \prod_{\substack{s=1\\ s\neq j}}^{n} \frac{s+1+i-1}{s-j} \prod_{\substack{s=1\\ s\neq i+1}}^{n} \frac{s+j-1}{s-1-i} \\
&= (i+j)\frac{1}{(i+j)^2} \frac{\prod_{s=2}^{n+1}(s+i-1)}{\prod_{\substack{s=1\\ s\neq j}}^{n}(s-j)} \frac{\prod_{s=1}^{n}(s+j-1)}{\prod_{\substack{s=0\\ s\neq i}}^{n-1}(s-i)} \\
&= \frac{(i+j-1)^2(n+i)(n-i)}{(i+j)i(-i)} \frac{\prod_{\substack{s=1\\ s\neq j}}^{n}(s+i-1)}{\prod_{\substack{s=1\\ s\neq j}}^{n}(s-j)} \frac{\prod_{\substack{s=1\\ s\neq i}}^{n}(s+j-1)}{\prod_{\substack{s=1\\ s\neq i}}^{n}(s-i)} \\
&= \frac{1}{i+j}\frac{i^2-n^2}{i^2}\left[(i+j-1)\prod_{\substack{s=1\\ s\neq j}}^{n}\frac{s+i-1}{s-j}\right]\left[(i+j-1)\prod_{\substack{s=1\\ s\neq i}}^{n}\frac{s+j-1}{s-i}\right] \\
&= \frac{1}{i+j}\frac{i^2-n^2}{i^2} f(i)f(j) \\
&= \frac{f(i+1)f(j)}{(i+1)+j-1},
\end{aligned}$$

so that (1.iv) holds for $(i+1, j)$. Carrying out a similar calculation for $(i, j+1)$, or using the symmetry of $\boldsymbol{T}_n$, we conclude by induction that (1.iv) holds for any i, j.

Chapter 2
Diagonally Dominant Tridiagonal Matrices; Three Examples

Exercises section 2.1

Exercise 2.1: The shifted power basis is a basis

Show that $\{(x - x_i)^j\}_{0 \le j \le n}$ is a basis for polynomials of degree n.

Hint.

Consider an arbitrary polynomial of degree n and expand it as a Taylor series around x_i.

Solution. We know that the set of polynomials of degree n is a vector space of dimension $n + 1$: It is spanned by $\{x^k\}_{k=0}^n$, and these are linearly independent (if a linear combination of these is zero, then it has in particular $n + 1$ zeros (since every x is a zero), and it follows from the fundamental theorem of algebra that the linear combination must be zero). Since the shifted power basis also has $n + 1$ vectors which are polynomials, all we need to show is that they are linearly independent. Suppose then that

$$\sum_{j=0}^{n} a_j (x - x_i)^j = 0.$$

In particular we can then pick $n + 1$ distinct values z_k for x so that this is zero. But then the polynomial $\sum_{j=0}^n a_j x^j$ has the $n + 1$ different zeros $z_k - x_i$. Since the $\{x^k\}_{k=0}^n$ are linearly independent, it follows that all $a_j = 0$, so that the shifted power basis also is a basis.

Exercise 2.2: The natural spline, $n = 1$

How can one define an N-spline when $n = 1$?

T. Lyche et al., *Exercises in Numerical Linear Algebra and Matrix Factorizations*, Texts in Computational Science and Engineering 23, https://doi.org/10.1007/978-3-030-59789-4_2

Solution. $n = 1$ implies that there is only segment, so that we need to find a third degree polynomial $p(x)$ so that $p(x_1) = y_1$, $p(x_2) = y_2$, $p''(x_1) = p''(x_2) = 0$. If $p(x) = ax^3 + bx^2 + cx + d$, then $p''(x) = 6ax + 2b$. If $a \neq 0$ this can be zero only for $x = -b/(2a)$, so that both x_1 and x_2 can't be zeros. Thus, a must be zero, but then $b = 0$ also, in order for the second derivative to have any zeros at all. We are thus left with finding a linear function $p(x) = cx + d$ so that $p(x_1) = y_1$, $p(x_2) = y_2$. An N-spline for $n = 1$ thus corresponds to connecting two points with a straight line.

Exercise 2.3: Bounding the moments

Show that for the N-spline the solution of the linear system (2.11) is bounded as follows:

$$\max_{2 \le j \le n} |\mu_j| \le \frac{3}{h^2} \max_{2 \le i \le n} |y_{i+1} - 2y_i + y_{i-1}|.$$

Hint.

Use Theorem 2.2.

Solution. We have that the σ_i from Theorem 2.2 are $\sigma_1 = \sigma_{n+1} = 3$, while all the other σ_i are 2. This means that the right hand side in this theorem is

$$\begin{aligned}\max\left(\frac{|b_i|}{\sigma_i}\right) &\le \frac{1}{\min(2,3)}\frac{6}{h^2}\max|y_{i+1} - 2y_i + y_{i-1}| \\ &= \frac{3}{h^2}\max|y_{i+1} - 2y_i + y_{i-1}|\end{aligned}$$

(since $\mu_1 = \mu_{n+1}$ for an N-spline). This is also the right hand side of the inequality, and the proof is done.

Exercise 2.4: Moment equations for 1. derivative boundary conditions

Suppose instead of the D_2 boundary conditions we use D_1 conditions given by $g'(a) = s_1$ and $g'(b) = s_{n+1}$ for some given numbers s_1 and s_{n+1}. Show that the linear system for the moments of a D_1-spline can be written

$$\begin{bmatrix} 2 & 1 & & & \\ 1 & 4 & 1 & & \\ & \ddots & \ddots & \ddots & \\ & & 1 & 4 & 1 \\ & & & 1 & 2 \end{bmatrix} \begin{bmatrix} \mu_1 \\ \mu_2 \\ \vdots \\ \mu_n \\ \mu_{n+1} \end{bmatrix} = \frac{6}{h^2} \begin{bmatrix} y_2 - y_1 - hs_1 \\ \delta^2 y_2 \\ \delta^2 y_3 \\ \vdots \\ \delta^2 y_{n-1} \\ \delta^2 y_n \\ hs_{n+1} - y_{n+1} + y_n \end{bmatrix},$$

where $\delta^2 y_i := y_{i+1} - 2y_i + y_{i-1}$, $i = 2, \ldots, n$. Is g unique?

Hint.
Equation (2.10) states the derivatives of the cubic spline interpolants at the knots (i.e., $p_{i-1}'(x_i)$ and $p_i'(x_i)$). Use these to compute $g'(x_1)$ and $g'(x_{n+1})$.

Solution. The first equation listed in (2.10) with $i = n + 1$ becomes

$$s_{n+1} = \frac{y_{n+1} - y_n}{h} + \frac{h}{6}\mu_n + \frac{h}{3}\mu_{n+1},$$

so that

$$\mu_n + 2\mu_{n+1} = \frac{6}{h^2}(hs_{n+1} - y_{n+1} + y_n),$$

which is the last equation of the stated system. The second equation in (2.10) with $i = 1$ gives

$$s_1 = \frac{y_2 - y_1}{h} - \frac{h}{3}\mu_1 - \frac{h}{6}\mu_2,$$

so that

$$2\mu_1 + \mu_2 = \frac{6}{h^2}(y_2 - y_1 - hs_1),$$

which is the first equation of the stated system. The remaining equations are the same. The system has a unique solution, because the matrix is strictly diagonally dominant.

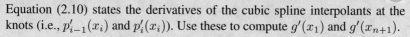

Exercise 2.5: Minimal norm property of the natural spline

Study proof of the following theorem[a].

Theorem 2.5 (Minimal norm property of a cubic spline)
Suppose g is an N-spline. Then

$$\int_a^b \big(g''(x)\big)^2 dx \le \int_a^b \big(h''(x)\big)^2 dx$$

for all $h \in C^2[a,b]$ such that $h(x_i) = g(x_i)$, $i = 1,\dots,n+1$.

Proof. Let h be any interpolant as in the theorem. We first show the orthogonality condition

$$\int_a^b g''e'' = 0, \qquad e := h - g. \tag{2.i}$$

Integration by parts gives $\int_a^b g''e'' = \big[g''e'\big]_a^b - \int_a^b g'''e'$. The first term is zero since g'' is continuous and $g''(b) = g''(a) = 0$. For the second term, since g''' is equal to a constant v_i on each subinterval (x_i, x_{i+1}) and $e(x_i) = 0$, for $i = 1,\dots,n+1$

$$\begin{aligned}\int_a^b g'''e' &= \sum_{i=1}^n \int_{x_i}^{x_{i+1}} g'''e' = \sum_{i=1}^n v_i \int_{x_i}^{x_{i+1}} e'\\ &= \sum_{i=1}^n v_i\big(e(x_{i+1}) - e(x_i)\big) = 0.\end{aligned}$$

Writing $h = g + e$ and using (2.i),

$$\begin{aligned}\int_a^b (h'')^2 &= \int_a^b (g'' + e'')^2\\ &= \int_a^b (g'')^2 + \int_a^b (e'')^2 + 2\int_a^b g''e''\\ &= \int_a^b (g'')^2 + \int_a^b (e'')^2 \ge \int_a^b (g'')^2,\end{aligned}$$

and the proof is complete. □

---

[a] The name spline is inherited from a "physical analogue", an elastic ruler that is used to draw smooth curves. Heavy weights, called *ducks*, are used to force the ruler to pass through, or near given locations. The ruler will take a shape that minimizes its potential energy. Since the potential energy is proportional to the integral of the square of the curvature, and the curvature can be approximated by the second derivative it follows from Theorem 2.5 that the mathematical N-spline approximately models the physical spline.

Exercise 2.6: Computing the D_2-spline

Let g be the D_2-spline corresponding to an interval $[a, b]$, a vector $\boldsymbol{y} \in \mathbb{R}^{n+1}$, and boundary conditions μ_1, μ_{n+1}. The vector $\boldsymbol{x} = [x_1, \ldots, x_n]$ and the coefficient matrix $\boldsymbol{C} \in \mathbb{R}^{n\times 4}$ in (2.7) are returned in Algorithm `splineint`. It uses Algorithms 2.1 and 2.2 to solve the tridiagonal linear system. Use the algorithm to compute the $c_{i,j}$ in Example 2.2.

code/splineint.m

```
function [x,C]=splineint(a,b,y,mu1,munp1)
y=y(:); n=length(y)-1;
h=(b-a)/n; x=a:h:b-h; c=ones(n-2,1);
[l,u]= trifactor(c,4*ones(n-1,1),c);
b1=6/h^2*(y(3:n+1)-2*y(2:n)+y(1:n-1));
b1(1)=b1(1)-mu1; b1(n-1)=b1(n-1)-munp1;
mu= [mu1;trisolve(l,u,c,b1);munp1];
C=zeros(4*n,1);
C(1:4:4*n-3)=y(1:n);
C(2:4:4*n-2)=(y(2:n+1)-y(1:n))/h-h*mu(1:n)/3 - h*mu(2:n+1)/6;
C(3:4:4*n-1)=mu(1:n)/2;
C(4:4:4*n)=(mu(2:n+1)-mu(1:n))/(6*h);
C=reshape(C,4,n)';
```

Listing 2.1: Compute the knot vector $\boldsymbol{x}$ and coefficient matrix $\boldsymbol{C}$ for the D_2-spline expressed as piecewise polynomials in the shifted monomial basis.

Solution. The $c_{i,j}$ can be computed as follows.

code/test_splineint.m

```
[x,C]=splineint(0,4,[0 1/6 2/3 1/6 0],0,0);
```

Listing 2.2: Use Listing 2.1 to compute the coefficients c_i, j in Example 2.2.

Exercise 2.7: Spline evaluation

To plot a piecewise polynomial g in the form (2.4) we need to compute values $g(r_j)$ at a number of sites $\boldsymbol{r} = [r_1, \ldots, r_m] \in \mathbb{R}^m$ for some reasonably large integer m. To determine $g(r_j)$ for some j we need to find an integer i_j so that $g(r_j) = p_{i_j}(r_j)$.

Given $k \in \mathbb{N}$, $\boldsymbol{t} = [t_1, \ldots, t_k]$ and a real number x. We consider the problem of computing an integer i so that $i = 1$ if $x < t_2$, $i = k$ if $x \geq t_k$, and $t_i \leq x < t_{i+1}$ otherwise. If $\boldsymbol{x} \in \mathbb{R}^m$ is a vector then an m-vector $\boldsymbol{i}$ should be computed, such that the jth component of $\boldsymbol{i}$ gives the location of the jth component of $\boldsymbol{x}$. The MATLAB function `findsubintervals` determines $\boldsymbol{i} = [i_1, \ldots, i_m]$. It uses the built-in MATLAB functions `length`, `min`, `sort`, `find`.

Use `findsubintervals` and the algorithm `splineval` below to make the plots in Figure 2.5 in the book. Given output $\boldsymbol{x}, \boldsymbol{C}$ of `splineint`, defining a cubic spline g, and a vector $\boldsymbol{X}$, `splineval` computes the vector $\boldsymbol{G} = g(\boldsymbol{X})$.

code/findsubintervals.m

```
function i = findsubintervals(t,x)
%i = findsubintervals(t,x)
k=length(t); m=length(x);
 if k<2
    i=ones(m,1);
 else
   t(1)=min(x(1),t(1))-1;
   [~,j]=sort([t(:)',x(:)']);
   i=(find(j>k)-(1:m))';
 end
```

Listing 2.3: Determine the index vector $\boldsymbol{i}$, for which the jth component gives the location of the jth component of $\boldsymbol{x}$.

code/splineval.m

```
function [X,G]=splineval(x,C,X)
m=length(X);
i=findsubintervals(x,X);
G=zeros(m,1);
for j=1:m
    k=i(j);
    t=X(j)-x(k);
   G(j)=[1,t,t^2,t^3]*C(k,:)';
end
```

Listing 2.4: Find the values $\boldsymbol{G}$ of the spline with knot vector $\boldsymbol{x}$ and coefficient matrix $\boldsymbol{C}$ at the points $\boldsymbol{X}$.

Solution. The following code reproduces Figure 2.5 in the book.

code/test2_splineint.m

```
x = linspace(-1,1,14);
y = atan(10*x)+pi/2;
[x,C]=splineint(-1,1,y,0,0);
X = linspace(-1,1,200);
[X,G]=splineval(x,C,X);
plot(X,G)
hold on;
plot(X,atan(10*X)+pi/2,'r--');
```

Listing 2.5: Use Listing 2.1 to reproduce Figure 2.5 in the book.

Exercises section 2.2

Exercise 2.8: Central difference approximation of 2. derivative

Consider

$$\delta^2 f(x) := \frac{f(x+h) - 2f(x) + f(x-h)}{h^2}, \quad h > 0, \quad f : [x-h, x+h] \to \mathbb{R}.$$

$\delta^2 f(x)$ is known as the *central difference approximation* to the second derivative at x.

a) Show using Taylor expansion that if $f \in C^2[x-h, x+h]$ then for some η_2

$$\delta^2 f(x) = f''(\eta_2), \quad x - h < \eta_2 < x + h.$$

Solution. The first order Taylor series with remainder gives

$$f(x-h) = f(x) - hf'(x) + \frac{h^2}{2} f''(\nu_0),$$
$$f(x+h) = f(x) + hf'(x) + \frac{h^2}{2} f''(\nu_1),$$

where $\nu_0 \in [x-h, x]$, $\nu_1 \in [x, x+h]$. Adding these and reorganizing we get

$$\frac{f(x+h) - 2f(x) + f(x-h)}{h^2} = \frac{1}{2}\big(f''(\nu_0) + f''(\nu_1)\big).$$

From the intermediate value theorem it follows that there exists a $\nu_2 \in [x-h, x+h]$ so that $f''(\nu_2) = \frac{1}{2}(f''(\nu_0) + f''(\nu_1))$.

b) Show that, if $f \in C^4[x-h, x+h]$ then for some η_4

$$\delta^2 f(x) = f''(x) + \frac{h^2}{12} f^{(4)}(\eta_4), \quad x - h < \eta_4 < x + h.$$

Solution. We add the two third order Taylor series with remainder

$$f(x-h) = f(x) - hf'(x) + \frac{h^2}{2} f''(x) - \frac{h^3}{6} f^{(3)}(x) + \frac{h^4}{24} f^{(4)}(\nu_0),$$
$$f(x+h) = f(x) + hf'(x) + \frac{h^2}{2} f''(x) + \frac{h^3}{6} f^{(3)}(x) + \frac{h^4}{24} f^{(4)}(\nu_1)$$

to obtain

$$\frac{f(x+h) - 2f(x) + f(x-h)}{h^2} = f''(x) + \frac{h^2}{12} \cdot \frac{1}{2}\big(f^{(4)}(\nu_0) + f^{(4)}(\nu_1)\big).$$

The result now follows from the intermediate value theorem in the same way.

Exercise 2.9: Two point boundary value problem

We consider a finite difference method for the two point boundary value problem

$$\begin{aligned} -u''(x) + r(x)u'(x) + q(x)u(x) &= f(x), \text{ for } x \in [a, b], \\ u(a) = g_0, \quad u(b) &= g_1. \end{aligned} \tag{2.36}$$

We assume that the given functions f, q and r are continuous on $[a, b]$ and that $q(x) \geq 0$ for $x \in [a, b]$. It can then be shown that (2.36) has a unique solution u.

To solve (2.36) numerically we choose $m \in \mathbb{N}$, $h = (b-a)/(m+1)$, $x_j = a + jh$ for $j = 0, 1, \ldots, m+1$ and solve the difference equation

$$\frac{-v_{j-1} + 2v_j - v_{j+1}}{h^2} + r(x_j)\frac{v_{j+1} - v_{j-1}}{2h} + q(x_j)v_j = f(x_j), \quad j = 1, \ldots, m, \tag{2.37}$$

with $v_0 = g_0$ and $v_{m+1} = g_1$.

a) Show that (2.37) leads to a tridiagonal linear system $\boldsymbol{A}\boldsymbol{v} = \boldsymbol{b}$, where $\boldsymbol{A} = \operatorname{tridiag}(a_j, d_j, c_j) \in \mathbb{R}^{m \times m}$ has elements

$$a_j = -1 - \frac{h}{2} r(x_j), \qquad c_j = -1 + \frac{h}{2} r(x_j), \qquad d_j = 2 + h^2 q(x_j),$$

and

$$b_j = \begin{cases} h^2 f(x_1) - a_1 g_0, & \text{if } j = 1, \\ h^2 f(x_j), & \text{if } 2 \leq j \leq m-1, \\ h^2 f(x_m) - c_m g_1, & \text{if } j = m. \end{cases}$$

Solution. For $j = 1, \ldots, m$, we get when we gather terms that

$$h^2 f(x_j) = \left(-1 - \frac{h}{2} r(x_j)\right) v_{j-1} + (2 + h^2 q(x_j)) v_j + \left(-1 + \frac{h}{2} r(x_j)\right) v_{j+1}$$

From this we get the desired formula for a_j, c_j, and d_j, and the right hand sides b_j for $2 \le j \le m-1$.
For $j = 1$, since v_0 is known we have to move $\left(-1 - \frac{h}{2} r(x_0)\right) v_0 = a_1 g_0$ over to the right hand side, so that we obtain $b_0 = h^2 f(x_1) - a_1 g_0$.
For $j = m$, since v_{m+1} is known we have to move $\left(-1 + \frac{h}{2} r(x_m)\right) v_{m+1} = c_m g_1$ over to the right hand side, so that we obtain $b_m = h^2 f(x_m) - c_m g_1$. This leads to the tridiagonal system $\boldsymbol{A}\boldsymbol{v} = \boldsymbol{b}$ in the exercise.

b) Show that the linear system satisfies the conditions in Theorem 2.4 if the spacing h is so small that $\frac{h}{2}|r(x)| < 1$ for all $x \in [a, b]$.

Solution. When $h|r(x)|/2 < 1$ for all $x \in [a, b]$, we see that $a_j, c_j \in (-2, 0)$. It follows that $|a_j| + |c_j| = 1 + \frac{h}{2} r(x_m) + 1 + \frac{h}{2} r(x_m) = 2$. Since $q(x_j) \ge 0$, $|d_j| = d_j \ge 2$, so that $\boldsymbol{A}$ is weakly diagonally dominant. Since $|c_j| = 1 + \frac{h}{2} r(x_j) < 2$, and $|d_j| > 2$ it follows in particular that $|d_1| > |c_1|$. Clearly also all $a_j > 0$ since $h|r(x)|/2 < 1$, and since also $|d_j| > 2$, in particular $d_n \neq 0$, so that all the conditions in the theorem are fulfilled.

c) Propose a method to find $v_1, \ldots, v_m$.

Solution. We can use the method `trisolve` to find the $v_1, \ldots, v_m$. Note that the indexing of the a_j should be shifted with one in this exercise, to be compatible with the notation used in `tridiag(`a_j`,` d_j`,` c_j`)` (a_j and d_j have the same index when they are in the same column of the matrix. In this exercise they have the same index when they are in the same row).

Exercise 2.10: Two point boundary value problem; computation

a) Consider the problem (2.36) with $r = 0$, $f = q = 1$ and boundary conditions $u(0) = 1$, $u(1) = 0$. The exact solution is $u(x) = 1 - \sinh x / \sinh 1$. Write a computer program to solve (2.37) for $h = 0.1, 0.05, 0.025, 0.0125$, and compute the "error" $\max_{1 \le j \le m} |u(x_j) - v_j|$ for each h.

Solution. The provided values for r, f, q give $a_j = c_j = -1$, $d_j = 2 + h^2$. The initial conditions are $g_0 = 1$, $g_1 = 0$, so that $\boldsymbol{b} = (h^2 + 1, h^2, \ldots, h^2)$. The code can look as follows.

code/exgtpbvpa.m

```
for m = [9 19 39 79, 159]
```

```
    h = 1/(m+1);
    x = h*(1:m)';
    [l, u] = trifactor( -ones(1, m - 1), ...
                     (2+h^2)*ones(1, m), ...
                     -ones(1, m - 1));
    b = h^2*ones(m, 1); b(1) = b(1) + 1;
    v = trisolve(l, u, -ones(1, m - 1), b);
    err = max(abs( (1-sinh(x)/sinh(1)) - v))
    log(err)/log(h)
end
```

Listing 2.6: Solve, for $h = 0.1, 0.05, 0.025, 0.0125$, the difference equation (2.36) with $r = 0$, $f = q = 1$ and boundary conditions $u(0) = 1$, $u(1) = 0$, and compute the "error" $\max_{1 \le j \le m} |u(x_j) - v_j|$ for each h.

The code also solves **c)**: If the error is proportional to h^p, then $err = Ch^p$ for some C. But then $p = (\log(err) - \log C)/\log h \approx \log(err)/\log h$ for small h, which is the quantity computed inside the `for`-loop. It seems that this converges to 3, so that one would guess that the error is proportional to h^3.

b) Make a combined plot of the solution u and the computed points v_j, $j = 0, \ldots, m+1$ for $h = 0.1$.

Solution. The code can look as follows.

code/exgtpbvpb.m

```
m = 9;
h = 1/(m+1);
x = h*(1:m)';
[l, u] = trifactor( -ones(1, m - 1), ...
                 (2+h^2)*ones(1, m), ...
                 -ones(1, m - 1));
b = h^2*ones(m, 1); b(1) = b(1) + 1;
v = trisolve(l, u, -ones(1, m - 1), b);
plot(x, (1-sinh(x)/sinh(1)), x, v)
legend('Exact solution', 'Estimated solution')
```

Listing 2.7: Plot the exact solution and estimated solution for the two-point boundary value problem.

The code produces the plot in Figure 2.1. The exact and estimated solution can't be distinguished.

c) One can show that the error is proportional to h^p for some integer p. Estimate p based on the error for $h = 0.1, 0.05, 0.025, 0.0125$.

Solution. This was solved by the code in **a)**.

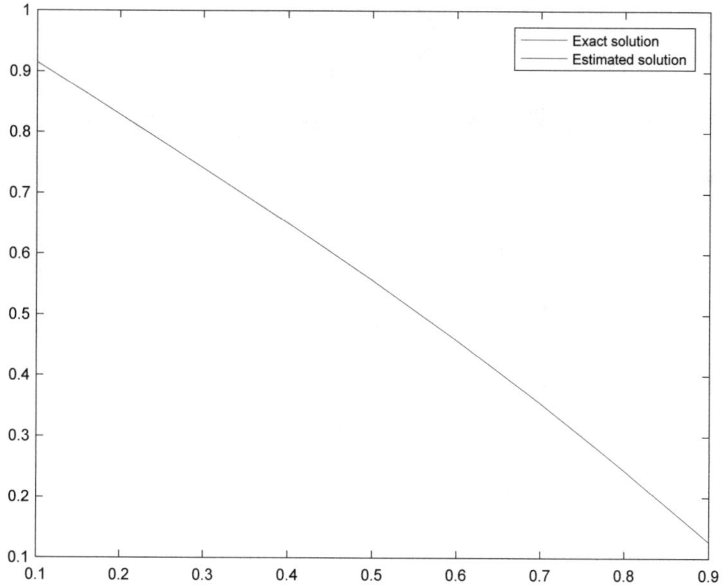

Figure 2.1: The exact solution and estimated solution for the two-point boundary value problem.

Exercises section 2.3

Exercise 2.11: Approximate force

Show that

$$F = \frac{4\sin^2(\pi h/2)R}{h^2L^2} = \frac{\pi^2 R}{L^2} + \mathcal{O}(h^2).$$

Solution. Taylor expansion around $x = 0$ yields

$$\sin x = x - x^3/3! + x^5/5! - \cdots,$$

implying $\sin(\pi h/2) = \pi h/2 + \mathcal{O}(h^3)$. Squaring both sides yields $\sin^2(\pi h/2) = \pi^2 h^2/4 + \mathcal{O}(h^4)$, and from this we obtain $4\sin^2(\pi h/2)R/(h^2L^2) = \pi^2 R/L^2 + \mathcal{O}(h^2)$.

Exercise 2.12: Symmetrize matrix (Exam exercise 1977-3)

Let $\boldsymbol{A} \in \mathbb{R}^{n\times n}$ be tridiagonal and suppose $a_{i,i+1}a_{i+1,i} > 0$ for $i = 1, \ldots, n-1$. Show that there exists a diagonal matrix $\boldsymbol{D} = \operatorname{diag}(d_1, \ldots, d_n)$ with $d_i > 0$ for all i such that $\boldsymbol{B} := \boldsymbol{DAD}^{-1}$ is symmetric.

Solution. We have

$$\boldsymbol{DAD}^{-1} = \begin{bmatrix} a_{11} & \frac{d_1}{d_2}a_{12} & 0 & \cdots & 0 \\ \frac{d_2}{d_1}a_{21} & a_{22} & \frac{d_2}{d_3}a_{23} & \ddots & \vdots \\ 0 & \frac{d_3}{d_2}a_{32} & \ddots & \ddots & 0 \\ \vdots & \ddots & \ddots & a_{n-1,n-1} & \frac{d_{n-1}}{d_n}a_{n-1,n} \\ 0 & \cdots & 0 & \frac{d_n}{d_{n-1}}a_{n,n-1} & a_{n,n} \end{bmatrix}$$

Symmetry would imply

$$\frac{d_i}{d_{i+1}}a_{i,i+1} = \frac{d_{i+1}}{d_i}a_{i+1,i}, \qquad i = 1, 2, \ldots, n-1.$$

We find $\left(\frac{d_i}{d_{i+1}}\right)^2 = \frac{a_{i+1,i}}{a_{i,i+1}}$, or

$$d_{i+1} = \sqrt{\frac{a_{i,i+1}}{a_{i+1,i}}}d_i, \qquad i = 1, 2, \ldots, n-1.$$

We can for example chose $d_1 = 1$.

Exercises section 2.4

Exercise 2.13: Eigenpairs $\boldsymbol{T}$ of order 2

Compute directly the eigenvalues and eigenvectors for $\boldsymbol{T}$ when $m = 2$ and thus verify Lemma 2.2 in this case.

Solution. When the order $m = 2$, one has $h := 1/(m+1) = 1/3$ and the 1D test matrix takes the form

$$\boldsymbol{T} = \begin{bmatrix} d & a \\ a & d \end{bmatrix}, \qquad a, d \in \mathbb{R}.$$

The eigenvalues λ of $\boldsymbol{T}$ satisfy

$$0 = \det(\boldsymbol{T} - \lambda \boldsymbol{I}) = \det \begin{bmatrix} d-\lambda & a \\ a & d-\lambda \end{bmatrix} = (d-\lambda)^2 - a^2 = \lambda^2 - 2d\lambda + d^2 - a^2,$$

and we find eigenvalues $\lambda_\pm := d \pm a$. The corresponding eigenvectors satisfy

$$\boldsymbol{v}_\pm \in \ker(\boldsymbol{T} - \lambda_\pm \boldsymbol{I}) = \ker \begin{bmatrix} \mp a & a \\ a & \mp a \end{bmatrix} = \operatorname{span} \begin{bmatrix} \pm 1 \\ 1 \end{bmatrix},$$

which agrees with Lemma 2.2 since

$$\begin{aligned} \boldsymbol{s}_1 &= [\sin(\pi h), \sin(2\pi h)]^{\mathrm{T}} = [\sin(\pi/3), \sin(2\pi/3)]^{\mathrm{T}} = \frac{\sqrt{3}}{2}[1,1]^{\mathrm{T}}, \\ \lambda_1 &= d + 2a\cos(\pi h) = d + 2a\cos(\pi/3) = d + a. \end{aligned}$$

and

$$\begin{aligned} \boldsymbol{s}_2 &= [\sin(2\pi h), \sin(4\pi h)]^{\mathrm{T}} = [\sin(2\pi/3), \sin(4\pi/3)]^{\mathrm{T}} = \frac{\sqrt{3}}{2}[1,-1]^{\mathrm{T}}, \\ \lambda_2 &= d + 2a\cos(2\pi h) = d + 2a\cos(2\pi/3) = d - a. \end{aligned}$$

Exercise 2.14: LU factorization of 2. derivative matrix

Show that $\boldsymbol{T} = \boldsymbol{L}\boldsymbol{U}$, where

$$\boldsymbol{L} = \begin{bmatrix} 1 & 0 & \cdots & \cdots & 0 \\ -\frac{1}{2} & 1 & \ddots & & \vdots \\ 0 & -\frac{2}{3} & 1 & \ddots & \vdots \\ \vdots & \ddots & \ddots & \ddots & 0 \\ 0 & \cdots & 0 & -\frac{m-1}{m} & 1 \end{bmatrix}, \qquad \boldsymbol{U} = \begin{bmatrix} 2 & -1 & 0 & \cdots & 0 \\ 0 & \frac{3}{2} & -1 & \ddots & \vdots \\ \vdots & \ddots & \ddots & \ddots & 0 \\ \vdots & & \ddots & \frac{m}{m-1} & -1 \\ 0 & \cdots & \cdots & t0 & \frac{m+1}{m} \end{bmatrix}.$$

This is the LU factorization of $\boldsymbol{T}$.

Solution. Let $\boldsymbol{L} = (l_{ij})_{ij}, \boldsymbol{U} = (r_{ij})_{ij}$ and $\boldsymbol{T}$ be as in the exercise. Clearly $\boldsymbol{L}$ is unit lower triangular and $\boldsymbol{U}$ is upper triangular. We compute the product $\boldsymbol{L}\boldsymbol{U}$ by separating cases for its entries. There are several ways to carry out and write down this computation, some more precise than others. For instance,

$$\begin{aligned} (\boldsymbol{LU})_{11} &= 1 \cdot 2 = 2, \\ (\boldsymbol{LU})_{ii} &= -\frac{i-1}{i} \cdot -1 + 1 \cdot \frac{i+1}{i} = 2, && \text{for } i = 2, \ldots, m, \\ (\boldsymbol{LU})_{i,i-1} &= -\frac{i-1}{i} \cdot \frac{i}{i-1} = -1, && \text{for } i = 2, \ldots, m, \\ (\boldsymbol{LU})_{i-1,i} &= 1 \cdot -1 = -1, && \text{for } i = 2, \ldots, m, \\ (\boldsymbol{LU})_{ij} &= 0, && \text{for } |i-j| \geq 2. \end{aligned}$$

It follows that $\boldsymbol{T} = \boldsymbol{L}\boldsymbol{U}$ is an LU factorization.

One can also show this by induction using the `trifactor`-algorithm. Since $\boldsymbol{T}$ and $\boldsymbol{U}$ have the same super-diagonal, we must have $c_m = -1$ for all m. Assume now that $\boldsymbol{L}_m\boldsymbol{U}_m = \boldsymbol{T}_m$, and that $l_{m-1} = -(m-1)/m$ and $u_m = (m+1)/m$. From the `trifactor`-algorithm,

$$\begin{aligned} l_m &= a_m/u_m = -1/((m+1)/m) = -m/(m+1) \\ u_{m+1} &= d_{m+1} - l_m c_m = 2 - m/(m+1) = (m+2)/(m+1). \end{aligned}$$

This shows that the `trifactor`-algorithm produces the desired terms in $\boldsymbol{L}_{m+1}$ and $\boldsymbol{U}_{m+1}$ as well.

Another way to show this is by induction. For $m = 1$, one has $\boldsymbol{L}_1\boldsymbol{U}_1 = 1 \cdot 2 = \boldsymbol{T}_1$. Now let $m > 1$ be arbitrary and assume that $\boldsymbol{L}_m\boldsymbol{U}_m = \boldsymbol{T}_m$. With

$$\boldsymbol{a} := \left[0, \dots, 0, -\frac{m}{m+1}\right]^{\mathrm{T}}, \qquad \boldsymbol{b} := [0, \dots, 0, -1]^{\mathrm{T}},$$

block multiplication yields

$$\begin{aligned} \boldsymbol{L}_{m+1}\boldsymbol{U}_{m+1} &= \begin{bmatrix} \boldsymbol{L}_m & \boldsymbol{0} \\ \boldsymbol{a}^{\mathrm{T}} & 1 \end{bmatrix} \begin{bmatrix} \boldsymbol{U}_m & \boldsymbol{b} \\ \boldsymbol{0} & \frac{m+2}{m+1} \end{bmatrix} \\ &= \begin{bmatrix} \boldsymbol{T}_m & \boldsymbol{L}_m\boldsymbol{b} \\ \boldsymbol{a}^{\mathrm{T}}\boldsymbol{U}_m & \boldsymbol{a}^{\mathrm{T}}\boldsymbol{b} + \frac{m+2}{m+1} \end{bmatrix} = \begin{bmatrix} \boldsymbol{T}_m & \boldsymbol{b} \\ \boldsymbol{b}^{\mathrm{T}} & 2 \end{bmatrix} = \boldsymbol{T}_{m+1}. \end{aligned}$$

By induction, we conclude that $\boldsymbol{T}_m = \boldsymbol{L}_m\boldsymbol{U}_m$ for all $m \geq 1$.

Exercise 2.15: Inverse of the 2. derivative matrix

Let $\boldsymbol{S} \in \mathbb{R}^{m\times m}$ have elements s_{ij} given by

$$s_{i,j} = s_{j,i} = \frac{1}{m+1} j(m+1-i), \qquad 1 \leq j \leq i \leq m.$$

Show that $\boldsymbol{S}\boldsymbol{T} = \boldsymbol{I}$ and conclude that $\boldsymbol{T}^{-1} = \boldsymbol{S}$.

Solution. Note that

$$s_{ij} = s_{ji} = \left(1 - \frac{i}{m+1}\right) j, \qquad \text{for } 1 \leq j \leq i \leq m.$$

In order to show that $\boldsymbol{S} = \boldsymbol{T}^{-1}$, we multiply $\boldsymbol{S}$ by $\boldsymbol{T}$ and show that the result is the identity matrix. To simplify notation we define $s_{ij} := 0$ whenever $i = 0$, $i = m+1$, $j = 0$, or $j = m+1$. With $1 \leq j < i \leq m$, we find

$$(\boldsymbol{S}\boldsymbol{T})_{i,j} = \sum_{k=1}^{m} s_{i,k}\boldsymbol{T}_{k,j} = -s_{i,j-1} + 2s_{i,j} - s_{i,j+1}$$

$$= \left(1 - \frac{i}{m+1}\right)(-j + 1 + 2j - j - 1) = 0,$$

$$(\boldsymbol{ST})_{j,i} = \sum_{k=1}^{m} s_{j,k}\boldsymbol{T}_{k,i} = -s_{j,i-1} + 2s_{j,i} - s_{j,i+1}$$

$$= -\left(1 - \frac{i-1}{m+1}\right)j + 2\left(1 - \frac{i}{m+1}\right)j - \left(1 - \frac{i+1}{m+1}\right)j$$

$$= -j + 2j - j + j \cdot \frac{i - 1 - 2i + i + 1}{m+1} = 0,$$

$$(\boldsymbol{ST})_{i,i} = \sum_{k=1}^{m} s_{i,k}\boldsymbol{T}_{k,i} = -s_{i,i-1} + 2s_{i,i} - s_{i,i+1}$$

$$= -\left(1 - \frac{i}{m+1}\right)(i-1) + 2\left(1 - \frac{i}{m+1}\right)i - \left(1 - \frac{i+1}{m+1}\right)i = 1,$$

which means that $\boldsymbol{ST} = \boldsymbol{I}$. Moreover, since $\boldsymbol{S}, \boldsymbol{T}$, and $\boldsymbol{I}$ are symmetric, transposing this equation yields $\boldsymbol{TS} = \boldsymbol{I}$. We conclude that $\boldsymbol{S} = \boldsymbol{T}^{-1}$.

Exercises section 2.5

Exercise 2.16: Matrix element as a quadratic form

For any matrix $\boldsymbol{A}$ show that $a_{ij} = \boldsymbol{e}_i^{\mathrm{T}}\boldsymbol{A}\boldsymbol{e}_j$ for all i, j.

Solution. Write $\boldsymbol{A} = (a_{ij})_{ij}$ and $\boldsymbol{e}_i = (\delta_{ki})_k$, where

$$\delta_{ki} = \begin{cases} 1 \text{ if } i = k, \\ 0 \text{ otherwise}, \end{cases}$$

is the Kronecker delta. Then, by the definition of the matrix product,

$$\boldsymbol{e}_i^{\mathrm{T}}\boldsymbol{A}\boldsymbol{e}_j = \boldsymbol{e}_i^{\mathrm{T}}(\boldsymbol{A}\boldsymbol{e}_j) = \boldsymbol{e}_i^{\mathrm{T}}\left(\sum_k a_{lk}\delta_{kj}\right)_l = \boldsymbol{e}_i^{\mathrm{T}}(a_{lj})_l = \sum_l \delta_{il}a_{lj} = a_{ij}.$$

Exercise 2.17: Outer product expansion of a matrix

For any matrix $\boldsymbol{A} \in \mathbb{C}^{m\times n}$ show that $\boldsymbol{A} = \sum_{i=1}^{m}\sum_{j=1}^{n} a_{ij}\boldsymbol{e}_i\boldsymbol{e}_j^{\mathrm{T}}$.

Solution. Clearly $\boldsymbol{e}_i\boldsymbol{e}_j^{\mathrm{T}}$ is the matrix $\boldsymbol{E}_{i,j}$ with 1 at entry (i, j) and zero elsewhere. Clearly also $\boldsymbol{A} = \sum_{i,j} a_{i,j}\boldsymbol{E}_{i,j} = \sum_{i,j} a_{i,j}\boldsymbol{e}_i\boldsymbol{e}_j^{\mathrm{T}}$.

Exercise 2.18: The product $A^{\mathrm{T}}A$

Let $\boldsymbol{B} = \boldsymbol{A}^{\mathrm{T}}\boldsymbol{A}$. Explain why this product is defined for any matrix $\boldsymbol{A}$. Show that $b_{ij} = \boldsymbol{a}_{:i}^{\mathrm{T}}\boldsymbol{a}_{:j}$ for all i, j.

Solution. A matrix product is defined as long as the dimensions of the matrices are compatible. More precisely, for the matrix product $\boldsymbol{AB}$ to be defined, the number of columns in $\boldsymbol{A}$ must equal the number of rows in $\boldsymbol{B}$.

Let now $\boldsymbol{A}$ be an $n \times m$ matrix. Then $\boldsymbol{A}^{\mathrm{T}}$ is an $m \times n$ matrix, and as a consequence the product $\boldsymbol{B} := \boldsymbol{A}^{\mathrm{T}}\boldsymbol{A}$ is well defined. Moreover, the (i, j)-th entry of $\boldsymbol{B}$ is given by

$$(\boldsymbol{B})_{ij} = \left(\boldsymbol{A}^{\mathrm{T}}\boldsymbol{A}\right)_{ij} = \sum_{k=1}^{n} a_{ki}a_{kj} = \boldsymbol{a}_{.i}^{\mathrm{T}}\boldsymbol{a}_{.j} = \langle \boldsymbol{a}_{.i}, \boldsymbol{a}_{.j} \rangle,$$

which is what needed to be shown.

Exercise 2.19: Outer product expansion

For $\boldsymbol{A} \in \mathbb{R}^{m \times n}$ and $\boldsymbol{B} \in \mathbb{R}^{p \times n}$ show that

$$\boldsymbol{AB}^{\mathrm{T}} = \boldsymbol{a}_{:1}\boldsymbol{b}_{:1}^{\mathrm{T}} + \boldsymbol{a}_{:2}\boldsymbol{b}_{:2}^{\mathrm{T}} + \cdots + \boldsymbol{a}_{:n}\boldsymbol{b}_{:n}^{\mathrm{T}}.$$

This is called the *outer product expansion* of the columns of $\boldsymbol{A}$ and $\boldsymbol{B}$.

Solution. Recall that the matrix product of $\boldsymbol{A} \in \mathbb{R}^{m,n}$ and $\boldsymbol{B}^{\mathrm{T}} = \boldsymbol{C} \in \mathbb{R}^{n,p}$ is defined by

$$(\boldsymbol{AC})_{ij} = \sum_{k=1}^{n} a_{ik}c_{kj} = \sum_{k=1}^{n} a_{ik}b_{jk}.$$

For the outer product expansion of the columns of $\boldsymbol{A}$ and $\boldsymbol{B}$, on the other hand, we find $\left(\boldsymbol{a}_{:k}\boldsymbol{b}_{:k}^{\mathrm{T}}\right)_{ij} = a_{ik}b_{jk}$. It follows that

$$\left(\boldsymbol{AB}^{\mathrm{T}}\right)_{ij} = \sum_{k=1}^{n} a_{ik}b_{jk} = \sum_{k=1}^{n} \left(\boldsymbol{a}_{:k}\boldsymbol{b}_{:k}^{\mathrm{T}}\right)_{ij}.$$

Exercise 2.20: System with many right hand sides; compact form

Suppose $\boldsymbol{A} \in \mathbb{R}^{m \times n}$, $\boldsymbol{B} \in \mathbb{R}^{m \times p}$, and $\boldsymbol{X} \in \mathbb{R}^{n \times p}$. Show that

$$\boldsymbol{AX} = \boldsymbol{B} \iff \boldsymbol{Ax}_{:j} = \boldsymbol{b}_{:j},\ j = 1, \ldots, p.$$

Solution. Let $\boldsymbol{A}, \boldsymbol{B}$, and $\boldsymbol{X}$ be as in the exercise.

($\Longrightarrow$): Suppose $\boldsymbol{AX} = \boldsymbol{B}$. Multiplying this equation from the right by $\boldsymbol{e}_j$ yields $\boldsymbol{Ax}_{:j} = \boldsymbol{b}_{:j}$ for $j = 1, \ldots, p$.

($\Longleftarrow$): Suppose $\boldsymbol{A}\boldsymbol{x}_{:j} = \boldsymbol{b}_{:j}$ for $j = 1, \ldots, p$. Let $\boldsymbol{I}$ denote the identity matrix of size p. Then

$$\begin{aligned}\boldsymbol{AX} &= \boldsymbol{AXI} = \boldsymbol{AX}[\boldsymbol{e}_1, \ldots, \boldsymbol{e}_p] = [\boldsymbol{AXe}_1, \ldots, \boldsymbol{AXe}_p] \\ &= [\boldsymbol{Ax}_{:1}, \ldots, \boldsymbol{Ax}_{:p}] = [\boldsymbol{b}_{:1}, \ldots, \boldsymbol{b}_{:p}] = \boldsymbol{B}.\end{aligned}$$

Exercise 2.21: Block multiplication example

Suppose $\boldsymbol{A} = [\boldsymbol{A}_1, \boldsymbol{A}_2]$ and $\boldsymbol{B} = \begin{bmatrix} \boldsymbol{B}_1 \\ \boldsymbol{0} \end{bmatrix}$. When is $\boldsymbol{AB} = \boldsymbol{A}_1\boldsymbol{B}_1$?

Solution. The product $\boldsymbol{AB}$ of two matrices $\boldsymbol{A}$ and $\boldsymbol{B}$ is defined precisely when the number of columns of $\boldsymbol{A}$ is equal to the number of rows of $\boldsymbol{B}$. For both sides in the equation $\boldsymbol{AB} = \boldsymbol{A}_1\boldsymbol{B}_1$ to make sense, both pairs $(\boldsymbol{A}, \boldsymbol{B})$ and $(\boldsymbol{A}_1, \boldsymbol{B}_1)$ need to be compatible in this way. Conversely, if the number of columns of $\boldsymbol{A}$ equals the number of rows of $\boldsymbol{B}$ and the number of columns of $\boldsymbol{A}_1$ equals the number of rows of $\boldsymbol{B}_1$, then there exists integers m, p, n, and s with $1 \leq s \leq p$ such that

$$\boldsymbol{A} \in \mathbb{C}^{m,p}, \ \boldsymbol{B} \in \mathbb{C}^{p,n}, \ \boldsymbol{A}_1 \in \mathbb{C}^{m,s}, \ \boldsymbol{A}_2 \in \mathbb{C}^{m,p-s}, \ \boldsymbol{B}_1 \in \mathbb{C}^{s,n}.$$

Then

$$(\boldsymbol{AB})_{ij} = \sum_{k=1}^{p} a_{ik}b_{kj} = \sum_{k=1}^{s} a_{ik}b_{kj} + \sum_{k=s+1}^{p} a_{ik} \cdot 0 = (\boldsymbol{A}_1\boldsymbol{B}_1)_{ij}.$$

Exercise 2.22: Another block multiplication example

Suppose $\boldsymbol{A}, \boldsymbol{B}, \boldsymbol{C} \in \mathbb{R}^{n \times n}$ are given in block form by

$$\boldsymbol{A} := \begin{bmatrix} \lambda & \boldsymbol{a}^{\mathrm{T}} \\ 0 & \boldsymbol{A}_1 \end{bmatrix}, \qquad \boldsymbol{B} := \begin{bmatrix} 1 & \boldsymbol{0}^{\mathrm{T}} \\ \boldsymbol{0} & \boldsymbol{B}_1 \end{bmatrix}, \qquad \boldsymbol{C} := \begin{bmatrix} 1 & \boldsymbol{0}^{\mathrm{T}} \\ \boldsymbol{0} & \boldsymbol{C}_1 \end{bmatrix},$$

where $\boldsymbol{A}_1, \boldsymbol{B}_1, \boldsymbol{C}_1 \in \mathbb{R}^{(n-1)\times(n-1)}$. Show that

$$\boldsymbol{CAB} = \begin{bmatrix} \lambda & \boldsymbol{a}^{\mathrm{T}}\boldsymbol{B}_1 \\ \boldsymbol{0} & \boldsymbol{C}_1\boldsymbol{A}_1\boldsymbol{B}_1 \end{bmatrix}.$$

Solution. Since the matrices have compatible dimensions, a direct computation gives

$$\boldsymbol{CAB} = \begin{bmatrix} 1 & \boldsymbol{0}^{\mathrm{T}} \\ \boldsymbol{0} & \boldsymbol{C}_1 \end{bmatrix} \begin{bmatrix} \lambda & \boldsymbol{a}^{\mathrm{T}} \\ \boldsymbol{0} & \boldsymbol{A}_1 \end{bmatrix} \begin{bmatrix} 1 & \boldsymbol{0}^{\mathrm{T}} \\ \boldsymbol{0} & \boldsymbol{B}_1 \end{bmatrix} = \begin{bmatrix} \lambda & \boldsymbol{a}^{\mathrm{T}} \\ \boldsymbol{0} & \boldsymbol{C}_1\boldsymbol{A}_1 \end{bmatrix} \begin{bmatrix} 1 & \boldsymbol{0}^{\mathrm{T}} \\ \boldsymbol{0} & \boldsymbol{B}_1 \end{bmatrix} = \begin{bmatrix} \lambda & \boldsymbol{a}^{\mathrm{T}}\boldsymbol{B}_1 \\ \boldsymbol{0} & \boldsymbol{C}_1\boldsymbol{A}_1\boldsymbol{B}_1 \end{bmatrix}.$$

Chapter 3
Gaussian Elimination and LU Factorizations

Exercises section 3.3

Exercise 3.1: Column oriented backsolve

Suppose $\boldsymbol{A} \in \mathbb{C}^{n\times n}$ is nonsingular, upper triangular, d-banded, and $\boldsymbol{b} \in \mathbb{C}^n$. Justify Algorithm 3.5 in the book for column oriented backward solving of $\boldsymbol{Ax} = \boldsymbol{b}$.

Solution. If $\boldsymbol{A}$ is upper triangular, suppose that we after $n-k$ steps of the algorithm have reduced our system to one of the form

$$\begin{bmatrix} a_{1,1} & a_{1,2} & \cdots & a_{1,k} \\ 0 & a_{2,1} & \cdots & a_{2,k} \\ \vdots & \vdots & \ddots & \vdots \\ 0 & 0 & \cdots & a_{k,k} \end{bmatrix} \begin{bmatrix} x_1 \\ x_2 \\ \vdots \\ x_k \end{bmatrix} = \begin{bmatrix} b_1 \\ b_2 \\ \vdots \\ b_k \end{bmatrix}$$

Then clearly $x_k = b_k/a_{k,k}$ (this explains the first statement inside the `for`-loop). Eliminating the x_k-variable we obtain the system

$$\begin{bmatrix} a_{1,1} & a_{1,2} & \cdots & a_{1,k-1} \\ 0 & a_{2,1} & \cdots & a_{2,k-1} \\ \vdots & \vdots & \ddots & \vdots \\ 0 & 0 & \cdots & a_{k-1,k-1} \end{bmatrix} \begin{bmatrix} x_1 \\ x_2 \\ \vdots \\ x_{k-1} \end{bmatrix} = \begin{bmatrix} b_1 \\ b_2 \\ \vdots \\ b_{k-1} \end{bmatrix} - x_k \begin{bmatrix} a_{1,k} \\ a_{2,k} \\ \vdots \\ a_{k-1,k} \end{bmatrix}.$$

This means that the right hand side $\boldsymbol{b}$ should be updated by subtracting `A(1:(k-1),k)*x(k)`. If $\boldsymbol{A}$ is d-banded, $A_{1,k} = \cdots = A_{k-d-1,k} = 0$, so that this is the same as subtracting `A(lk:(k-1),k)*x(k)` with `lk` being the maximum of 1 and $k-d$. This explains the second part inside the `for`-loop. Finally we end up with a 1×1-matrix, so to find x_1 we only need to divide by $a_{1,1}$.

T. Lyche et al., *Exercises in Numerical Linear Algebra and Matrix Factorizations*, Texts in Computational Science and Engineering 23, https://doi.org/10.1007/978-3-030-59789-4_3

Exercise 3.2: Computing the inverse of a triangular matrix

Suppose $\boldsymbol{A} \in \mathbb{C}^{n\times n}$ is a nonsingular lower triangular matrix. By Lemma 2.5 the inverse $\boldsymbol{B} = [\boldsymbol{b}_1, \ldots, \boldsymbol{b}_n]$ is also lower triangular. The kth column $\boldsymbol{b}_k$ of $\boldsymbol{B}$ is the solution of the linear systems $\boldsymbol{A}\boldsymbol{b}_k = \boldsymbol{e}_k$. Show that $\boldsymbol{b}_k(k) = 1/a(k,k)$ for $k = 1, \ldots, n$, and explain why we can find $\boldsymbol{b}_k$ by solving the linear systems

$$\boldsymbol{A}\big((k+1){:}n, (k+1){:}n\big)\boldsymbol{b}_k\big((k+1){:}n\big) = -\boldsymbol{A}\big((k+1){:}n, k\big)b_k(k) \quad (3.22)$$

for $k = 1, \ldots, n-1$. Is it possible to store the interesting part of $\boldsymbol{b}_k$ in $\boldsymbol{A}$ as soon as it is computed?

When $\boldsymbol{A}$ instead is upper triangular, show also that we can find $\boldsymbol{b}_k$ by solving the linear systems

$$\boldsymbol{A}(1{:}k, 1{:}k)\boldsymbol{b}_k(1{:}k) = \boldsymbol{I}(1{:}k, k), \quad k = n, n-1, \ldots, 1, \quad (3.23)$$

for $k = n, n-1, \ldots, 1$.

Solution. This exercise introduces an efficient method for computing the inverse $\boldsymbol{B}$ of a triangular matrix $\boldsymbol{A}$. Let $\boldsymbol{b}_k = [b_{1k}, \ldots, b_{nk}]^{\mathrm{T}}$ be column k in $\boldsymbol{B}$.

Let us solve the exercise for an upper triangular matrix (the lower triangular case is similar). By the rules of block multiplication,

$$[\boldsymbol{A}\boldsymbol{b}_1, \ldots, \boldsymbol{A}\boldsymbol{b}_n] = \boldsymbol{A}[\boldsymbol{b}_1, \ldots, \boldsymbol{b}_n] = \boldsymbol{A}\boldsymbol{B} = \boldsymbol{I} = [\boldsymbol{e}_1, \ldots, \boldsymbol{e}_n].$$

The kth column in this matrix equation says that $\boldsymbol{A}\boldsymbol{b}_k = \boldsymbol{e}_k$.

Splitting $\boldsymbol{A}$ into blocks $\begin{bmatrix} \boldsymbol{A}_{11} & \boldsymbol{A}_{12} \\ 0 & \boldsymbol{A}_{22} \end{bmatrix}$ where $\boldsymbol{A}_{11}$ has size $k \times k$ ($\boldsymbol{A}_{11}$ and $\boldsymbol{A}_{22}$ are then upper triangular), we get

$$\begin{bmatrix} \boldsymbol{A}_{11} & \boldsymbol{A}_{12} \\ 0 & \boldsymbol{A}_{22} \end{bmatrix} \left[\begin{array}{c} b_{1k} \\ \vdots \\ b_{kk} \\ \hline b_{(k+1)k} \\ \vdots \\ b_{nk} \end{array}\right] = \left[\begin{array}{c} \boldsymbol{A}_{11} \begin{bmatrix} b_{1k} \\ \vdots \\ b_{kk} \end{bmatrix} + \boldsymbol{A}_{12} \begin{bmatrix} b_{(k+1)k} \\ \vdots \\ b_{nk} \end{bmatrix} \\ \hline \boldsymbol{A}_{22} \begin{bmatrix} b_{(k+1)k} \\ \vdots \\ b_{nk} \end{bmatrix} \end{array}\right] = \begin{bmatrix} \boldsymbol{e}_k \\ 0 \end{bmatrix}.$$

In particular,

$$\boldsymbol{A}_{22} \begin{bmatrix} b_{(k+1)k} \\ \vdots \\ b_{nk} \end{bmatrix} = 0.$$

Since $\boldsymbol{A}_{22}$ is invertible (or use back substitution), it follows that $b_{(k+1)k} = \cdots = b_{nk} = 0$, so that $\boldsymbol{B}$ is upper triangular (this is also stated in Lemma 2.5). It follows that we need to solve

$$
\boldsymbol{A}_{11}\begin{bmatrix} b_{1k} \\ b_{2k} \\ \vdots \\ b_{kk} \end{bmatrix} = \begin{bmatrix} a_{11} & a_{12} & \cdots & a_{1k} \\ 0 & a_{22} & \cdots & a_{2k} \\ \vdots & \vdots & \ddots & \vdots \\ 0 & 0 & \cdots & a_{kk} \end{bmatrix} \begin{bmatrix} b_{1k} \\ b_{2k} \\ \vdots \\ b_{kk} \end{bmatrix} = \boldsymbol{e}_k. \tag{3.i}
$$

This yields (3.23) for solving for the kth column of $\boldsymbol{B}$ (note that the MATLAB notation $\boldsymbol{I}(1:k,k)$ yields $\boldsymbol{e}_k$).

Let us consider the number of arithmetic operations needed to compute the inverse. In finding $\boldsymbol{b}_k$ we need to solve a $k \times k$ triangular system of the form $\boldsymbol{A}_{11}\boldsymbol{x} = \boldsymbol{e}_k$. Solving for x_1 we need to compute $k-1$ multiplications, $k-2$ additions, and one division. This gives a total number of $2k-2$ arithmetic operations. Solving for x_2 needs $2k-4$ operations, and so on, all the way down to x_{k-1} which needs 2 operations. Solving for $x_k = 1/a_{k,k}$ needs an additional division, so that we need to perform

$$
1 + \sum_{r=1}^{k-1} 2r = 1 + (k-1)k
$$

operations. Since we solve a triangular system for every $1 \le k \le n$, we end up with a total of

$$
\sum_{k=1}^{n}(1+(k-1)k) = n + \sum_{k=1}^{n}(k-1)k = n + \frac{1}{3}(n-1)n(n+1) = \frac{1}{3}n(n^2+2).
$$

arithmetic operations. Here we used the formulas that are deduced in Exercise 3.3.

Usually we are just interesting in the "leading term" for the number of operations (here $n^3/3$). This can be obtained more simply by approximating the sums with integrals as in the book: solving the $k \times k$ triangular system can be solved in $1 + \sum_{r=1}^{k-1} 2r \approx \int_{r=1}^{k-1} 2r \approx (k-1)^2 \approx k^2$ operations, and adding together the number of operations for all k we obtain $\sum_{k=1}^{n} k^2 \approx \int_{k=1}^{n} k^2 dk \approx n^3/3$ operations.

Performing this block multiplication for $k = n, n-1, \ldots, 1$, we see that the computations after step k only use the first $k-1$ leading principal submatrices of $\boldsymbol{A}$. It follows that the column $\boldsymbol{b}_k$ computed at step k can be stored in row (or column) k of $\boldsymbol{A}$ without altering the remaining computations. An implementation which stores the inverse (in-place) in $\boldsymbol{A}$ can thus look as follows:

code/exlowertrianginv.m

```
n = 8;
A = rand(n);
A = triu(A);
U=A;
for k=n:-1:1
    U(k,k) = 1/U(k,k);
```

```
    for r=k-1:-1:1
        U(r, k) = -U(r,r+1:k)*U(r+1:k,k)/U(r,r);
    end
end
U*A
```

Listing 3.1: Compute the inverse of a triangular matrix.

In the code, `r` and `k` are row- and column indices, respectively. Inside the `for`-loop we compute x_r for the system (3.i). The contribution from $x_{r+1}, \ldots, x_k$ can be written as a dot product, which here is computed as a matrix product (the minus sign comes from that we isolate x_r on the left hand side). Note that k goes from n and downwards. If we did this the other way we would overwrite matrix entries needed for later calculations.

Exercise 3.3: Finite sums of integers

Use induction on m, or some other method, to show that

$$1 + 2 + \cdots + m = \frac{1}{2}m(m+1), \tag{3.24}$$

$$1^2 + 2^2 + \cdots + m^2 = \frac{1}{3}m(m+\frac{1}{2})(m+1), \tag{3.25}$$

$$1 + 3 + 5 + \cdots + 2m - 1 = m^2, \tag{3.26}$$

$$1 \cdot 2 + 2 \cdot 3 + 3 \cdot 4 + \cdots + (m-1)m = \frac{1}{3}(m-1)m(m+1). \tag{3.27}$$

Solution. There are many ways to prove these identities. The quickest is perhaps by induction. We choose instead an approach based on what is called a generating function. This approach does not assume knowledge of the sum-expressions we want to derive, and the approach also works in a wide range of other circumstances.

It is easily checked that the identities hold for $m = 1, 2, 3$. So let $m \geq 4$ and define

$$P_m(x) := 1 + x + \cdots + x^m = \frac{1 - x^{m+1}}{1 - x}.$$

Then

$$P'_m(x) = \frac{1 - (m+1)x^m + mx^{m+1}}{(x-1)^2},$$

$$P''_m(x) = \frac{-2 + (m^2+m)x^{m-1} + 2(1-m^2)x^m + (m^2-m)x^{m+1}}{(x-1)^3}.$$

Applying l'Hôpital's rule twice, we find

$$\begin{aligned} 1 + 2 + \cdots + m &= P'_m(1) \\ &= \lim_{x \to 1} \frac{1 - (m+1)x^m + mx^{m+1}}{(x-1)^2} \end{aligned}$$

$$= \lim_{x \to 1} \frac{-m(m+1)x^{m-1} + m(m+1)x^m}{2(x-1)}$$
$$= \frac{1}{2}m(m+1),$$

establishing Equation (3.24). In addition it follows that

$$1+3+\cdots+2m-1 = \sum_{k=1}^{m}(2k-1) = -m + 2\sum_{k=1}^{m} k = -m + m(m+1) = m^2,$$

which establishes Equation (3.26). Next, applying l'Hôpital's rule three times, we find that

$$1 \cdot 2 + 2 \cdot 3 + \cdots + (m-1) \cdot m = P_m''(1)$$

is equal to

$$\lim_{x \to 1} \frac{-2 + (m^2+m)x^{m-1} + 2(1-m^2)x^m + (m^2-m)x^{m+1}}{(x-1)^3}$$
$$= \lim_{x \to 1} \frac{(m-1)(m^2+m)x^{m-2} + 2m(1-m^2)x^{m-1} + (m+1)(m^2-m)x^m}{3(x-1)^2}$$
$$= \lim_{x \to 1} (m-1)m(m+1)\frac{(m-2)x^{m-3} - 2(m-1)x^{m-2} + mx^{m-1}}{6(x-1)}$$
$$= \frac{1}{3}(m-1)m(m+1),$$

establishing Equation (3.27). Finally,

$$1^2 + 2^2 + \cdots + m^2 = \sum_{k=1}^{m} k^2 = \sum_{k=1}^{m} ((k-1)k + k) = \sum_{k=1}^{m}(k-1)k + \sum_{k=1}^{m} k$$
$$= \frac{1}{3}(m-1)m(m+1) + \frac{1}{2}m(m+1)$$
$$= \frac{1}{3}(m+1)(m+\frac{1}{2})m,$$

which establishes Equation (3.25).

Exercise 3.4: Multiplying triangular matrices

Show that the matrix multiplication $\boldsymbol{AB}$ can be done in $\frac{1}{3}n(2n^2+1) \approx G_n$ arithmetic operations when $\boldsymbol{A} \in \mathbb{R}^{n\times n}$ is lower triangular and $\boldsymbol{B} \in \mathbb{R}^{n\times n}$ is upper triangular. What about $\boldsymbol{BA}$?

Solution. Computing the (i,j)-th entry of the matrix $\boldsymbol{AB}$ amounts to computing the inner product of the ith row $\boldsymbol{a}_{i:}^{\mathrm{T}}$ of $\boldsymbol{A}$ and the jth column $\boldsymbol{b}_{:j}$ of $\boldsymbol{B}$. Because of the triangular nature of $\boldsymbol{A}$ and $\boldsymbol{B}$, only the first i entries of $\boldsymbol{a}_{i:}^{\mathrm{T}}$ can be nonzero and only the first j entries of $\boldsymbol{b}_{:j}$ can be nonzero. The computation $\boldsymbol{a}_{i:}^{\mathrm{T}}\boldsymbol{b}_{:j}$ therefore involves $\min\{i,j\}$ multiplications and $\min\{i,j\}-1$ additions. Carrying out this calculation for all i and j, and using the identities from Exercise 3.3, yields a total number of

$$\begin{aligned}
&\sum_{i=1}^{n}\sum_{j=1}^{n}(2\min\{i,j\}-1)\\
&=\sum_{i=1}^{n}\left(\sum_{j=1}^{i}(2j-1)+\sum_{j=i+1}^{n}(2i-1)\right)=\sum_{i=1}^{n}\left(i^2+(n-i)(2i-1)\right)\\
&=\sum_{i=1}^{n}\left(-i^2+2ni-n+i\right)=-n^2+(2n+1)\sum_{i=1}^{n}i-\sum_{i=1}^{n}i^2\\
&=-n^2+\frac{1}{2}n(n+1)(2n+1)-\frac{1}{6}n(n+1)(2n+1)\\
&=-n^2+\frac{1}{3}n(n+1)(2n+1)=\frac{2}{3}n^3+\frac{1}{3}n=\frac{1}{3}n(2n^2+1)
\end{aligned}$$

arithmetic operations. A similar calculation gives the same result for the product $\boldsymbol{BA}$.

Exercises section 3.4

Exercise 3.5: Using PLU for $\boldsymbol{A}^*$

Suppose we know the PLU factors $\boldsymbol{P},\boldsymbol{L},\boldsymbol{U}$ in a PLU factorization $\boldsymbol{A}=\boldsymbol{PLU}$ of $\boldsymbol{A}\in\mathbb{C}^{n\times n}$. Explain how we can solve the system $\boldsymbol{A}^*\boldsymbol{x}=\boldsymbol{b}$ economically.

Solution. If $\boldsymbol{A}=\boldsymbol{PLR}$, then $\boldsymbol{A}^*=\boldsymbol{R}^*\boldsymbol{L}^*\boldsymbol{P}^*$. The matrix $\boldsymbol{L}^*$ is upper triangular and the matrix $\boldsymbol{R}^*$ is lower triangular, implying that $\boldsymbol{R}^*\boldsymbol{L}^*$ is an LU factorization of $\boldsymbol{A}^*\boldsymbol{P}$. Since $\boldsymbol{A}$ is nonsingular, the matrix $\boldsymbol{R}^*$ must be nonsingular, and we can apply Algorithms 3.1 and 3.2 to economically solve the systems $\boldsymbol{R}^*\boldsymbol{z}=\boldsymbol{b}$, $\boldsymbol{L}^*\boldsymbol{y}=\boldsymbol{z}$, and $\boldsymbol{P}^*\boldsymbol{x}=\boldsymbol{y}$, to find a solution $\boldsymbol{x}$ to the system $\boldsymbol{R}^*\boldsymbol{L}^*\boldsymbol{P}^*\boldsymbol{x}=\boldsymbol{A}^*\boldsymbol{x}=\boldsymbol{b}$.

Exercise 3.6: Using PLU for determinant

Suppose we know the PLU factors $\boldsymbol{P},\boldsymbol{L},\boldsymbol{U}$ in a PLU factorization $\boldsymbol{A}=\boldsymbol{PLU}$ of $\boldsymbol{A}\in\mathbb{C}^{n\times n}$. Explain how we can use this to compute the determinant of $\boldsymbol{A}$.

Solution. If $\boldsymbol{A}=\boldsymbol{PLU}$, then

$$\det(\boldsymbol{A})=\det(\boldsymbol{PLU})=\det(\boldsymbol{P})\det(\boldsymbol{L})\det(\boldsymbol{U})$$

and the determinant of $\boldsymbol{A}$ can be computed from the determinants of $\boldsymbol{P}$, $\boldsymbol{L}$, and $\boldsymbol{U}$. Since the latter two matrices are triangular, their determinants are simply the products of their diagonal entries. The matrix $\boldsymbol{P}$, on the other hand, is a permutation matrix, so that every row and column is everywhere 0, except for a single entry (where it is 1). Its determinant is therefore quickly computed by cofactor expansion.

Exercise 3.7: Using PLU for $\boldsymbol{A}^{-1}$

Suppose the factors $\boldsymbol{P}, \boldsymbol{L}, \boldsymbol{U}$ in a PLU factorization of $\boldsymbol{A} \in \mathbb{C}^{n\times n}$ are known. Use Exercise 3.4 to show that it takes approximately $2G_n$ arithmetic operations to compute $\boldsymbol{A}^{-1} = \boldsymbol{U}^{-1}\boldsymbol{L}^{-1}\boldsymbol{P}^{\mathrm{T}}$. Here we have not counted the final multiplication with $\boldsymbol{P}^{\mathrm{T}}$ which amounts to n row interchanges.

Solution. It is clear from the `rforwardsolve` and `rbacksolve` algorithms that solving an $n \times n$-triangular system takes n^2 operations. From Exercise 3.2 it is thus clear that inverting an upper/lower triangular matrix takes $\sum_{k=1}^{n} k^2 \approx n^3/3$ operations (see Exercise 3.3). Inverting both $\boldsymbol{L}$ and $\boldsymbol{U}$ thus takes $2n^3/3 \approx G_n$ operations. According to Exercise 3.4, it takes approximately G_n arithmetic operations to multiply an upper and a lower triangular matrix. It thus takes approximately $G_n + G_n = 2G_n$ operations to compute $\boldsymbol{U}^{-1}\boldsymbol{L}^{-1}$.

Exercise 3.8: Upper Hessenberg system (Exam exercise (1994-2)

Gaussian elimination with row pivoting can be written in the following form if for each k we exchange rows k and $k+1$:

Algorithm 1

1. for $k = 1, 2, \ldots, n-1$
   (a) exchange $a_{k,j}$ and $a_{k+1,j}$ for $j = k, k+1, \ldots, n$
   (b) for $i = k+1, k+2, \ldots, n$
      i. $a_{i,k} = m_{i,k} := a_{i,k}/a_{k,k}$
      ii. $a_{i,j} = a_{i,j} - m_{i,k}a_{k,j}$ for $j = k+1, k+2, \ldots, n$

To solve the set of equations $\boldsymbol{Ax} = \boldsymbol{b}$ we have the following algorithm:

Algorithm 2

1. for $k = 1, 2, \ldots, n-1$
   (a) exchange b_k and b_{k+1}
   (b) $b_i = b_i - a_{i,k}b_k$ for $i = k+1, k+2, \ldots, n$
2. $x_n = b_n/a_{n,n}$
3. for $k = n-1, n-2, \ldots, 1$
   (a) $sum = 0$
   (b) $sum = sum + a_{k,j}x_j$ for $j = k+1, k+2, \ldots, n$
   (c) $x_k = (b_k - sum)/a_{k,k}$

We say that $\boldsymbol{H} \in \mathbb{R}^{n\times n}$ is *unreduced upper Hessenberg* if it is upper Hessenberg and the subdiagonal elements $h_{i,i-1} \neq 0$ for $i = 2, \ldots, n$.

a) Let $\boldsymbol{H} \in \mathbb{R}^{n\times n}$ be unreduced upper Hessenberg. Give an $\mathcal{O}(n^2)$ algorithm for solving the linear system $\boldsymbol{Hx} = \boldsymbol{b}$ using suitable specializations of Algorithms 1 and 2.

Solution. Note that the `rbacksolve` algorithm can be used for the second part, and there is no need for storing the multipliers: They can simply be applied directly to the augmented matrix. MATLAB code corresponding to the pseudocode above can thus take the following simplified form.

code/solve_row_pivot.m

```
function x = solve_row_pivot(A,b)
  n = size(A,1);
  A = [A b];
  for k=1:(n-1)
    A([k k+1],k:end) = A([k+1 k],k:end);
```

```
        for i=(k+1):n
            m = A(i,k)/A(k,k);
            A(i,(k+1):end) = A(i,(k+1):end) - m*A(k,(k+1):end);
        end
    end
    x = rbacksolve(A(:,1:n),A(:,n+1),n);
end
```

Listing 3.2: Solving a linear system with row pivoting.

MATLAB code when an unreduced upper Hessenberg matrix is assumed can be as follows.

code/solve_row_pivot_hess.m

```
function x = solve_row_pivot_hess(H,b)
    n = size(H, 1);
    H = [H b];
    for k=1:(n-1)
        H([k k+1],k:end) = H([k+1 k],k:end);
        m = H(k+1,k)/H(k,k);
        H(k+1,(k+1):end)=H(k+1,(k+1):end) - m*H(k,(k+1):end);
    end
    x = rbacksolve(H(:,1:n),H(:,n+1),n);
end
```

Listing 3.3: Solving a linear system with row pivoting, assuming the matrix has the Hessenberg form.

We can test both functions as follows.

code/solve_row_pivot_test.m

```
n = 10;
A = rand(n);
b = rand(n,1);
x = solve_row_pivot(A,b);
max(abs(A*x-b))

for k=3:n
    for l=1:(k-2)
        A(k,l) = 0;
    end
end
x = solve_row_pivot_hess(A,b);
max(abs(A*x-b))
```

Listing 3.4: Test code for Listings 3.2 and 3.3.

b) Find the number of multiplications/divisions in the algorithm you developed in exercise **a**). Is division by zero possible?

Solution. Number of multiplications/divisions:

$$\text{Algorithm 1H: } \sum_{k=1}^{n-1}\left(1+\sum_{j=k+1}^{n}1\right)=\sum_{k=1}^{n-1}(n-k+1)=\frac{1}{2}(n-1)(n+2),$$

$$\text{Algorithm 2H: } \sum_{k=1}^{n-1}1+1+\sum_{k=1}^{n-1}\left(\sum_{j=k+1}^{n}1+1\right)=\frac{1}{2}n^2+\frac{3}{2}n-1.$$

The total for the two algorithms is n^2+2n-2.

In Algorithm 1H division by zero will not happen because the element $h_{k+1,k}$ in 1(b), which becomes $h_{k,k}$ in 1(a), is the original element in $\boldsymbol{H}$ and therefore nonzero by definition. If $\boldsymbol{H}$ is nonsingular division by zero will not happen in Algorithm 2H. If $\boldsymbol{H}$ is singular then $h_{n,n}=0$, and we get division by zero in statement 2 of Algorithm 2H.

c) Let $\boldsymbol{U}\in\mathbb{R}^{n\times n}$ be upper triangular and nonsingular. We define

$$\boldsymbol{C}:=\boldsymbol{U}+\boldsymbol{v}\boldsymbol{e}_1^{\mathrm{T}}, \tag{3.28}$$

where $\boldsymbol{v}\in\mathbb{R}^n$ and $\boldsymbol{e}_1$ is the first unit vector in $\mathbb{R}^n$. We also let

$$\boldsymbol{P}:=\boldsymbol{I}_{1,2}\boldsymbol{I}_{2,3}\cdots\boldsymbol{I}_{n-1,n}, \tag{3.29}$$

where the $\boldsymbol{I}_{i,j}$ are obtained from the identity matrix by interchanging rows i and j. Explain why the matrix $\boldsymbol{E}:=\boldsymbol{C}\boldsymbol{P}$ is unreduced upper Hessenberg.

Solution. The operation $\boldsymbol{A}\boldsymbol{I}_{k,k+1}$ implies that column k and $k+1$ in $\boldsymbol{A}$ are exchanged. The operation $\boldsymbol{A}\boldsymbol{P}$ means that column j is replaced by column $j-1$ for $j=2,\dots,n$, while column 1 is replaced by column n. We illustrate what happens by a Wilkinson diagram of size 4. Using the symbol $\otimes$ for the diagonal elements in $\boldsymbol{U}$ we find

$$\boldsymbol{C}=\begin{bmatrix}\otimes & x & x & x\\ 0 & \otimes & x & x\\ 0 & 0 & \otimes & x\\ 0 & 0 & 0 & \otimes\end{bmatrix}+\begin{bmatrix}x&0&0&0\\x&0&0&0\\x&0&0&0\\x&0&0&0\end{bmatrix}=\begin{bmatrix}x & x & x & x\\ x & \otimes & x & x\\ x & 0 & \otimes & x\\ x & 0 & 0 & \otimes\end{bmatrix},$$

$$\boldsymbol{C}\boldsymbol{I}_{12}=\begin{bmatrix}x & x & x & x\\ \otimes & x & x & x\\ 0 & x & \otimes & x\\ 0 & x & 0 & \otimes\end{bmatrix},\quad \boldsymbol{C}\boldsymbol{I}_{12}\boldsymbol{I}_{23}=\begin{bmatrix}x & x & x & x\\ \otimes & x & x & x\\ 0 & \otimes & x & x\\ 0 & 0 & x & \otimes\end{bmatrix}$$

Finally,

$$\boldsymbol{C}\boldsymbol{I}_{12}\boldsymbol{I}_{23}\boldsymbol{I}_{34}=\begin{bmatrix}x & x & x & x\\ \otimes & x & x & x\\ 0 & \otimes & x & x\\ 0 & 0 & \otimes & x\end{bmatrix}$$

is upper Hessenberg. The subdiagonal elements $e_{k+1,k}$ in $\boldsymbol{E}$ are the original diagonal elements in $\boldsymbol{U}$ and therefore nonzero.

d) Let $\boldsymbol{A} \in \mathbb{R}^{n \times n}$ be nonsingular. We assume that $\boldsymbol{A}$ has a unique L1U factorization $\boldsymbol{A} = \boldsymbol{LU}$. For a given $\boldsymbol{W} \in \mathbb{R}^n$ we define a rank one modification of $\boldsymbol{A}$ by

$$\boldsymbol{B} := \boldsymbol{A} + \boldsymbol{w}\boldsymbol{e}_1^{\mathrm{T}}. \tag{3.30}$$

Show that $\boldsymbol{B}$ has the factorization $\boldsymbol{B} = \boldsymbol{LHP}^{\mathrm{T}}$, where $\boldsymbol{L}$ is unit lower triangular, $\boldsymbol{P}$ is given by (3.29) and $\boldsymbol{H}$ is unreduced upper Hessenberg.

Solution. We have

$$\boldsymbol{B} = \boldsymbol{A} + \boldsymbol{w}\boldsymbol{e}_1^{\mathrm{T}} = \boldsymbol{LU} + \boldsymbol{w}\boldsymbol{e}_1^{\mathrm{T}} = \boldsymbol{L}(\boldsymbol{U} + (\boldsymbol{L}^{-1}\boldsymbol{w})\boldsymbol{e}_1^{\mathrm{T}})$$

and therefore $\boldsymbol{B} = \boldsymbol{LC}$, where $\boldsymbol{C}$ is of the form (3.28) with $\boldsymbol{v} = \boldsymbol{L}^{-1}\boldsymbol{w}$. Since $\boldsymbol{A}$ is nonsingular it follows that $\boldsymbol{U}$ is nonsingular. We define $\boldsymbol{H} := \boldsymbol{CP}$. **c)** implies that $\boldsymbol{H}$ is unreduced upper Hessenberg. Since $\boldsymbol{P} = [\boldsymbol{e}_2, \boldsymbol{e}_3, \ldots, \boldsymbol{e}_n, \boldsymbol{e}_1]$ is a permutation matrix it is orthogonal. Therefore $\boldsymbol{C} = \boldsymbol{HP}^{\mathrm{T}}$ and $\boldsymbol{B} = \boldsymbol{LC} = \boldsymbol{LHP}^{\mathrm{T}}$, where $\boldsymbol{H}$ is unreduced upper Hessenberg.

e) Use the results above to sketch an $\mathcal{O}(n^2)$ algorithm for solving the linear system $\boldsymbol{Bx} = \boldsymbol{b}$, where $\boldsymbol{B}$ is given by (3.30). We assume that the matrices $\boldsymbol{L}$ and $\boldsymbol{U}$ in the L1U factorization of $\boldsymbol{A}$ have already been computed.

Solution. We have $\boldsymbol{Bx} = \boldsymbol{b} \iff \boldsymbol{LHP}^{\mathrm{T}}\boldsymbol{x} = \boldsymbol{b}$ giving the following algorithm with the indicated number of multiplications/divisions:

1. Solve $\boldsymbol{Ly} = \boldsymbol{b}$ for $\boldsymbol{y}$ (forward substitution $\mathcal{O}(n^2)$)
2. Solve $\boldsymbol{Lv} = \boldsymbol{w}$ for $\boldsymbol{v}$ (forward substitution $\mathcal{O}(n^2)$)
3. $\boldsymbol{H} := (\boldsymbol{U} + \boldsymbol{v}\boldsymbol{e}_1^{\mathrm{T}})\boldsymbol{P}$ (a number of exchanges)
4. Solve $\boldsymbol{Hz} = \boldsymbol{y}$ for $\boldsymbol{z}$ (from 2b) $\mathcal{O}(n^2)$)
5. $\boldsymbol{x} = \boldsymbol{I}_{1,2} \cdots \boldsymbol{I}_{n-1,n}\boldsymbol{z}$ (a number of exchanges)

Exercises section 3.5

Exercise 3.9: # operations for banded triangular systems

Show that for $1 \le d \le n$ Algorithm 3.4, with `A(k,k)=1` for $k = 1, \ldots, n$ in Algorithm 3.1, requires exactly $N_{LU}(n,d) := (2d^2+d)n - (d^2+d)(8d+1)/6 = \mathcal{O}(d^2n)$ operations. In particular, for a full matrix $d = n-1$ and we find $N_{LU}(n,n) = \frac{2}{3}n^3 - \frac{1}{2}n^2 - \frac{1}{6}n \approx G_n$ in agreement with the exact count (3.9) for Gaussian elimination, while for a tridiagonal matrix $N_{LU}(n,1) = 3n-3 = \mathcal{O}(n)$.

Hint.
Consider the cases $2 \le k \le d$ and $d+1 \le k \le n$ separately.

Solution. Assume first that $\boldsymbol{A}$ is a full (i.e., with no assumption on the bandwidth), lower triangular matrix. According to Equation (3.8) the number of operations for a given k is $2k-1$ (since $l_k = 1$ in the case of a full matrix), and the total number of operations in `rforwardsolve` is thus

$$1 + \sum_{k=2}^{n}(2k-1) = n^2.$$

If $\boldsymbol{A}$ also has ones on the diagonal, as in the text of the exercise, the number of operations reduces to

$$\sum_{k=2}^{n}(2k-2) = 2\sum_{k=1}^{n-1} k = n^2 - n,$$

since divisions with one are not counted.

Algorithm 3.4 reduces the problem to smaller, full, triangular systems, and can be split into the cases

- $2 \le k \le d$: forwardsolves of size $k-1$ and k (the last with ones on the diagonal).
- $d+1 \le k \le n$: forwardsolves of size d ad $d+1$ (the last with ones on the diagonal).

Using the above counts n^2 and n^2-n, these total a number of operations of

$$\begin{aligned}
&\sum_{k=2}^{d}((k-1)^2 + k^2 - k) + \sum_{k=d+1}^{n}(d^2 + (d+1)^2 - (d+1)) \\
&= 2\sum_{k=2}^{d} k^2 - 3\sum_{k=2}^{d} k + d - 1 + (n-d)(2d^2+d) \\
&= \frac{d(d+1)(2d+1)}{3} - 2 - \frac{3}{2}d(d+1) + 3 + d - 1 - d^2(2d+1) + n(2d^2+d)
\end{aligned}$$

$$= \frac{d(d+1)(2d+1)}{3} - \frac{3}{2}d(d+1) + d - d^2(2d+1) + n(2d^2+d)$$
$$= \frac{d(d+1)(2d+1)}{3} - \frac{3}{2}d(d+1) - d(d+1)(2d-1) + n(2d^2+d)$$
$$= \frac{(d^2+d)(4d+2-9-6(2d-1))}{6} + n(2d^2+d)$$
$$= \frac{(d^2+d)(-8d-1)}{6} + n(2d^2+d) = (2d^2+d)n - (d^2+d)(8d+1)/6.$$

Exercise 3.10: L1U and LU1

Show that the matrix $\boldsymbol{A}_3$ in Example 3.5 has no LU1 or LDU factorization. Give an example of a matrix that has an LU1 factorization, but no LDU or L1U factorization.

Solution. Suppose the matrix $\boldsymbol{A}_3$ has an LU1 factorization, i.e.,

$$\boldsymbol{A}_3 = \begin{bmatrix} 0 & 1 \\ 0 & 2 \end{bmatrix} = \begin{bmatrix} l_1 & 0 \\ l_2 & l_3 \end{bmatrix} \begin{bmatrix} 1 & u_1 \\ 0 & 1 \end{bmatrix} = \begin{bmatrix} l_1 & l_1 u_1 \\ l_2 & l_2 u_1 + l_3 \end{bmatrix}.$$

Comparing the first column gives $l_1 = l_2 = 0$. This implies that $l_1 u_1 = 0$. Since $a_{12} \neq 0$, this gives a contradiction. If $\boldsymbol{LDU}$ was an LDU factorization, then $(\boldsymbol{LD})\boldsymbol{U}$ is an LU1 factorization, so we can't have an LDU factorization either.

Thus $\boldsymbol{A}_3$ has an L1U factorization, but no LU1 factorization. If $\boldsymbol{A}_3 = \boldsymbol{LU}$ is an L1U factorization, then $\boldsymbol{A}_3^{\mathrm{T}} = \boldsymbol{U}^{\mathrm{T}}\boldsymbol{L}^{\mathrm{T}}$ is an LU1 factorization of $\boldsymbol{A}_3^{\mathrm{T}}$. It follows that $\boldsymbol{A}_3^{\mathrm{T}}$ has an LU1 factorization. It clearly can have no L1U factorization, since otherwise $\boldsymbol{A}_3$ would have an LU1 factorization, which we have have proved is not the case.

Exercise 3.11: LU of nonsingular matrix

Show that the following are equivalent for a nonsingular matrix $\boldsymbol{A} \in \mathbb{C}^{n \times n}$.

1. $\boldsymbol{A}$ has an LDU factorization.
2. $\boldsymbol{A}$ has an L1U factorization.
3. $\boldsymbol{A}$ has an LU1 factorization.

Solution. Note first that, since $\boldsymbol{A}$ is nonsingular, any diagonal entries in $\boldsymbol{L}$, $\boldsymbol{D}$, and $\boldsymbol{U}$ in such factorizations must be nonzero. If $\boldsymbol{L}$ is lower triangular we can write $\boldsymbol{L} = \boldsymbol{L}'\boldsymbol{D}'$ where $(\boldsymbol{L}')_{ij} = l_{ij}/l_{jj}$, and $(\boldsymbol{D}')_{ii}$ the diagonal matrix with entries $(\boldsymbol{D}')_{ii} = l_{ii}$. Since $\boldsymbol{L}'$ is lower triangular with ones on the diagonal, it follows that an LU1 factorization implies an LDU factorization. It follows similarly that an LU1 factorization implies an LDU factorization.

If $\boldsymbol{LDU}$ is an LDU factorization, $(\boldsymbol{LD})\boldsymbol{U}$ is an LU1 factorization, and $\boldsymbol{L}(\boldsymbol{DU})$ is an L1U factorization. The result follows.

Exercise 3.12: Row interchange

Show that $\boldsymbol{A} = \left[\begin{smallmatrix}1&1\\0&1\end{smallmatrix}\right]$ has a unique L1U factorization. Note that we have only interchanged rows in Example 3.5.

Solution. Suppose we are given an LU factorization

$$\begin{bmatrix}1&1\\0&1\end{bmatrix} = \begin{bmatrix}1&0\\l_{21}&1\end{bmatrix}\begin{bmatrix}u_{11}&u_{12}\\0&u_{22}\end{bmatrix}.$$

Carrying out the matrix multiplication on the right hand side, one finds that

$$\begin{bmatrix}1&1\\0&1\end{bmatrix} = \begin{bmatrix}u_{11}&u_{12}\\l_{21}u_{11}&l_{21}u_{12}+u_{22}\end{bmatrix},$$

implying that $u_{11} = u_{12} = 1$. It follows that necessarily $l_{21} = 0$ and $u_{22} = 1$, and the pair

$$\boldsymbol{L} = \begin{bmatrix}1&0\\0&1\end{bmatrix}, \qquad \boldsymbol{U} = \begin{bmatrix}1&1\\0&1\end{bmatrix}$$

is the only possible LU factorization of the matrix $\begin{bmatrix}1&1\\0&1\end{bmatrix}$. One directly checks that this is indeed an LU factorization.

Exercise 3.13: LU and determinant

Suppose $\boldsymbol{A}$ has an L1U factorization $\boldsymbol{A} = \boldsymbol{L}\boldsymbol{U}$. Show that $\det(\boldsymbol{A}_{[k]}) = u_{11}u_{22}\cdots u_{kk}$ for $k = 1,\ldots,n$.

Solution. Suppose $\boldsymbol{A}$ has an LU factorization $\boldsymbol{A} = \boldsymbol{L}\boldsymbol{U}$. Then, by Lemma 3.1, $\boldsymbol{A}_{[k]} = \boldsymbol{L}_{[k]}\boldsymbol{U}_{[k]}$ is an LU factorization of the leading principal submatrices of size $k = 1,\ldots,n$. By induction, the cofactor expansion of the determinant yields that the determinant of a triangular matrix is the product of its diagonal entries. One therefore finds that $\det(\boldsymbol{L}_{[k]}) = 1$, $\det(\boldsymbol{U}_{[k]}) = u_{11}\cdots u_{kk}$ and

$$\det(\boldsymbol{A}_{[k]}) = \det(\boldsymbol{L}_{[k]}\boldsymbol{U}_{[k]}) = \det(\boldsymbol{L}_{[k]})\det(\boldsymbol{U}_{[k]}) = u_{11}\cdots u_{kk}$$

for $k = 1,\ldots,n$.

Exercise 3.14: Diagonal elements in $\boldsymbol{U}$

Suppose $\boldsymbol{A} \in \mathbb{C}^{n\times n}$ and $\boldsymbol{A}_{[k]}$ is nonsingular for $k = 1,\ldots,n-1$. Use Exercise 3.13 to show that the diagonal elements u_{kk} in the L1U factorization are

$$u_{11} = a_{11}, \qquad u_{kk} = \frac{\det(\boldsymbol{A}_{[k]})}{\det(\boldsymbol{A}_{[k-1]})}, \qquad \text{for } k = 2,\ldots,n.$$

Solution. From Exercise 3.13, we know that $\det(\boldsymbol{A}_{[k]}) = u_{11}\cdots u_{kk}$ for $k = 1,\ldots,n$. Since $\boldsymbol{A}$ is nonsingular, its determinant $\det(\boldsymbol{A}) = u_{11}\cdots u_{nn}$ is nonzero. This implies that $\det(\boldsymbol{A}_{[k]}) = u_{11}\cdots u_{kk} \neq 0$ for $k = 1,\ldots,n$, yielding $a_{11} = u_{11}$ for $k = 1$ and a well-defined quotient

$$\frac{\det(\boldsymbol{A}_{[k]})}{\det(\boldsymbol{A}_{[k-1]})} = \frac{u_{1,1}\cdots u_{k-1,k-1}u_{k,k}}{u_{1,1}\cdots u_{k-1,k-1}} = u_{k,k},$$

for $k = 2,\ldots,n$.

Exercise 3.15: Proof of LDU theorem

Give a proof of the LU theorem (Theorem 3.4) for the LDU case.

Solution. Suppose first that the $\{\boldsymbol{A}_{[k]}\}_{k=1}^{n-1}$ are nonsingular. Assume we have shown by induction that all these have unique LDU factorizations $\boldsymbol{A}_{[k]} = \boldsymbol{L}_k\boldsymbol{D}_k\boldsymbol{U}_k$. The $\{\boldsymbol{L}_k, \boldsymbol{D}_k, \boldsymbol{U}_k\}_{k=1}^{n-1}$ must then also be nonsingular. We need to solve

$$\begin{aligned}\begin{bmatrix}\boldsymbol{A}_{[n-1]} & \boldsymbol{c}_n\\ \boldsymbol{r}_n^{\mathrm{T}} & a_{nn}\end{bmatrix} &= \begin{bmatrix}\boldsymbol{L}_{n-1} & \boldsymbol{0}\\ \boldsymbol{l}_n^{\mathrm{T}} & 1\end{bmatrix}\begin{bmatrix}\boldsymbol{D}_{n-1} & \boldsymbol{0}\\ \boldsymbol{0} & d_{nn}\end{bmatrix}\begin{bmatrix}\boldsymbol{U}_{n-1} & \boldsymbol{u}_n\\ \boldsymbol{0} & 1\end{bmatrix}\\ &= \begin{bmatrix}\boldsymbol{L}_{n-1}\boldsymbol{D}_{n-1} & \boldsymbol{0}\\ \boldsymbol{l}_n^{\mathrm{T}}\boldsymbol{D}_{n-1} & d_{nn}\end{bmatrix}\begin{bmatrix}\boldsymbol{U}_{n-1} & \boldsymbol{u}_n\\ \boldsymbol{0} & 1\end{bmatrix}\end{aligned}$$

for $\boldsymbol{l}_n$, d_{nn}, and $\boldsymbol{u}_n$. This is equivalent to solving

$$\begin{aligned}\boldsymbol{c}_n &= \boldsymbol{L}_{n-1}\boldsymbol{D}_{n-1}\boldsymbol{u}_n,\\ \boldsymbol{r}_n^{\mathrm{T}} &= \boldsymbol{l}_n^{\mathrm{T}}\boldsymbol{D}_{n-1}\boldsymbol{U}_{n-1},\\ a_{nn} &= \boldsymbol{l}_n^{\mathrm{T}}\boldsymbol{D}_{n-1}\boldsymbol{u}_n + d_{nn}.\end{aligned}$$

These have unique solutions, so that we obtain a unique LDU factorization (since $\boldsymbol{A}_{[n-1]} = \boldsymbol{L}_{[n-1]}\boldsymbol{D}_{[n-1]}\boldsymbol{U}_{[n-1]}$ gives uniqueness for the $\boldsymbol{L}_{n-1}$, $\boldsymbol{D}_{n-1}$, $\boldsymbol{U}_{n-1}$).

The other way, suppose that $\boldsymbol{A} = \boldsymbol{LDU}$ is a unique LDU factorization, and let k be the smallest number so that $\boldsymbol{A}_{[k]}$ is singular. It follows from the above that $\boldsymbol{A}_{[k]}$ has a unique LDU factorization, which must be $\boldsymbol{L}_{[k]}\boldsymbol{D}_{[k]}\boldsymbol{U}_{[k]}$. In this factorization $\boldsymbol{D}_{[k]}$ must be singular. Copying Lemma 3.1 for an LDU factorization, we get an equation of the form $\boldsymbol{U}_{[k]}^{\mathrm{T}}\boldsymbol{D}_{[k]}\boldsymbol{M}_k^{\mathrm{T}} = \boldsymbol{C}_k^{\mathrm{T}}$ for the lower $(n-k)\times k$ matrix $\boldsymbol{M}_k$ of $\boldsymbol{L}$. Since $\boldsymbol{D}_{[k]}$ is singular there is no unique solution for $\boldsymbol{M}_k$, so that the LDU factorization can not be unique, which is a contradiction.

Exercise 3.16: Proof of LU1 theorem

Give a proof of the LU theorem (Theorem 3.4) for the LU1 case.

Solution. Rather than copying the proof of Theorem 3.4 as we did for the LDU case above, let us deduce the LU1 case from the L1U case directly. Clearly $\boldsymbol{A}$ has a unique LU1 factorization if and only if $\boldsymbol{A}^{\mathrm{T}}$ has a unique L1U factorization, which, according to Theorem 3.4, is the case if and only if the first $n-1$ leading principal submatrices of $\boldsymbol{A}^{\mathrm{T}}$ are nonsingular. But these are $(\boldsymbol{A}_{[k]})^{\mathrm{T}}$, which are nonsingular if and only if the $\boldsymbol{A}_k$ are nonsingular. The result follows.

Exercise 3.17: Computing the inverse (Exam exercise 1978-1)

Let $\boldsymbol{A} \in \mathbb{R}^{n\times n}$ be nonsingular and with a unique L1U factorization $\boldsymbol{A} = \boldsymbol{L}\boldsymbol{U}$. We partition $\boldsymbol{L}$ and $\boldsymbol{U}$ as follows

$$\boldsymbol{L} = \begin{bmatrix} 1 & 0 \\ \boldsymbol{\ell}_1 & \boldsymbol{L}_{2,2} \end{bmatrix}, \qquad \boldsymbol{U} = \begin{bmatrix} u_{1,1} & \boldsymbol{u}_1^{\mathrm{T}} \\ 0 & \boldsymbol{U}_{2,2} \end{bmatrix}, \tag{3.32}$$

where $\boldsymbol{L}_{2,2}, \boldsymbol{U}_{2,2} \in \mathbb{R}^{(n-1)\times(n-1)}$. Define $\boldsymbol{A}_{2,2} := \boldsymbol{L}_{2,2}\boldsymbol{U}_{2,2}$ and $\boldsymbol{B}_{2,2} := \boldsymbol{A}_{2,2}^{-1}$.

a) Show that $\boldsymbol{A}^{-1} = \boldsymbol{B}$, where

$$\boldsymbol{B} := \begin{bmatrix} (1 + \boldsymbol{u}_1^{\mathrm{T}}\boldsymbol{B}_{2,2}\boldsymbol{\ell}_1)/u_{1,1} & -\boldsymbol{u}_1^{\mathrm{T}}\boldsymbol{B}_{2,2}/u_{1,1} \\ -\boldsymbol{B}_{2,2}\boldsymbol{\ell}_1 & \boldsymbol{B}_{2,2} \end{bmatrix}. \tag{3.33}$$

Solution. Since $\boldsymbol{A}$ is nonsingular, $\boldsymbol{L}$ and $\boldsymbol{U}$ must be as well. But then $\boldsymbol{L}_{2,2}$ and $\boldsymbol{U}_{2,2}$ must also be nonsingular. It follows that $\boldsymbol{A}_{2,2} = \boldsymbol{L}_{2,2}\boldsymbol{U}_{2,2}$ also is nonsingular, with $\boldsymbol{B}_{2,2} := \boldsymbol{A}_{2,2}^{-1} = \boldsymbol{U}_{2,2}^{-1}\boldsymbol{L}_{2,2}^{-1}$. From Lemma 2.4 it follows that

$$\begin{bmatrix} \boldsymbol{B} & \boldsymbol{C} \\ \boldsymbol{0} & \boldsymbol{D} \end{bmatrix}^{-1} = \begin{bmatrix} \boldsymbol{B}^{-1} & -\boldsymbol{B}^{-1}\boldsymbol{C}\boldsymbol{D}^{-1} \\ 0 & \boldsymbol{D}^{-1} \end{bmatrix}, \qquad \begin{bmatrix} \boldsymbol{B} & \boldsymbol{0} \\ \boldsymbol{C} & \boldsymbol{D} \end{bmatrix}^{-1} = \begin{bmatrix} \boldsymbol{B}^{-1} & 0 \\ -\boldsymbol{D}^{-1}\boldsymbol{C}\boldsymbol{B}^{-1} & \boldsymbol{D}^{-1} \end{bmatrix}.$$

But then

$$\boldsymbol{U}^{-1} = \begin{bmatrix} u_{1,1}^{-1} & -u_{1,1}^{-1}\boldsymbol{u}_1^{\mathrm{T}}\boldsymbol{U}_{2,2}^{-1} \\ \boldsymbol{0} & \boldsymbol{U}_{2,2}^{-1} \end{bmatrix}, \qquad \boldsymbol{L}^{-1} = \begin{bmatrix} 1 & \boldsymbol{0} \\ -\boldsymbol{L}_{2,2}^{-1}\boldsymbol{\ell}_1 & \boldsymbol{L}_{2,2}^{-1} \end{bmatrix},$$

$$\begin{aligned} \boldsymbol{A}^{-1} = \boldsymbol{U}^{-1}\boldsymbol{L}^{-1} &= \begin{bmatrix} (1 + \boldsymbol{u}_1^{\mathrm{T}}\boldsymbol{U}_{2,2}^{-1}\boldsymbol{L}_{2,2}^{-1}\boldsymbol{\ell}_1)/u_{1,1} & -\boldsymbol{u}_1^{\mathrm{T}}\boldsymbol{U}_{2,2}^{-1}\boldsymbol{L}_{2,2}^{-1}/u_{1,1} \\ -\boldsymbol{U}_{2,2}^{-1}\boldsymbol{L}_{2,2}^{-1}\boldsymbol{\ell}_1 & \boldsymbol{U}_{2,2}^{-1}\boldsymbol{L}_{2,2}^{-1} \end{bmatrix} \\ &= \begin{bmatrix} (1 + \boldsymbol{u}_1^{\mathrm{T}}\boldsymbol{A}_{2,2}^{-1}\boldsymbol{\ell}_1)/u_{1,1} & -\boldsymbol{u}_1^{\mathrm{T}}\boldsymbol{A}_{2,2}^{-1}/u_{1,1} \\ -\boldsymbol{A}_{2,2}^{-1}\boldsymbol{\ell}_1 & \boldsymbol{A}_{2,2}^{-1} \end{bmatrix}. \end{aligned} \tag{3.ii}$$

Inserting $\boldsymbol{B}_{2,2} = \boldsymbol{A}_{2,2}^{-1}$, Equation (3.33) follows.

b) Suppose that the elements $l_{i,j}$, $i > j$ in $\boldsymbol{L}$ and $u_{i,j}$, $j \geq i$ in $\boldsymbol{U}$ are stored in $\boldsymbol{A}$ with elements $a_{i,j}$. Write an algorithm that overwrites the elements in $\boldsymbol{A}$ with ones in $\boldsymbol{A}^{-1}$. Only one extra vector $\boldsymbol{s} \in \mathbb{R}^n$ should be used.

Solution. Define more generally the matrices $\boldsymbol{L}_{k,k}$ and $\boldsymbol{U}_{k,k}$ as the lower right $(n-k+1)\times(n-k+1)$ corners of $\boldsymbol{L}$ and $\boldsymbol{U}$, respectively. Define also $\boldsymbol{A}_{k,k} := \boldsymbol{L}_{k,k}\boldsymbol{U}_{k,k}$. Assume now that `A` is a MATLAB matrix where $\boldsymbol{L}$ and $\boldsymbol{U}$ are stored as the lower and upper triangular parts. Since $\boldsymbol{L}_{k+1,k+1}$ and $\boldsymbol{U}_{k+1,k+1}$ are the lower right corners of dimension one less than $\boldsymbol{L}_{k,k}$ and $\boldsymbol{U}_{k,k}$, respectively, Equation (3.ii) says that

$$\boldsymbol{A}_{k,k}^{-1} = \begin{bmatrix} (1+\boldsymbol{u}_k^{\mathrm{T}}\boldsymbol{A}_{k+1,k+1}^{-1}\boldsymbol{\ell}_k)/u_{k,1} & -\boldsymbol{u}_k^{\mathrm{T}}\boldsymbol{A}_{k+1,k+1}^{-1}/u_{k,1} \\ -\boldsymbol{A}_{k+1,k+1}^{-1}\boldsymbol{\ell}_k & \boldsymbol{A}_{k+1,k+1}^{-1} \end{bmatrix},$$

where $\boldsymbol{\ell}_k$ = `A(k+1:n,k)` are the entries below the main diagonal in the first column of $\boldsymbol{L}_{k,k}$, and $(u_{k,1}, \boldsymbol{u}_k)^{\mathrm{T}}$ = `A(k,k:n)` is the first row in $\boldsymbol{U}_{k,k}$. Note that $\boldsymbol{u}_k^{\mathrm{T}}\boldsymbol{A}_{k+1,k+1}^{-1}/u_{k,1}$ is a common term in the first row, and an algorithm can exploit this with the following two steps:

$$\begin{aligned} \boldsymbol{c}^{\mathrm{T}} &= -\texttt{A(k,(k+1):n)}\,\boldsymbol{A}_{k+1,k+1}^{-1}/\texttt{A(k,k)} \\ \boldsymbol{A}_{k,k}^{-1} &= \begin{bmatrix} 1/\texttt{A(k,k)} - \boldsymbol{c}^{\mathrm{T}}\texttt{A((k+1):n,k)} & \boldsymbol{c}^{\mathrm{T}} \\ -\boldsymbol{A}_{k+1,k+1}^{-1}\texttt{A((k+1):n,k)} & \boldsymbol{A}_{k+1,k+1}^{-1} \end{bmatrix} \end{aligned}$$

Note that an algorithm can store $\boldsymbol{c}^{\mathrm{T}}$ directly into `A(k,(k+1):n)`, since this part is not needed in the remaining steps, and can therefore be overwritten. An algorithm can then overwrite `A(k,k)` with $1/$`A(k,k)` $-\ \boldsymbol{c}^{\mathrm{T}}$`A((k+1):n,k)`, since `A(k,k)` is not needed in the remaining step. Finally we can compute the lower left corner of the block matrix, and we then have filled in all the entries of $\boldsymbol{A}_{k,k}^{-1}$ directly in `A` (the entries corresponding to $\boldsymbol{A}_{k+1,k+1}^{-1}$ require no changes). Since $\boldsymbol{A}_{n,n} = \boldsymbol{L}_{n,n}\boldsymbol{U}_{n,n} = \boldsymbol{U}_{n,n}$, we have that $\boldsymbol{A}_{n,n}$ =`A(n,n)`, so that $\boldsymbol{A}_{n,n}^{-1}$ =`1/A(n,n)`. This can be used as the starting point of the algorithm, which thus can look as follows (we replace $\boldsymbol{A}_{k+1,k+1}^{-1}$ with `A((k+1):n,(k+1):n)`):

code/compute_inverse.m

```
function A=compute_inverse(A)
  n = size(A,1);
  A(n,n) = 1/A(n,n);
  for k=(n-1):(-1):1
    A(k,(k+1):n) = -A(k,(k+1):n)*A((k+1):n,(k+1):n)/A(k,k);
    A(k,k) = 1/A(k,k) - A(k,(k+1):n)*A((k+1):n,k);
    A((k+1):n,k) = -A((k+1):n,(k+1):n)*A((k+1):n,k);
  end
end
```

Listing 3.5: Computing the inverse of a matrix.

The code returns $\boldsymbol{A}_{1,1}^{-1} = (\boldsymbol{L}_{1,1}\boldsymbol{U}_{1,1})^{-1} = (\boldsymbol{L}\boldsymbol{U})^{-1} = \boldsymbol{A}^{-1}$. The following code tests our algorithm for a random 10×10-matrix.

code/test_compute_inverse.m

```
n = 10;
A = rand(n);
realA = (tril(A,-1) + eye(n))*triu(A);
realA*compute_inverse(A)
```

Listing 3.6: Test computing the inverse of a matrix.

An alternative method is to compute $\boldsymbol{L}^{-1}$, $\boldsymbol{U}^{-1}$ and then $\boldsymbol{U}^{-1}\boldsymbol{L}^{-1}$ (see Exercise 3.2).

Exercise 3.18: Solving $\boldsymbol{TH}\boldsymbol{x} = \boldsymbol{b}$ (Exam exercise 1981-3)

In this exercise we consider nonsingular matrices $\boldsymbol{T}, \boldsymbol{H}, \boldsymbol{S} \in \mathbb{R}^{n\times n}$ with $\boldsymbol{T} = (t_{ij})$ upper triangular, $\boldsymbol{H} = (h_{ij})$ upper Hessenberg and $\boldsymbol{S} := \boldsymbol{TH}$. We assume that $\boldsymbol{H}$ has a unique LU1 factorization $\boldsymbol{H} = \boldsymbol{LU}$ with $\|\boldsymbol{L}\|_\infty \|\boldsymbol{U}\|_\infty \le K\|\boldsymbol{H}\|_\infty$ for a constant K not too large. In this exercise the number of operations is the highest order term in the number of multiplications and divisions.

a) Give an algorithm which computes $\boldsymbol{S}$ from $\boldsymbol{T}$ and $\boldsymbol{H}$ without using the lower parts $(t_{ij}, i > j)$ of $\boldsymbol{T}$ and $(h_{ij}, i > j+1)$ of $\boldsymbol{H}$. In what order should the elements in $\boldsymbol{S}$ be computed if $\boldsymbol{S}$ overwrites the elements in $\boldsymbol{H}$? What is the number of operations of the algorithm?

Solution. We find

$$s_{ij} = \sum_{r=i}^{\min(j+1,n)} t_{ir}h_{rj}, \qquad j = \max(1, i-1), \ldots, n, \qquad i = 1, 2, \ldots, n.$$

We can store $\boldsymbol{S}$ in $\boldsymbol{H}$ if we compute from top to bottom line by line or column by column. A row-based version is shown in Listing 3.7, while a column-based version is shown in Listing 3.8.

code/mult_tri_hess_rows.m

```
for i = 1:n
   for j = (max(1,i-1)):n
      s = min(j+1,n);
      H(i,j) = T(i,i:s)*H(i:s,j);
   end
end
```

Listing 3.7: Multiplying an upper triangular and a Hessenberg matrix — row-based version

code/mult_tri_hess_columns.m

```
for j = 1:n
   s = min(n,j+1);
   for i = 1:s
      H(i,j) = T(i,i:s)*H(i:s,j);
   end
end
```

Listing 3.8: Multiplying an upper triangular and a Hessenberg matrix — column-based version

The number of operations is of the order

$$\sum_{i=1}^{n}\sum_{j=i-1}^{n}\sum_{r=i}^{j+1} 1 \approx \int_0^n \int_i^n \int_i^j drdjdi = \frac{1}{6}n^3.$$

b) Show that $\boldsymbol{L}$ is upper Hessenberg.

Solution. We will need the following fact, which is easily proved: If $\boldsymbol{L}$ is lower triangular and nonsingular, and $\boldsymbol{b}$ starts with k zeros, then the solution $\boldsymbol{x}$ to $\boldsymbol{L}\boldsymbol{x} = \boldsymbol{b}$ also must start with k zeros. Now, denote rows i in $\boldsymbol{L}$ and $\boldsymbol{H}$ by $\boldsymbol{l}_i^{\mathrm{T}}$ and $\boldsymbol{h}_i^{\mathrm{T}}$, respectively. We have that $\boldsymbol{l}_i^{\mathrm{T}}\boldsymbol{U} = \boldsymbol{h}_i^{\mathrm{T}}$, so that $\boldsymbol{U}^{\mathrm{T}}\boldsymbol{l}_i = \boldsymbol{h}_i$. That $\boldsymbol{H}$ is upper Hessenberg is equivalent to $\boldsymbol{h}_i$ starting with $i-2$ zeros for any i. By the fact above, $\boldsymbol{l}_i$ starts with $i-2$ zeros as well, so that $\boldsymbol{L}$ too is upper Hessenberg.

c) Give a detailed algorithm for finding the LU1-factorization of $\boldsymbol{H}$ stored in $\boldsymbol{H}$. Determine the number of operations in the algorithm.

Solution. Since $\boldsymbol{H}$ has a unique LU1 factorization and $\|\boldsymbol{L}\|_\infty\|\boldsymbol{U}\|_\infty \le K\|\boldsymbol{H}\|_\infty$ for a constant K not too large, we do the factorization without row interchanges. This leads to Algorithm 3.9.

code/lu1fact_hess.m

```
for i=1:(n-1)
   H(i,(i+1):n) = H(i,(i+1):n)/H(i,i);
   H(i+1,(i+1):n) = H(i+1,(i+1):n) - H(i+1,i)*H(i,(i+1):n);
end
```

Listing 3.9: LU1 factorization of a Hessenberg matrix.

Here the nonzero elements of $\boldsymbol{L}$ are stored in the diagonal and subdiagonal of $\boldsymbol{H}$ and the elements of $\boldsymbol{U}$ are stored above the diagonal in $\boldsymbol{H}$. The code applies two elementary matrices $\boldsymbol{A}_i$ and $\boldsymbol{B}_i$ for column i:

1. $\boldsymbol{A}_i$ multiplies row i with $1/H_{i,i}$,

2. $\boldsymbol{B}_i$ subtracts $H_{i+1,i}$ times row i from row $i+1$.

The matrices $\boldsymbol{A}_i$ and $\boldsymbol{B}_i$ thus equal the identity matrix, except for one entry each: $(\boldsymbol{A}_i)_{i,i} = 1/H_{i,i}$ and $(\boldsymbol{B}_i)_{i+1,i} = -H_{i+1,i}$. Clearly

$$\boldsymbol{B}_{n-1}\boldsymbol{A}_{n-1}\cdots\boldsymbol{B}_2\boldsymbol{A}_2\boldsymbol{B}_1\boldsymbol{A}_1\boldsymbol{H} = \boldsymbol{U}$$

is upper triangular with ones on the diagonal, and we then have

$$\boldsymbol{H} = \boldsymbol{A}_1^{-1}\boldsymbol{B}_1^{-1}\cdots\boldsymbol{A}_{n-1}^{-1}\boldsymbol{B}_{n-1}^{-1}\boldsymbol{U}.$$

Here $\boldsymbol{A}_i^{-1}$ and $\boldsymbol{B}_i^{-1}$ also equal the identity matrix, except for one entry each: $(\boldsymbol{A}_i^{-1})_{i,i} = H_{i,i}$ and $(\boldsymbol{B}_i^{-1})_{i+1,i} = H_{i+1,i}$. Note also that $\boldsymbol{A}_i^{-1}$ and $\boldsymbol{B}_i^{-1}$ do not alter columns $i+1,\ldots,n$, since the i-th rows of these matrices have only zeros in these columns. It is easily seen that the multiplication with $\boldsymbol{A}_i^{-1}\boldsymbol{B}_i^{-1}$ above simply puts $H_{i,i}$, and $H_{i+1,i}$ at the locations with the same indices. It follows that $\boldsymbol{A}_1^{-1}\boldsymbol{B}_1^{-1}\cdots\boldsymbol{A}_{n-1}^{-1}\boldsymbol{B}_{n-1}^{-1}$ is upper Hessenberg and lower triangular, so that the LU1-factorization is stored correctly in $\boldsymbol{H}$ in the above code.

The number of operations is of the order

$$\sum_{k=2}^{n}\sum_{i=2}^{k} 2 \approx 2\int_0^n\int_0^k di dk = n^2.$$

d) Given $\boldsymbol{b} \in \mathbb{R}^n$ and $\boldsymbol{T}$,$\boldsymbol{H}$ as before. Suppose $\boldsymbol{S}$ and the LU1-factorization are not computed. We want to find $\boldsymbol{x} \in \mathbb{R}^n$ such that $\boldsymbol{S}\boldsymbol{x} = \boldsymbol{b}$. We have the 2 following methods

Method 1:

1. $\boldsymbol{S} = \boldsymbol{T}\boldsymbol{H}$
2. Solve $\boldsymbol{S}\boldsymbol{x} = \boldsymbol{b}$

Method 2:

2. Solve $\boldsymbol{T}\boldsymbol{z} = \boldsymbol{b}$
2. Solve $\boldsymbol{H}\boldsymbol{x} = \boldsymbol{z}$

What method would you prefer? Give reasons for your answer.

Solution. Table 3.1 presents a breakdown of the operation complexity of Methods 1 and 2.

Method 2 has fewer arithmetic operations than Method 1. In addition row interchanges might be necessary for the LU-factorization of $\boldsymbol{S}$. These are not necessary in method 2.

Method 1	Operation	Complexity	Method 2	Operation	Complexity
	$\boldsymbol{S} = \boldsymbol{T}\boldsymbol{H}$	$\frac{1}{6}n^3$		$\boldsymbol{T}\boldsymbol{z} = \boldsymbol{b}$	$\frac{1}{2}n^2$
	$\boldsymbol{S} = \boldsymbol{L}\boldsymbol{U}$	n^2		$\boldsymbol{H} = \boldsymbol{L}\boldsymbol{U}$	n^2
	$\boldsymbol{L}\boldsymbol{y} = \boldsymbol{b}$	$2n$		$\boldsymbol{L}\boldsymbol{y} = \boldsymbol{b}$	$2n$
	$\boldsymbol{U}\boldsymbol{x} = \boldsymbol{y}$	$\frac{1}{2}n^2$		$\boldsymbol{U}\boldsymbol{x} = \boldsymbol{y}$	$\frac{1}{2}n^2$
	Total	$\frac{1}{6}n^3 + \mathcal{O}(n^2)$		Total	$2n^2 + \mathcal{O}(n)$

Table 3.1: Breakdown of the operation complexity of Methods 1 and 2.

Exercise 3.19: L1U factorization update (Exam exercise 1983-1)

Let $\boldsymbol{A} \in \mathbb{R}^{n\times n}$ be nonsingular with columns $\boldsymbol{a}_1, \boldsymbol{a}_2, \ldots, \boldsymbol{a}_n$. We assume that $\boldsymbol{A}$ has a unique L1U factorization $\boldsymbol{A} = \boldsymbol{L}\boldsymbol{U}$.

For a positive integer $p \le n$ and $\boldsymbol{b} \in \mathbb{R}^n$ we define

$$\boldsymbol{B} := [\boldsymbol{a}_1, \ldots, \boldsymbol{a}_{p-1}, \boldsymbol{a}_{p+1}, \ldots, \boldsymbol{a}_n, \boldsymbol{b}] \in \mathbb{R}^{n\times n}.$$

a) Show that $\boldsymbol{H} := \boldsymbol{L}^{-1}\boldsymbol{B}$ is upper Hessenberg. We assume that $\boldsymbol{H}$ has a unique L1U factorization. $\boldsymbol{H} = \boldsymbol{L}_H\boldsymbol{U}_H$.

Solution. Since $\boldsymbol{L}^{-1}\boldsymbol{A} = \boldsymbol{U} = [\boldsymbol{u}_1, \ldots, \boldsymbol{u}_n]$ we have $\boldsymbol{L}^{-1}\boldsymbol{a}_j = \boldsymbol{u}_j$, $j = 1, \ldots, n$. We find

$$\boldsymbol{H} = \boldsymbol{L}^{-1}\boldsymbol{B} = [\boldsymbol{u}_1, \ldots, \boldsymbol{u}_{p-1}, \boldsymbol{u}_{p+1}, \ldots, \boldsymbol{u}_n, \boldsymbol{L}^{-1}\boldsymbol{b}].$$

Since $h_{i,j} = u_{i,j} = 0$ for $i > j$, $j = 1, \ldots, p-1$ and $h_{i,j} = u_{i,j+1} = 0$ for $i > j+1$, $j = p, \ldots, n-1$, we see that $h_{i,j} = 0$ for $i > j+1$ so that $\boldsymbol{H}$ is upper Hessenberg.

b) Describe briefly how many multiplications/divisions are required to find the L1U factorization of $\boldsymbol{H}$?

Solution. The number of operations is $\mathcal{O}(n^2)$, both to find $\boldsymbol{L}^{-1}\boldsymbol{b}$ and for Gaussian elimination on $\boldsymbol{H}$.

c) Suppose we have found the L1U factorization $\boldsymbol{H} := \boldsymbol{L}_H\boldsymbol{U}_H$ of $\boldsymbol{H}$. Explain how we can find the L1U factorization of $\boldsymbol{B}$ from $\boldsymbol{L}_H$ and $\boldsymbol{U}_H$.

Solution. Since $\boldsymbol{B} = \boldsymbol{L}\boldsymbol{H} = \boldsymbol{L}\boldsymbol{L}_H\boldsymbol{U}_H$ we only need to compute the product $\boldsymbol{L}\boldsymbol{L}_H$.

Exercise 3.20: U1L factorization (Exam exercise 1990-1)

We say that $\boldsymbol{A} \in \mathbb{R}^{n\times n}$ has a U1L factorization if $\boldsymbol{A} = \boldsymbol{U}\boldsymbol{L}$ for an upper triangular matrix $\boldsymbol{U} \in \mathbb{R}^{n\times n}$ with ones on the diagonal and a lower triangular $\boldsymbol{L} \in \mathbb{R}^{n\times n}$. A UL and the more common LU factorization are analogous, but normally not the same.

a) Find a U1L factorization of the matrix

$$\boldsymbol{A} := \begin{bmatrix} -3 & -2 \\ 4 & 2 \end{bmatrix}.$$

Solution. Comparing elements in

$$\begin{bmatrix} -3 & -2 \\ 4 & 2 \end{bmatrix} = \begin{bmatrix} 1 & u_{12} \\ 0 & 1 \end{bmatrix} \begin{bmatrix} l_{11} & 0 \\ l_{21} & l_{22} \end{bmatrix} = \begin{bmatrix} l_{11} + u_{12}l_{21} & u_{12}l_{22} \\ l_{21} & l_{22} \end{bmatrix}$$

we find $l_{21} = a_{21} = 4$, $l_{22} = a_{22} = 2$, $u_{12} = a_{12}/l_{22} = -1$ and $l_{11} = a_{11} - u_{12}l_{21} = 1$. We obtain

$$\begin{bmatrix} -3 & -2 \\ 4 & 2 \end{bmatrix} = \begin{bmatrix} 1 & -1 \\ 0 & 1 \end{bmatrix} \begin{bmatrix} 1 & 0 \\ 4 & 2 \end{bmatrix}.$$

b) Let the columns of $\boldsymbol{P} \in \mathbb{R}^{n\times n}$ be the unit vectors in reverse order, i.e.,

$$\boldsymbol{P} := [\boldsymbol{e}_n, \boldsymbol{e}_{n-1}, \ldots, \boldsymbol{e}_1].$$

Show that $\boldsymbol{P}^{\mathrm{T}} = \boldsymbol{P}$ and $\boldsymbol{P}^2 = \boldsymbol{I}$. What is the connection between the elements in $\boldsymbol{A}$ and $\boldsymbol{P}\boldsymbol{A}$?

Solution. We have $p_{i,n-i+1} = 1$ for $i = 1, \ldots, n$ and $p_{i,j} = 0$ otherwise. Clearly, $p_{i,j} = p_{j,i}$ for all i, j and $\boldsymbol{P}^{\mathrm{T}} = \boldsymbol{P}$. Now

$$\boldsymbol{P}^2 = \boldsymbol{P}^{\mathrm{T}}\boldsymbol{P} = \begin{bmatrix} \boldsymbol{e}_n^{\mathrm{T}} \\ \vdots \\ \boldsymbol{e}_1^{\mathrm{T}} \end{bmatrix} [\boldsymbol{e}_n \cdots \boldsymbol{e}_1] = \left[\boldsymbol{e}_{n-i+1}^{\mathrm{T}}\boldsymbol{e}_{n-j+1}\right]_{i,j} = [\delta_{ij}]_{i,j} = \boldsymbol{I}.$$

The rows of $\boldsymbol{P}\boldsymbol{A}$ consists of the rows in $\boldsymbol{A}$ in reverse order.

c) Let $\boldsymbol{B} := \boldsymbol{P}\boldsymbol{A}\boldsymbol{P}$. Find integers r, s, depending on i, j, n, such that $b_{i,j} = a_{r,s}$.

Solution. We find

$$(\boldsymbol{P}\boldsymbol{A}\boldsymbol{P})(i,j) = \boldsymbol{e}_{n-i+1}^{\mathrm{T}}\boldsymbol{A}\boldsymbol{e}_{n-j+1} = a_{n-i+1,n-j+1}$$

so that $r = n - i + 1$ and $s = n - j + 1$. If $\boldsymbol{A}$ is lower triangular then $\boldsymbol{B}$ is upper triangular.

d) Make a detailed algorithm which to given $\boldsymbol{A} \in \mathbb{R}^{n\times n}$ determines $\boldsymbol{B} := \boldsymbol{PAP}$. The elements $b_{i,j}$ in $\boldsymbol{B}$ should be stored in position i, j in $\boldsymbol{A}$. You should not use other matrices than $\boldsymbol{A}$ and a scalar $w \in \mathbb{R}$.

Solution. This is shown in Algorithm 3.10.

code/pap_compute.m

```
A = A(end:(-1):1,end:(-1):1)
```

Listing 3.10: Reversing the order in a matrix in both axes.

e) Let $\boldsymbol{PAP} = \boldsymbol{MR}$ be an L1U factorization of $\boldsymbol{PAP}$, i.e., $\boldsymbol{M}$ is lower triangular with ones on the diagonal and $\boldsymbol{R}$ is upper triangular. Express the matrices $\boldsymbol{U}$ and $\boldsymbol{L}$ in a U1L factorization of $\boldsymbol{A}$ in terms of $\boldsymbol{M}$, $\boldsymbol{R}$ and $\boldsymbol{P}$.

Solution. Since $\boldsymbol{PAP} = \boldsymbol{MR}$ and $\boldsymbol{P}^{-1} = \boldsymbol{P}$ we obtain

$$\boldsymbol{A} = (\boldsymbol{PMP})(\boldsymbol{PRP}) =: \boldsymbol{UL},$$

where $\boldsymbol{U}$ is upper triangular with ones on the diagonal and $\boldsymbol{L}$ is lower triangular.

f) Give necessary and sufficient conditions for a matrix to have a unique U1L factorization.

Solution. We first show that $\boldsymbol{A}$ has a unique U1L factorization if and only if $\boldsymbol{B} := \boldsymbol{PAP}$ has a unique L1U factorization. Suppose $\boldsymbol{A}$ has a unique U1L factorization $\boldsymbol{A} = \boldsymbol{UL}$. Then $\boldsymbol{B} = \boldsymbol{MR}$, where $\boldsymbol{M} := \boldsymbol{PUP}$ and $\boldsymbol{R} := \boldsymbol{PLP}$, is an L1U factorization of $\boldsymbol{B}$. We need to show that $\boldsymbol{M}$ and $\boldsymbol{R}$ are unique. Suppose $\boldsymbol{B} = \boldsymbol{M}_1\boldsymbol{R}_1 = \boldsymbol{M}_2\boldsymbol{R}_2$ are two L1U factorizations of $\boldsymbol{B}$. Then

$$\boldsymbol{A} = \boldsymbol{U}_1\boldsymbol{L}_1 = \boldsymbol{U}_2\boldsymbol{L}_2, \qquad \boldsymbol{U}_i = \boldsymbol{PM}_i\boldsymbol{P}, \qquad \boldsymbol{L}_i = \boldsymbol{PR}_i\boldsymbol{P}, \qquad i = 1, 2.$$

Since $\boldsymbol{A}$ has a unique U1L factorization and $\boldsymbol{P}^{-1} = \boldsymbol{P}$ we have

$$\boldsymbol{M}_1 = \boldsymbol{PU}_1\boldsymbol{P} = \boldsymbol{PU}_2\boldsymbol{P} = \boldsymbol{M}_2, \qquad \boldsymbol{R}_1 = \boldsymbol{PL}_1\boldsymbol{P} = \boldsymbol{PL}_2\boldsymbol{P} = \boldsymbol{R}_2,$$

and uniqueness follows. The converse is analogous.

We can now show that a matrix has a unique U1L factorization if and only if the matrices

$$\boldsymbol{P}_k\boldsymbol{B}_{[k]}\boldsymbol{P}_k := \begin{bmatrix} a_{n-k+1,n-k+1} & \cdots & a_{n-k+1,n} \\ \vdots & & \vdots \\ a_{n,n-k+1} & \cdots & a_{n,n} \end{bmatrix}$$

are nonsingular for $k = 1, \ldots, n-1$. Here $\boldsymbol{P}_k := [\boldsymbol{e}_k, \ldots, \boldsymbol{e}_1]$ with $\boldsymbol{e}_j$ the unit vectors in $\mathbb{R}^k$, $j = 1, \ldots, k$ and

$$\boldsymbol{B}_{[k]} := \begin{bmatrix} b_{11} & \cdots & b_{1k} \\ \vdots & & \vdots \\ b_{k1} & \cdots & b_{kk} \end{bmatrix}.$$

Indeed, $\boldsymbol{A}$ has a unique U1L factorization if and only if

$$\begin{aligned} & \boldsymbol{B} = \boldsymbol{PAP} \text{ has a unique L1U factorization} \\ \overset{\text{LU-theorem}}{\Longleftrightarrow} \; & \boldsymbol{B}_{[k]} \text{ nonsingular for } k = 1, \ldots, n-1 \\ \Longleftrightarrow \; & \boldsymbol{P}_k \boldsymbol{B}_{[k]} \boldsymbol{P}_k \text{ nonsingular for } k = 1, \ldots, n-1. \end{aligned}$$

Exercises section 3.6

Exercise 3.21: Making block LU into LU

Show that $\hat{\boldsymbol{L}}$ is unit lower triangular and $\hat{\boldsymbol{U}}$ is upper triangular.

Solution. We can write a block LU factorization of $\boldsymbol{A}$ as

$$\boldsymbol{A} = \boldsymbol{LU} = \begin{bmatrix} \boldsymbol{I} & 0 & \cdots & 0 \\ \boldsymbol{L}_{21} & \boldsymbol{I} & \cdots & 0 \\ \vdots & \vdots & \ddots & \vdots \\ \boldsymbol{L}_{m1} & \boldsymbol{L}_{m2} & \cdots & \boldsymbol{I} \end{bmatrix} \begin{bmatrix} \boldsymbol{U}_{11} & \boldsymbol{U}_{12} & \cdots & \boldsymbol{U}_{1m} \\ 0 & \boldsymbol{U}_{22} & \cdots & \boldsymbol{U}_{2m} \\ \vdots & \vdots & \ddots & \vdots \\ 0 & 0 & \cdots & \boldsymbol{U}_{mm} \end{bmatrix}$$

(i.e., the blocks are denoted by $\boldsymbol{L}_{ij}$, $\boldsymbol{U}_{ij}$). We now assume that $\boldsymbol{U}_{ii}$ has an LU factorization $\tilde{\boldsymbol{L}}_{ii}\tilde{\boldsymbol{U}}_{ii}$ ($\tilde{\boldsymbol{L}}_{ii}$ unit lower triangular, $\tilde{\boldsymbol{U}}_{ii}$ upper triangular), and define $\hat{\boldsymbol{L}} := \boldsymbol{L}\,\mathrm{diag}(\tilde{\boldsymbol{L}}_{ii})$, $\hat{\boldsymbol{U}} := \mathrm{diag}(\tilde{\boldsymbol{L}}_{ii}^{-1})\boldsymbol{U}$. We get that

$$\hat{\boldsymbol{L}} = \begin{bmatrix} \boldsymbol{I} & 0 & \cdots & 0 \\ \boldsymbol{L}_{21} & \boldsymbol{I} & \cdots & 0 \\ \vdots & \vdots & \ddots & \vdots \\ \boldsymbol{L}_{m1} & \boldsymbol{L}_{m2} & \cdots & \boldsymbol{I} \end{bmatrix} \begin{bmatrix} \tilde{\boldsymbol{L}}_{11} & 0 & \cdots & 0 \\ 0 & \tilde{\boldsymbol{L}}_{22} & \cdots & 0 \\ \vdots & \vdots & \ddots & \vdots \\ 0 & 0 & \cdots & \tilde{\boldsymbol{L}}_{mm} \end{bmatrix} = \begin{bmatrix} \tilde{\boldsymbol{L}}_{11} & 0 & \cdots & 0 \\ \boldsymbol{L}_{21}\tilde{\boldsymbol{L}}_{11} & \tilde{\boldsymbol{L}}_{22} & \cdots & 0 \\ \vdots & \vdots & \ddots & \vdots \\ \boldsymbol{L}_{m1}\tilde{\boldsymbol{L}}_{11} & \boldsymbol{L}_{m2}\tilde{\boldsymbol{L}}_{22} & \cdots & \tilde{\boldsymbol{L}}_{mm} \end{bmatrix}.$$

This shows that $\hat{\boldsymbol{L}}$ has the blocks $\tilde{\boldsymbol{L}}_{ii}$ on the diagonal, and since these are unit lower triangular, it follows that also $\hat{\boldsymbol{L}}$ is unit lower triangular. Also,

$$\hat{\boldsymbol{U}} := \mathrm{diag}(\tilde{\boldsymbol{L}}_{ii}^{-1})\boldsymbol{U} = \begin{bmatrix} \tilde{\boldsymbol{L}}_{11}^{-1} & 0 & \cdots & 0 \\ 0 & \tilde{\boldsymbol{L}}_{22}^{-1} & \cdots & 0 \\ \vdots & \vdots & \ddots & \vdots \\ 0 & 0 & \cdots & \tilde{\boldsymbol{L}}_{mm}^{-1} \end{bmatrix} \begin{bmatrix} \boldsymbol{U}_{11} & \boldsymbol{U}_{12} & \cdots & \boldsymbol{U}_{1m} \\ 0 & \boldsymbol{U}_{22} & \cdots & \boldsymbol{U}_{2m} \\ \vdots & \vdots & \ddots & \vdots \\ 0 & 0 & \cdots & \boldsymbol{U}_{mm} \end{bmatrix}$$

$$
\begin{aligned}
&= \begin{bmatrix} \tilde{\boldsymbol{L}}_{11}^{-1}\boldsymbol{U}_{11} & \tilde{\boldsymbol{L}}_{11}^{-1}\boldsymbol{U}_{12} & \cdots & \tilde{\boldsymbol{L}}_{11}^{-1}\boldsymbol{U}_{1m} \\ 0 & \tilde{\boldsymbol{L}}_{22}^{-1}\boldsymbol{U}_{22} & \cdots & \tilde{\boldsymbol{L}}_{22}^{-1}\boldsymbol{U}_{2m} \\ \vdots & \vdots & \ddots & \vdots \\ 0 & 0 & \cdots & \tilde{\boldsymbol{L}}_{mm}^{-1}\boldsymbol{U}_{mm} \end{bmatrix} \\
&= \begin{bmatrix} \tilde{\boldsymbol{L}}_{11}^{-1}\tilde{\boldsymbol{L}}_{11}\tilde{\boldsymbol{U}}_{11} & \tilde{\boldsymbol{L}}_{11}^{-1}\boldsymbol{U}_{12} & \cdots & \tilde{\boldsymbol{L}}_{11}^{-1}\boldsymbol{U}_{1m} \\ 0 & \tilde{\boldsymbol{L}}_{22}^{-1}\tilde{\boldsymbol{L}}_{22}\tilde{\boldsymbol{U}}_{22} & \cdots & \tilde{\boldsymbol{L}}_{22}^{-1}\boldsymbol{U}_{2m} \\ \vdots & \vdots & \ddots & \vdots \\ 0 & 0 & \cdots & \tilde{\boldsymbol{L}}_{mm}^{-1}\tilde{\boldsymbol{L}}_{mm}\tilde{\boldsymbol{U}}_{mm} \end{bmatrix} \\
&= \begin{bmatrix} \tilde{\boldsymbol{U}}_{11} & \tilde{\boldsymbol{L}}_{11}^{-1}\boldsymbol{U}_{12} & \cdots & \tilde{\boldsymbol{L}}_{11}^{-1}\boldsymbol{U}_{1m} \\ 0 & \tilde{\boldsymbol{U}}_{22} & \cdots & \tilde{\boldsymbol{L}}_{22}^{-1}\boldsymbol{U}_{2m} \\ \vdots & \vdots & \ddots & \vdots \\ 0 & 0 & \cdots & \tilde{\boldsymbol{U}}_{mm} \end{bmatrix},
\end{aligned}
$$

where we inserted $\boldsymbol{U}_{ii} = \tilde{\boldsymbol{L}}_{ii}\tilde{\boldsymbol{U}}_{ii}$. This shows $\hat{\boldsymbol{U}}$ has the blocks $\tilde{\boldsymbol{U}}_{ii}$ on the diagonal, and since these are upper triangular, it follows that also $\hat{\boldsymbol{U}}$ is upper triangular.

Chapter 4
LDL* Factorization and Positive Definite Matrices

Exercises section 4.2

Exercise 4.1: Positive definite characterizations

Show directly that all 4 characterizations in Theorem 4.4 hold for the matrix

$$\boldsymbol{A} = \begin{bmatrix} 2 & 1 \\ 1 & 2 \end{bmatrix}.$$

Solution. We check the equivalent statements of Theorem 4.4 for the matrix $\boldsymbol{A}$.

1. Obviously $\boldsymbol{A}$ is symmetric. In addition $\boldsymbol{A}$ is positive definite, because

   $$[x\ y] \begin{bmatrix} 2 & 1 \\ 1 & 2 \end{bmatrix} \begin{bmatrix} x \\ y \end{bmatrix} = 2x^2 + 2xy + 2y^2 = (x+y)^2 + x^2 + y^2 > 0$$

   for any nonzero vector $[x, y]^{\mathrm{T}} \in \mathbb{R}^2$.
2. The eigenvalues of $\boldsymbol{A}$ are the roots of the characteristic equation

   $$0 = \det(\boldsymbol{A} - \lambda \boldsymbol{I}) = (2 - \lambda)^2 - 1 = (\lambda - 1)(\lambda - 3).$$

   Hence the eigenvalues are $\lambda = 1$ and $\lambda = 3$, which are both positive.
3. The leading principal submatrices of $\boldsymbol{A}$ are $[2]$ and $\boldsymbol{A}$ itself, which have positive determinants 2 and 3.
4. If we assume as in a Cholesky factorization that $\boldsymbol{B}$ is lower triangular, then

   $$\boldsymbol{B}\boldsymbol{B}^{\mathrm{T}} = \begin{bmatrix} b_{11} & 0 \\ b_{21} & b_{22} \end{bmatrix} \begin{bmatrix} b_{11} & b_{21} \\ 0 & b_{22} \end{bmatrix} = \begin{bmatrix} b_{11}^2 & b_{11}b_{21} \\ b_{21}b_{11} & b_{21}^2 + b_{22}^2 \end{bmatrix} = \begin{bmatrix} 2 & 1 \\ 1 & 2 \end{bmatrix}.$$

   Since $b_{11}^2 = 2$ we can choose $b_{11} = \sqrt{2}$. Using this, $b_{11}b_{21} = 1$ gives that $b_{21} = 1/\sqrt{2}$, and $b_{21}^2 + b_{22}^2 = 2$ finally gives $b_{22} = \sqrt{2 - 1/2} = \sqrt{3/2}$ (we chose the positive square root). This means that we can choose

T. Lyche et al., *Exercises in Numerical Linear Algebra and Matrix Factorizations*, Texts in Computational Science and Engineering 23, https://doi.org/10.1007/978-3-030-59789-4_4

$$B = \begin{bmatrix} \sqrt{2} & 0 \\ 1/\sqrt{2} & \sqrt{3/2} \end{bmatrix}.$$

This could also have been obtained by writing down an LDL-factorization (as in the proof for its existence), and then multiplying in the square root of the diagonal matrix.

Exercise 4.2: L1U factorization (Exam exercise 1982-1)

Find the L1U factorization of the following matrix $A \in \mathbb{R}^{n\times n}$

$$A = \begin{bmatrix} 1 & -1 & 0 & \cdots & 0 \\ -1 & 2 & -1 & \ddots & \vdots \\ 0 & \ddots & \ddots & \ddots & 0 \\ \vdots & \ddots & -1 & 2 & -1 \\ 0 & \cdots & 0 & -1 & 2 \end{bmatrix}.$$

Is A positive definite?

Solution. By (2.16) we find

$$u_1 = 1, \qquad l_k = -1, \qquad u_{k+1} = 1, \qquad k = 1, \ldots, n-1,$$

so that

$$L = \begin{bmatrix} 1 & & & \\ -1 & 1 & & \\ & \ddots & \ddots & \\ & & -1 & 1 \end{bmatrix}, \qquad U = \begin{bmatrix} 1 & -1 & & \\ & \ddots & \ddots & \\ & & 1 & -1 \\ & & & 1 \end{bmatrix}.$$

Since $U^{\mathrm{T}} = L$ and L has positive diagonal elelements, the L1U factorization is also a Cholesky factorization. Then it follows from Theorem 4.2 that A is positive definite.

Exercise 4.3: A counterexample

In the non-symmetric case a nonsingular positive semidefinite matrix is not necessarily positive definite. Show this by considering the matrix $A := \begin{bmatrix} 1 & 0 \\ -2 & 1 \end{bmatrix}$.

Solution. For any vector $x = [x, y]^{\mathrm{T}} \in \mathbb{C}^2$, one obtains

$$x^{\mathrm{T}} A x = [x\ y] \begin{bmatrix} 1 & 0 \\ -2 & 1 \end{bmatrix} \begin{bmatrix} x \\ y \end{bmatrix} = [x\ y] \begin{bmatrix} x \\ y - 2x \end{bmatrix} = x^2 - 2xy + y^2 = (x-y)^2 \geq 0,$$

showing that $\boldsymbol{A}$ is positive semi-definite. Equality holds precisely when $x = y$, and hence there exist nonzero vectors $\boldsymbol{x}$ satisfying $\boldsymbol{x}^{\mathrm{T}}\boldsymbol{A}\boldsymbol{x} = 0$, showing that $\boldsymbol{A}$ is not positive definite.

Exercise 4.4: Cholesky update (Exam exercise 2015-2)

a) Let $\boldsymbol{E} \in \mathbb{R}^{n\times n}$ be of the form $\boldsymbol{E} = \boldsymbol{I} + \boldsymbol{u}\boldsymbol{u}^{\mathrm{T}}$, where $\boldsymbol{u} \in \mathbb{R}^n$. Show that $\boldsymbol{E}$ is symmetric and positive definite, and find an expression for $\boldsymbol{E}^{-1}$.

Hint.
The matrix $\boldsymbol{E}^{-1}$ is of the form $\boldsymbol{E}^{-1} = \boldsymbol{I} + a\boldsymbol{u}\boldsymbol{u}^{\mathrm{T}}$ for some $a \in \mathbb{R}$.

Solution. We have that

$$\boldsymbol{E}^{\mathrm{T}} = (\boldsymbol{I} + \boldsymbol{u}\boldsymbol{u}^{\mathrm{T}})^{\mathrm{T}} = \boldsymbol{I}^{\mathrm{T}} + (\boldsymbol{u}\boldsymbol{u}^{\mathrm{T}})^{\mathrm{T}} = \boldsymbol{I} + \boldsymbol{u}\boldsymbol{u}^{\mathrm{T}} = \boldsymbol{E},$$

and

$$\boldsymbol{x}^{\mathrm{T}}\boldsymbol{E}\boldsymbol{x} = \boldsymbol{x}^{\mathrm{T}}(\boldsymbol{I} + \boldsymbol{u}\boldsymbol{u}^{\mathrm{T}})\boldsymbol{x} = \boldsymbol{x}^{\mathrm{T}}\boldsymbol{x} + \boldsymbol{x}^{\mathrm{T}}\boldsymbol{u}\boldsymbol{u}^{\mathrm{T}}\boldsymbol{x} = \|\boldsymbol{x}\|^2 + (\boldsymbol{x}^{\mathrm{T}}\boldsymbol{u})^2 > 0,$$

so that $\boldsymbol{E}$ is symmetric and positive definite. Using the hint we compute

$$(\boldsymbol{I}+a\boldsymbol{u}\boldsymbol{u}^{\mathrm{T}})(\boldsymbol{I}+\boldsymbol{u}\boldsymbol{u}^{\mathrm{T}}) = \boldsymbol{I}+(1+a)\boldsymbol{u}\boldsymbol{u}^{\mathrm{T}}+a\boldsymbol{u}\boldsymbol{u}^{\mathrm{T}}\boldsymbol{u}\boldsymbol{u}^{\mathrm{T}} = \boldsymbol{I}+(1+a+a\|\boldsymbol{u}\|^2)\boldsymbol{u}\boldsymbol{u}^{\mathrm{T}}.$$

This equals $\boldsymbol{I}$ if $1 + a + a\|\boldsymbol{u}\|^2 = 0$, i.e., if $a = -1/(1 + \|\boldsymbol{u}\|^2)$. This shows that

$$\boldsymbol{E}^{-1} = \boldsymbol{I} - \frac{1}{1 + \|\boldsymbol{u}\|^2}\boldsymbol{u}\boldsymbol{u}^{\mathrm{T}}.$$

b) Let $\boldsymbol{A} \in \mathbb{R}^{n\times n}$ be of the form $\boldsymbol{A} = \boldsymbol{B} + \boldsymbol{u}\boldsymbol{u}^{\mathrm{T}}$, where $\boldsymbol{B} \in \mathbb{R}^{n\times n}$ is symmetric and positive definite, and $\boldsymbol{u} \in \mathbb{R}^n$. Show that $\boldsymbol{A}$ can be decomposed as

$$\boldsymbol{A} = \boldsymbol{L}(\boldsymbol{I} + \boldsymbol{v}\boldsymbol{v}^{\mathrm{T}})\boldsymbol{L}^{\mathrm{T}},$$

where $\boldsymbol{L}$ is nonsingular and lower triangular, and $\boldsymbol{v} \in \mathbb{R}^n$.

Solution. Since $\boldsymbol{B}$ is symmetric and positive definite it has a Cholesky factorization $\boldsymbol{B} = \boldsymbol{L}\boldsymbol{L}^{\mathrm{T}}$. We have that

$$\boldsymbol{L}(\boldsymbol{I} + \boldsymbol{v}\boldsymbol{v}^{\mathrm{T}})\boldsymbol{L}^{\mathrm{T}} = \boldsymbol{L}\boldsymbol{L}^{\mathrm{T}} + \boldsymbol{L}\boldsymbol{v}\boldsymbol{v}^{\mathrm{T}}\boldsymbol{L}^{\mathrm{T}} = \boldsymbol{B} + \boldsymbol{L}\boldsymbol{v}(\boldsymbol{L}\boldsymbol{v})^{\mathrm{T}}.$$

If we now choose $\boldsymbol{v}$ so that $\boldsymbol{L}\boldsymbol{v} = \boldsymbol{u}$ (this is possible since $\boldsymbol{L}$ is nonsingular), this equals $\boldsymbol{B} + \boldsymbol{u}\boldsymbol{u}^{\mathrm{T}} = \boldsymbol{A}$, and this shows that $\boldsymbol{A}$ can be written in the desired form.

c) Assume that the Cholesky decomposition of $\boldsymbol{B}$ is already computed. Outline a procedure to solve the system $\boldsymbol{A}\boldsymbol{x} = \boldsymbol{b}$, where $\boldsymbol{A}$ is on the form above.

Solution. We first find a $\boldsymbol{v}$ so that $\boldsymbol{A} = \boldsymbol{L}(\boldsymbol{I}+\boldsymbol{v}\boldsymbol{v}^{\mathrm{T}})\boldsymbol{L}^{\mathrm{T}}$ (by solving $\boldsymbol{L}\boldsymbol{v} = \boldsymbol{u}$, which is a lower triangular system). Then we solve $\boldsymbol{L}\boldsymbol{z} = \boldsymbol{b}$ (lower triangular system), then $(\boldsymbol{I}+\boldsymbol{v}\boldsymbol{v}^{\mathrm{T}})\boldsymbol{w} = \boldsymbol{z}$ (where we can use **a**), where we found an expression for $(\boldsymbol{I}+\boldsymbol{v}\boldsymbol{v}^{\mathrm{T}})^{-1}$), and finally $\boldsymbol{L}^{\mathrm{T}}\boldsymbol{x} = \boldsymbol{w}$ (upper triangular system).

Exercise 4.5: Cholesky update (Exam exercise 2016-2)

Let $\boldsymbol{A} \in \mathbb{R}^{n\times n}$ be a symmetric positive definite matrix with a known Cholesky factorization $\boldsymbol{A} = \boldsymbol{L}\boldsymbol{L}^{\mathrm{T}}$. Furthermore, let $\boldsymbol{A}_+$ be a corresponding $(n+1)\times(n+1)$ matrix of the form

$$\boldsymbol{A}_+ = \begin{bmatrix} \boldsymbol{A} & \boldsymbol{a} \\ \boldsymbol{a}^{\mathrm{T}} & \alpha \end{bmatrix},$$

where $\boldsymbol{a}$ is a vector in $\mathbb{R}^n$, and α is a real number. We assume that the matrix $\boldsymbol{A}_+$ is symmetric positive definite.

a) Show that if $\boldsymbol{A}_+ = \boldsymbol{L}_+\boldsymbol{L}_+^{\mathrm{T}}$ is the Cholesky factorization of $\boldsymbol{A}_+$, then $\boldsymbol{L}_+$ is of the form

$$\boldsymbol{L}_+ = \begin{bmatrix} \boldsymbol{L} & 0 \\ \boldsymbol{y}^{\mathrm{T}} & \lambda \end{bmatrix},$$

i.e., that the leading principal $n\times n$ submatrix of $\boldsymbol{L}_+$ is $\boldsymbol{L}$.

Solution. Write $\boldsymbol{L}_+ = \begin{bmatrix} \boldsymbol{L}_0 & 0 \\ \boldsymbol{y}^{\mathrm{T}} & \lambda \end{bmatrix}$. We have that

$$\boldsymbol{A}_+ = \boldsymbol{L}_+\boldsymbol{L}_+^{\mathrm{T}} = \begin{bmatrix} \boldsymbol{L}_0 & 0 \\ \boldsymbol{y}^{\mathrm{T}} & \lambda \end{bmatrix}\begin{bmatrix} \boldsymbol{L}_0^{\mathrm{T}} & \boldsymbol{y} \\ 0 & \lambda \end{bmatrix} = \begin{bmatrix} \boldsymbol{L}_0\boldsymbol{L}_0^{\mathrm{T}} & \boldsymbol{L}_0\boldsymbol{y} \\ \boldsymbol{y}^{\mathrm{T}}\boldsymbol{L}_0^{\mathrm{T}} & \boldsymbol{y}^{\mathrm{T}}\boldsymbol{y}+\lambda^2 \end{bmatrix}.$$

Since $\boldsymbol{L}_0\boldsymbol{L}_0^{\mathrm{T}} = \boldsymbol{A}$ we must have that $\boldsymbol{L}_0 = \boldsymbol{L}$ by uniqueness of the Cholesky factorization, and it follows that $\boldsymbol{L}_+$ is on the given form.

b) Explain why $\alpha > \|\boldsymbol{L}^{-1}\boldsymbol{a}\|_2^2$.

Solution. From what we computed above it follows that $\boldsymbol{y}$ and α must satisfy $\boldsymbol{L}\boldsymbol{y} = \boldsymbol{a}$ (i.e., $\boldsymbol{y} = \boldsymbol{L}^{-1}\boldsymbol{a}$), and $\boldsymbol{y}^{\mathrm{T}}\boldsymbol{y}+\lambda^2 = \|\boldsymbol{L}^{-1}\boldsymbol{a}\|_2^2+\lambda^2 = \alpha$. Since $\boldsymbol{L}_+$ has positive diagonal elements, we must have that $\lambda > 0$, and it follows that $\alpha > \|\boldsymbol{L}^{-1}\boldsymbol{a}\|_2^2$.

Let us show that $\alpha > \|\boldsymbol{L}^{-1}\boldsymbol{a}\|_2^2$ implies that $\boldsymbol{A}_+$ is positive definite. This will be the case if $[\boldsymbol{z}^{\mathrm{T}}\ 1]\,\boldsymbol{A}_+\begin{bmatrix}\boldsymbol{z}\\1\end{bmatrix} > 0$ for any vector $\boldsymbol{z}\in\mathbb{R}^n$. We have that

$$\begin{aligned}
[\boldsymbol{z}^{\mathrm{T}}\ 1]\,\boldsymbol{A}_+\begin{bmatrix}\boldsymbol{z}\\1\end{bmatrix} &= [\boldsymbol{z}^{\mathrm{T}}\ 1]\begin{bmatrix} \boldsymbol{L}\boldsymbol{L}^{\mathrm{T}} & \boldsymbol{a} \\ \boldsymbol{a}^{\mathrm{T}} & \alpha \end{bmatrix}\begin{bmatrix}\boldsymbol{z}\\1\end{bmatrix} \\
&= \left[\boldsymbol{z}^{\mathrm{T}}\boldsymbol{L}\boldsymbol{L}^{\mathrm{T}}+\boldsymbol{a}^{\mathrm{T}},\ \boldsymbol{z}^{\mathrm{T}}\boldsymbol{a}+\alpha\right]\begin{bmatrix}\boldsymbol{z}\\1\end{bmatrix}
\end{aligned}$$

$$= \boldsymbol{z}^{\mathrm{T}}\boldsymbol{L}\boldsymbol{L}^{\mathrm{T}}\boldsymbol{z} + 2\boldsymbol{a}^{\mathrm{T}}\boldsymbol{z} + \alpha$$
$$= \boldsymbol{z}^{\mathrm{T}}\boldsymbol{A}\boldsymbol{z} + 2\boldsymbol{a}^{\mathrm{T}}\boldsymbol{z} + \alpha.$$

The minimum of this (as a function of $\boldsymbol{z}$) is attained when $2\boldsymbol{A}\boldsymbol{z} + 2\boldsymbol{a} = 0$, i.e., when $\boldsymbol{z} = -\boldsymbol{A}^{-1}\boldsymbol{a}$. Inserting this in the expression above we get

$$\begin{aligned}\boldsymbol{z}^{\mathrm{T}}\boldsymbol{A}\boldsymbol{z} + 2\boldsymbol{a}^{\mathrm{T}}\boldsymbol{z} + \alpha &= \boldsymbol{a}^{\mathrm{T}}(\boldsymbol{A}^{-1})^{\mathrm{T}}\boldsymbol{a} - 2\boldsymbol{a}^{\mathrm{T}}\boldsymbol{A}^{-1}\boldsymbol{a} + \alpha \\ &= -\boldsymbol{a}^{\mathrm{T}}\boldsymbol{A}^{-1}\boldsymbol{a} + \alpha \\ &= -\boldsymbol{a}^{\mathrm{T}}(\boldsymbol{L}\boldsymbol{L}^{\mathrm{T}})^{-1}\boldsymbol{a} + \alpha \\ &= -\boldsymbol{a}^{\mathrm{T}}(\boldsymbol{L}^{\mathrm{T}})^{-1}\boldsymbol{L}^{-1}\boldsymbol{a} + \alpha \\ &= -\|\boldsymbol{L}^{-1}\boldsymbol{a}\|_2^2 + \alpha,\end{aligned}$$

and this is > 0 whenever $\alpha > \|\boldsymbol{L}^{-1}\boldsymbol{a}\|_2^2$.

c) Explain how you can compute $\boldsymbol{L}_+$ when $\boldsymbol{L}$ is known.

Solution. The vector $\boldsymbol{y}$ must first be found by solving $\boldsymbol{L}\boldsymbol{y} = \boldsymbol{a}$. Then one obtains

$$\lambda = \sqrt{\alpha - \|\boldsymbol{y}\|_2^2} = \sqrt{\alpha - \|\boldsymbol{L}^{-1}\boldsymbol{a}\|_2^2}.$$

Chapter 5
Orthonormal and Unitary Transformations

Exercises section 5.1

Exercise 5.1: The A^*A inner product

Suppose $\boldsymbol{A} \in \mathbb{C}^{m\times n}$ has linearly independent columns. Show that $\langle \boldsymbol{x}, \boldsymbol{y}\rangle :=$ $\boldsymbol{y}^*\boldsymbol{A}^*\boldsymbol{A}\boldsymbol{x}$ defines an inner product on $\mathbb{C}^n$.

Solution. Assume that $\boldsymbol{A} \in \mathbb{C}^{m\times n}$ has linearly independent columns. We show that

$$\langle\cdot,\cdot\rangle_{\boldsymbol{A}} : (\boldsymbol{x}, \boldsymbol{y}) \longmapsto \boldsymbol{y}^*\boldsymbol{A}^*\boldsymbol{A}\boldsymbol{x}$$

satisfies the axioms of an inner product on a complex vector space $\mathcal{V}$, as described in Definition 5.1. Let $\boldsymbol{x}, \boldsymbol{y}, \boldsymbol{z} \in \mathcal{V}$ and $a, b \in \mathbb{C}$, and let $\langle\cdot,\cdot\rangle$ be the standard inner product on $\mathcal{V}$.

Positivity. One has

$$\langle \boldsymbol{x}, \boldsymbol{x}\rangle_{\boldsymbol{A}} = \boldsymbol{x}^*\boldsymbol{A}^*\boldsymbol{A}\boldsymbol{x} = (\boldsymbol{A}\boldsymbol{x})^*\boldsymbol{A}\boldsymbol{x} = \langle \boldsymbol{A}\boldsymbol{x}, \boldsymbol{A}\boldsymbol{x}\rangle \geq 0,$$

with equality holding if and only if $\boldsymbol{A}\boldsymbol{x} = 0$. Since $\boldsymbol{A}\boldsymbol{x}$ is a linearly combination of the columns of $\boldsymbol{A}$ with coefficients the entries of $\boldsymbol{x}$, and since the columns of $\boldsymbol{A}$ are assumed to be linearly independent, one has $\boldsymbol{A}\boldsymbol{x} = 0$ if and only if $\boldsymbol{x} = 0$.

Skew symmetry. One has

$$\langle \boldsymbol{x}, \boldsymbol{y}\rangle_{\boldsymbol{A}} = \boldsymbol{y}^*\boldsymbol{A}^*\boldsymbol{A}\boldsymbol{x} = \overline{(\boldsymbol{y}^*\boldsymbol{A}^*\boldsymbol{A}\boldsymbol{x})^*} = \overline{\boldsymbol{x}^*\boldsymbol{A}^*\boldsymbol{A}\boldsymbol{y}} = \overline{\langle \boldsymbol{y}, \boldsymbol{x}\rangle}_{\boldsymbol{A}}.$$

Linearity. One has

$$\begin{aligned}\langle a\boldsymbol{x} + b\boldsymbol{y}, \boldsymbol{z}\rangle_{\boldsymbol{A}} &= \boldsymbol{z}^*\boldsymbol{A}^*\boldsymbol{A}(a\boldsymbol{x} + b\boldsymbol{y}) = a\boldsymbol{z}^*\boldsymbol{A}^*\boldsymbol{A}\boldsymbol{x} + b\boldsymbol{z}^*\boldsymbol{A}^*\boldsymbol{A}\boldsymbol{y}\\ &= a\langle \boldsymbol{x}, \boldsymbol{z}\rangle_{\boldsymbol{A}} + b\langle \boldsymbol{y}, \boldsymbol{z}\rangle_{\boldsymbol{A}}.\end{aligned}$$

T. Lyche et al., *Exercises in Numerical Linear Algebra and Matrix Factorizations*, Texts in Computational Science and Engineering 23, https://doi.org/10.1007/978-3-030-59789-4_5

Exercise 5.2: Angle between vectors in complex case

Show that in the complex case there is a unique angle θ in $[0, \pi/2]$ such that

$$\cos\theta = \frac{|\langle \boldsymbol{x}, \boldsymbol{y}\rangle|}{\|\boldsymbol{x}\|\|\boldsymbol{y}\|}.$$

Solution. By the Cauchy-Schwarz inequality for a complex inner product space,

$$0 \leq \frac{|\langle \boldsymbol{x}, \boldsymbol{y}\rangle|}{\|\boldsymbol{x}\|\|\boldsymbol{y}\|} \leq 1.$$

Note that taking $\boldsymbol{x}$ and $\boldsymbol{y}$ perpendicular yields zero, taking $\boldsymbol{x}$ and $\boldsymbol{y}$ equal yields one, and any value in between can be obtained by picking an appropriate affine combination of these two cases.

Since the cosine decreases monotonously from one to zero on the interval $[0, \pi/2]$, there is a unique argument $\theta \in [0, \pi/2]$ such that

$$\cos\theta = \frac{|\langle \boldsymbol{x}, \boldsymbol{y}\rangle|}{\|\boldsymbol{x}\|\|\boldsymbol{y}\|}. \tag{5.i}$$

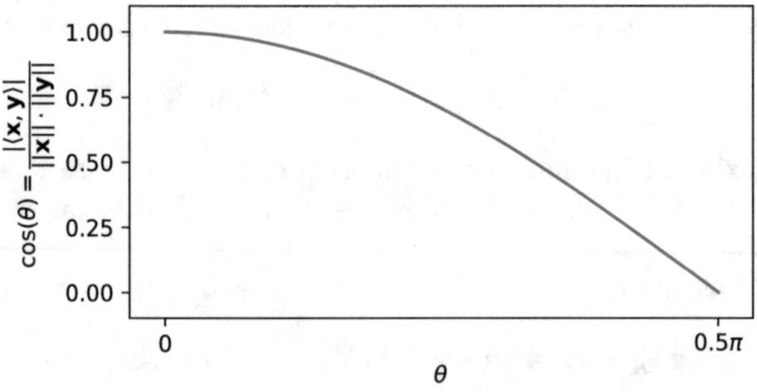

Figure 5.1: There is a unique angle $\theta \in [0, \pi/2]$ satisfying (5.i).

Exercise 5.3: $\boldsymbol{x}^{\mathrm{T}}\boldsymbol{A}\boldsymbol{y}$ inequality (Exam exercise 1979-3)

Suppose $\boldsymbol{A} \in \mathbb{R}^{n\times n}$ is symmetric positive definite. Show that

$$|\boldsymbol{x}^{\mathrm{T}}\boldsymbol{A}\boldsymbol{y}|^2 \leq \boldsymbol{x}^{\mathrm{T}}\boldsymbol{A}\boldsymbol{x}\,\boldsymbol{y}^{\mathrm{T}}\boldsymbol{A}\boldsymbol{y}$$

for all $\boldsymbol{x}, \boldsymbol{y} \in \mathbb{R}^n$, with equality if and only if $\boldsymbol{x}$ and $\boldsymbol{y}$ are linearly dependent.

Solution. Since $\boldsymbol{A}$ is symmetric positive definite $\langle \boldsymbol{x}, \boldsymbol{y} \rangle := \boldsymbol{x}^{\mathrm{T}} \boldsymbol{A} \boldsymbol{y}$ is an inner product on $\mathbb{R}^n$. Indeed, positivity, symmetry and linearity holds. The results then follows from the Cauchy-Schwarz inequality.

Exercises section 5.2

Exercise 5.4: What does algorithm housegen do when $\boldsymbol{x} = \boldsymbol{e}_1$?
Determine $\boldsymbol{H}$ in Algorithm 5.1 when $\boldsymbol{x} = \boldsymbol{e}_1$.

Solution. If $\boldsymbol{x} = \boldsymbol{e}_1$, then the algorithm yields $\rho = 1$, and $a = -\|\boldsymbol{e}_1\|_2 = -1$. We then get $\boldsymbol{z} = \boldsymbol{e}_1$, and

$$\boldsymbol{u} = \frac{\boldsymbol{z} + \boldsymbol{e}_1}{\sqrt{1 + z_1}} = \frac{2\boldsymbol{e}_1}{\sqrt{2}} = \sqrt{2}\boldsymbol{e}_1$$

and

$$\boldsymbol{H} = \boldsymbol{I} - \boldsymbol{u}\boldsymbol{u}^{\mathrm{T}} = \begin{bmatrix} -1 & 0 & \cdots & 0 \\ 0 & 1 & \cdots & 0 \\ \vdots & \vdots & \ddots & \vdots \\ 0 & 0 & \cdots & 1 \end{bmatrix}.$$

Exercise 5.5: Examples of Householder transformations
If $\boldsymbol{x}, \boldsymbol{y} \in \mathbb{R}^n$ with $\|\boldsymbol{x}\|_2 = \|\boldsymbol{y}\|_2$ and $\boldsymbol{v} := \boldsymbol{x} - \boldsymbol{y} \neq \boldsymbol{0}$ then it follows from Example 5.1 that $\left(\boldsymbol{I} - 2\frac{\boldsymbol{v}\boldsymbol{v}^{\mathrm{T}}}{\boldsymbol{v}^{\mathrm{T}}\boldsymbol{v}}\right)\boldsymbol{x} = \boldsymbol{y}$. Use this to construct a Householder transformation $\boldsymbol{H}$ such that $\boldsymbol{H}\boldsymbol{x} = \boldsymbol{y}$ in the following cases.

a) $\boldsymbol{x} = \begin{bmatrix} 3 \\ 4 \end{bmatrix}, \quad \boldsymbol{y} = \begin{bmatrix} 5 \\ 0 \end{bmatrix}.$

Solution. Let $\boldsymbol{x}$ and $\boldsymbol{y}$ be as in the exercise. As $\|\boldsymbol{x}\|_2 = \|\boldsymbol{y}\|_2$, we can apply what we did in Example 5.1 to obtain a vector $\boldsymbol{v}$ and a matrix $\boldsymbol{H}$,

$$\boldsymbol{v} = \boldsymbol{x} - \boldsymbol{y} = \begin{bmatrix} -2 \\ 4 \end{bmatrix}, \qquad \boldsymbol{H} = \boldsymbol{I} - 2\frac{\boldsymbol{v}\boldsymbol{v}^{\mathrm{T}}}{\boldsymbol{v}^{\mathrm{T}}\boldsymbol{v}} = \frac{1}{5}\begin{bmatrix} 3 & 4 \\ 4 & -3 \end{bmatrix},$$

such that $\boldsymbol{H}\boldsymbol{x} = \boldsymbol{y}$. As explained in the text above Example 5.1, $\boldsymbol{H}$ is a Householder transformation with $\boldsymbol{u} := \sqrt{2}\boldsymbol{v}/\|\boldsymbol{v}\|_2$.

b) $\boldsymbol{x} = \begin{bmatrix} 2 \\ 2 \\ 1 \end{bmatrix}, \quad \boldsymbol{y} = \begin{bmatrix} 0 \\ 3 \\ 0 \end{bmatrix}.$

Solution. Let $\boldsymbol{x}$ and $\boldsymbol{y}$ be as in the exercise. As $\|\boldsymbol{x}\|_2 = \|\boldsymbol{y}\|_2$, we can apply what we did in Example 5.1 to obtain a vector $\boldsymbol{v}$ and a Householder transformation $\boldsymbol{H}$,

$$\boldsymbol{v} = \boldsymbol{x} - \boldsymbol{y} = \begin{bmatrix} 2 \\ -1 \\ 1 \end{bmatrix}, \qquad \boldsymbol{H} = \boldsymbol{I} - 2\frac{\boldsymbol{v}\boldsymbol{v}^{\mathrm{T}}}{\boldsymbol{v}^{\mathrm{T}}\boldsymbol{v}} = \frac{1}{3}\begin{bmatrix} -1 & 2 & -2 \\ 2 & 2 & 1 \\ -2 & 1 & 2 \end{bmatrix},$$

such that $\boldsymbol{H}\boldsymbol{x} = \boldsymbol{y}$.

Exercise 5.6: 2 × 2 Householder transformation

Show that a real 2×2 Householder transformation can be written in the form

$$\boldsymbol{H} = \begin{bmatrix} -\cos\phi & \sin\phi \\ \sin\phi & \cos\phi \end{bmatrix}.$$

Find $\boldsymbol{H}\boldsymbol{x}$ if $\boldsymbol{x} = [\cos\phi, \sin\phi]^{\mathrm{T}}$.

Solution. Let $\boldsymbol{H} = \boldsymbol{I} - \boldsymbol{u}\boldsymbol{u}^{\mathrm{T}} \in \mathbb{R}^{2,2}$ be any Householder transformation. Then $\boldsymbol{u} = [u_1, u_2]^{\mathrm{T}} \in \mathbb{R}^2$ is a vector satisfying $u_1^2 + u_2^2 = \|\boldsymbol{u}\|_2^2 = 2$, implying that the components of $\boldsymbol{u}$ are related via $u_1^2 - 1 = 1 - u_2^2$. Moreover, as $0 \le u_1^2, u_2^2 \le \|\boldsymbol{u}\|^2 = 2$, one has $-1 \le u_1^2 - 1 = 1 - u_2^2 \le 1$, and there exists an angle $\phi' \in [0, 2\pi)$ such that $\cos(\phi') = u_1^2 - 1 = 1 - u_2^2$. For such an angle ϕ', one has

$$-u_1u_2 = \pm\sqrt{1+\cos\phi'}\sqrt{1-\cos\phi'} = \pm\sqrt{1-\cos^2\phi'} = \sin(\pm\phi').$$

We thus find an angle $\phi := \pm\phi'$ for which

$$\boldsymbol{H} = \begin{bmatrix} 1-u_1^2 & -u_1u_1 \\ -u_1u_2 & 1-u_2^2 \end{bmatrix} = \begin{bmatrix} -\cos(\phi') & \sin(\pm\phi') \\ \sin(\pm\phi') & \cos(\phi') \end{bmatrix} = \begin{bmatrix} -\cos(\phi) & \sin(\phi) \\ \sin(\phi) & \cos(\phi) \end{bmatrix}.$$

Furthermore, we find

$$\boldsymbol{H}\begin{bmatrix} \cos\phi \\ \sin\phi \end{bmatrix} = \begin{bmatrix} -\cos\phi & \sin\phi \\ \sin\phi & \cos\phi \end{bmatrix}\begin{bmatrix} \cos\phi \\ \sin\phi \end{bmatrix} = \begin{bmatrix} \sin^2\phi - \cos^2\phi \\ 2\sin\phi\cos\phi \end{bmatrix} = \begin{bmatrix} -\cos(2\phi) \\ \sin(2\phi) \end{bmatrix}.$$

When applied to the vector $[\cos\phi, \sin\phi]^{\mathrm{T}}$, therefore, $\boldsymbol{H}$ doubles the angle and reflects the result in the y-axis.

Exercise 5.7: Householder transformation (Exam exercise 2010-1)

a) Suppose $\boldsymbol{x}, \boldsymbol{y} \in \mathbb{R}^n$ with $\|\boldsymbol{x}\|_2 = \|\boldsymbol{y}\|_2$ and $\boldsymbol{v} := \boldsymbol{x} - \boldsymbol{y} \neq 0$. Show that

$$\boldsymbol{H}\boldsymbol{x} = \boldsymbol{y}, \quad \text{where} \quad \boldsymbol{H} := \boldsymbol{I} - 2\frac{\boldsymbol{v}\boldsymbol{v}^{\mathrm{T}}}{\boldsymbol{v}^{\mathrm{T}}\boldsymbol{v}}.$$

Solution. Since

$$\begin{aligned}\boldsymbol{v}^{\mathrm{T}}\boldsymbol{v} &= (\boldsymbol{x}-\boldsymbol{y})^{\mathrm{T}}(\boldsymbol{x}-\boldsymbol{y}) = \boldsymbol{x}^{\mathrm{T}}\boldsymbol{x} - 2\boldsymbol{x}^{\mathrm{T}}\boldsymbol{y} + \boldsymbol{y}^{\mathrm{T}}\boldsymbol{y} = 2\boldsymbol{x}^{\mathrm{T}}\boldsymbol{x} - 2\boldsymbol{x}^{\mathrm{T}}\boldsymbol{y}\\ &= 2\boldsymbol{x}^{\mathrm{T}}\boldsymbol{v} = 2\boldsymbol{v}^{\mathrm{T}}\boldsymbol{x},\end{aligned}$$

we find

$$\boldsymbol{H}\boldsymbol{x} = \left(\boldsymbol{I} - 2\frac{\boldsymbol{v}\boldsymbol{v}^{\mathrm{T}}}{\boldsymbol{v}^{\mathrm{T}}\boldsymbol{v}}\right)\boldsymbol{x} = \boldsymbol{x} - \frac{2\boldsymbol{v}^{\mathrm{T}}\boldsymbol{x}}{2\boldsymbol{v}^{\mathrm{T}}\boldsymbol{x}}\boldsymbol{v} = \boldsymbol{x} - (\boldsymbol{x}-\boldsymbol{y}) = \boldsymbol{y}.$$

b) Let $\boldsymbol{B} \in \mathbb{R}^{4,4}$ be given by

$$\boldsymbol{B} := \begin{bmatrix} 0 & 1 & 0 & 0\\ 0 & 0 & 1 & 0\\ 0 & 0 & 0 & 1\\ \epsilon & 0 & 0 & 0\end{bmatrix},$$

where $0 < \epsilon < 1$. Compute a Householder transformation $\boldsymbol{H}$ and a matrix $\boldsymbol{B}_1$ such that the first column of $\boldsymbol{B}_1 := \boldsymbol{H}\boldsymbol{B}\boldsymbol{H}$ has a zero in the last two positions.

Solution. Let $\boldsymbol{V} \in \mathbb{R}^{3,3}$ be a Householder transformation mapping $\boldsymbol{x} := [0, 0, \epsilon]^{\mathrm{T}}$ into $\boldsymbol{y} = [\epsilon, 0, 0]^{\mathrm{T}}$. With $\boldsymbol{v} = \boldsymbol{x} - \boldsymbol{y} = \epsilon[-1, 0, 1]^{\mathrm{T}}$ we find

$$\boldsymbol{V} = \boldsymbol{I} - 2\frac{\boldsymbol{v}\boldsymbol{v}^{\mathrm{T}}}{\boldsymbol{v}^{\mathrm{T}}\boldsymbol{v}} = \boldsymbol{I} - 2\epsilon^2\begin{bmatrix}-1\\ 0\\ 1\end{bmatrix}\begin{bmatrix}-1 & 0 & 1\end{bmatrix}/(2\epsilon^2) = \begin{bmatrix}0 & 0 & 1\\ 0 & 1 & 0\\ 1 & 0 & 0\end{bmatrix}$$

Let

$$\boldsymbol{H} = \begin{bmatrix}1 & \boldsymbol{0}^{\mathrm{T}}\\ \boldsymbol{0} & \boldsymbol{V}\end{bmatrix} = \begin{bmatrix}1 & 0 & 0 & 0\\ 0 & 0 & 0 & 1\\ 0 & 0 & 1 & 0\\ 0 & 1 & 0 & 0\end{bmatrix}.$$

Then

$$\boldsymbol{B}_1 = \boldsymbol{H}\boldsymbol{B}\boldsymbol{H} = \begin{bmatrix}0 & 0 & 0 & 1\\ \epsilon & 0 & 0 & 0\\ 0 & 1 & 0 & 0\\ 0 & 0 & 1 & 0\end{bmatrix}.$$

This matrix is upper Hessenberg and has the same eigenvalues as $\boldsymbol{B}$ since it is similar to $\boldsymbol{B}$.

Exercises section 5.4

Exercise 5.8: QR decomposition

$$A=\begin{bmatrix}1&2\\1&2\\1&0\\1&0\end{bmatrix},\qquad Q=\frac{1}{2}\begin{bmatrix}1&1&1&1\\1&1&-1&-1\\1&-1&-1&1\\1&-1&1&-1\end{bmatrix},\qquad R=\begin{bmatrix}2&2\\0&2\\0&0\\0&0\end{bmatrix}.$$

Show that Q is orthonormal and that QR is a QR decomposition of A. Find a QR factorization of A.

Solution. That Q is orthonormal, and therefore unitary, can be shown directly by verifying that $Q^{\mathrm{T}}Q = I$. A direct computation shows that $QR = A$. Moreover,

$$R=\begin{bmatrix}2&2\\0&2\\0&0\\0&0\end{bmatrix}=:\begin{bmatrix}R_1\\0_{2,2}\end{bmatrix},$$

where R_1 is upper triangular. It follows that $A = QR$ is a QR decomposition.

A QR factorization is obtained by removing the parts of Q and R that don't contribute anything to the product QR. Thus we find a QR factorization

$$A=Q_1R_1,\qquad Q_1:=\frac{1}{2}\begin{bmatrix}1&1\\1&1\\1&-1\\1&-1\end{bmatrix},\qquad R_1:=\begin{bmatrix}2&2\\0&2\end{bmatrix}.$$

Exercise 5.9: Householder triangulation

a) Let

$$A:=\begin{bmatrix}1&0&1\\-2&-1&0\\2&2&1\end{bmatrix}.$$

Find Householder transformations $H_1, H_2 \in \mathbb{R}^{3\times 3}$ such that H_2H_1A is upper triangular.

Solution. Write $A = [a_1, a_2, a_3]$ as in the exercise. We wish to find Householder transformations H_1, H_2 that produce zeros in the columns a_1, a_2 of A. Applying Algorithm 5.1 in the book to the first column of A, we find first that $a = -3$, $z = [1/3, -2/3, 2/3]^{\mathrm{T}}$, and then

$$u_1 = \frac{1}{\sqrt{3}}\begin{bmatrix} 2 \\ -1 \\ 1 \end{bmatrix}, \qquad H_1A := (I - u_1u_1^{\mathrm{T}})A = \begin{bmatrix} -3 & -2 & -1 \\ 0 & 0 & 1 \\ 0 & 1 & 0 \end{bmatrix}.$$

Next we need to map the bottom element $(H_1A)_{3,2}$ of the second column to zero, without changing the first row of H_1A. For this, we apply Algorithm 5.1 to the vector $[0,1]^{\mathrm{T}}$ to find $a = -1$, $z = [0,1]^{\mathrm{T}}$, and then

$$u_2 = \begin{bmatrix} 1 \\ 1 \end{bmatrix} \quad \text{and} \quad \hat{H}_2 := I - u_2u_2^{\mathrm{T}} = \begin{bmatrix} 0 & -1 \\ -1 & 0 \end{bmatrix},$$

which is a Householder transformation of size 2×2. Since

$$H_2H_1A = \begin{bmatrix} -3 & -2 & -1 \\ 0 & -1 & 0 \\ 0 & 0 & -1 \end{bmatrix}, \qquad H_2 := \begin{bmatrix} 1 & \mathbf{0} \\ \mathbf{0} & \hat{H}_2 \end{bmatrix},$$

it follows that the Householder transformations H_1 and H_2 bring A into upper triangular form.

b) Find the QR factorization of A, when R has positive diagonal elements.

Solution. Clearly the matrix $H_3 := -I$ is orthogonal and $R := H_3H_2H_1A$ is upper triangular with positive diagonal elements. It follows that

$$A = QR, \qquad Q := H_1^{\mathrm{T}}H_2^{\mathrm{T}}H_3^{\mathrm{T}} = H_1H_2H_3,$$

is a QR factorization of A of the required form.

Exercise 5.10: Hadamard's inequality

In this exercise we use the QR factorization to prove a classical determinant inequality. For any $A = [a_1, \ldots, a_n] \in \mathbb{C}^{n\times n}$ we have

$$|\det(A)| \le \prod_{j=1}^{n} \|a_j\|_2.$$

Equality holds if and only if A has a zero column or the columns of A are orthogonal.

a) Show that, if Q is unitary, then $|\det(Q)| = 1$.

Solution. Since

$$1 = \det(I) = \det(Q^*Q) = \det(Q^*)\det(Q) = \det(Q)^*\det(Q) = |\det(Q)|^2,$$

we have $|\det(Q)| = 1$.

b) Let $\boldsymbol{A} = \boldsymbol{QR}$ be a QR factorization of $\boldsymbol{A}$ and let $\boldsymbol{R} = [\boldsymbol{r}_1, \dots, \boldsymbol{r}_n]$. Show that $(\boldsymbol{A}^*\boldsymbol{A})_{jj} = \|\boldsymbol{a}_j\|_2^2 = (\boldsymbol{R}^*\boldsymbol{R})_{jj} = \|\boldsymbol{r}_j\|_2^2$.

Solution. Since $\boldsymbol{Q}$ is orthogonal,

$$\begin{aligned}\|\boldsymbol{a}_j\|_2^2 &= \sum_{k=1}^{n} \overline{a}_{kj} a_{kj} = (\boldsymbol{A}^*\boldsymbol{A})_{jj} = (\boldsymbol{R}^*\boldsymbol{Q}^*\boldsymbol{Q}\boldsymbol{R})_{jj} = (\boldsymbol{R}^*\boldsymbol{R})_{jj} \\ &= \sum_{k=1}^{n} \overline{r}_{kj} r_{kj} = \|\boldsymbol{r}_j\|_2^2.\end{aligned}$$

c) Show that $|\det(\boldsymbol{A})| = \prod_{j=1}^{n} |r_{jj}| \le \prod_{j=1}^{n} \|\boldsymbol{a}_j\|_2$.

Solution. From **b)** follows

$$|\det(\boldsymbol{A})| = |\det(\boldsymbol{Q})\det(\boldsymbol{R})| = |\det(\boldsymbol{R})| = \prod_{j=1}^{n} |r_{jj}| \le \prod_{j=1}^{n} \|\boldsymbol{r}_j\|_2 = \prod_{j=1}^{n} \|\boldsymbol{a}_j\|_2.$$

d) Show that we have equality if $\boldsymbol{A}$ has a zero column.

Solution. If one of the columns $\boldsymbol{a}_j$ of $\boldsymbol{A}$ is zero, then both the left hand side and right hand side in Hadamard's inequality (5.22) are zero, and equality holds.

e) Suppose the columns of $\boldsymbol{A}$ are nonzero. Show that we have equality if and only if the columns of $\boldsymbol{A}$ are orthogonal

Hint.
Show that we have equality $\iff$ $\boldsymbol{R}$ is diagonal $\iff$ $\boldsymbol{A}^*\boldsymbol{A}$ is diagonal.

Solution. Suppose the columns are nonzero. We have equality if and only if $|r_{jj}| = \|\boldsymbol{r}_j\|_2$ for $j = 1, \dots, n$. This happens if and only if $\boldsymbol{R}$ is diagonal. But then $\boldsymbol{A}^*\boldsymbol{A} = \boldsymbol{R}^*\boldsymbol{R}$ is diagonal, which means that the columns of $\boldsymbol{A}$ are orthogonal.

Exercise 5.11: QL factorization (Exam exercise 1982-2)

Suppose $\boldsymbol{B} \in \mathbb{R}^{n\times n}$ is symmetric and positive definite. It can be shown that $\boldsymbol{B}$ has a factorization of the form $\boldsymbol{B} = \boldsymbol{L}^{\mathrm{T}}\boldsymbol{L}$, where $\boldsymbol{L}$ is lower triangular with positive diagonal elements (you should not show this). Note that this is different from the Cholesky factorization $\boldsymbol{B} = \boldsymbol{L}\boldsymbol{L}^{\mathrm{T}}$.

a) Suppose $\boldsymbol{B} = \boldsymbol{L}^{\mathrm{T}}\boldsymbol{L}$. Write down the equations to determine the elements $l_{i,j}$ of $\boldsymbol{L}$, in the order $i = n, n-1, \dots, 1$ and $j = i, 1, 2 \dots, i-1$.

Solution. We have

$$b_{i,j} = \sum_{k=\max(i,j)}^{n} l_{k,i} l_{k,j}, \quad i,j = 1,\ldots,n.$$

Computing the entries in $\boldsymbol{L}$ in the order described in the exercise gives the following algorithm when we isolate $l_{i,j}$ on the left hand side.

code/ltlfact.m

```
function L=ltlfact(B)
   n=size(B,1);
   L=zeros(n);
   for i=n:(-1):1
      L(i,i) = sqrt(B(i,i)- sum(L((i+1):n,i).^2));
      L(i,1:(i-1)) = (B(i,1:(i-1))-(L((i+1):n,i))'*L((i+1):n
          ,1:(i-1)))/L(i,i);
   end
end
```

Listing 5.1: Compute the matrix $\boldsymbol{L}$ in the $\boldsymbol{L}^{\mathrm{T}}\boldsymbol{L}$ factorization of a symmetric and positive definite matrix $\boldsymbol{B}$.

b) Explain (without making a detailed algorithm) how the $\boldsymbol{L}^{\mathrm{T}}\boldsymbol{L}$ factorization can be used to solve the linear system $\boldsymbol{B}\boldsymbol{x} = \boldsymbol{c}$. Compute $\|\boldsymbol{L}\|_F$. Is the algorithm stable?

Solution.

First solve the upper triangular system $\boldsymbol{L}^{\mathrm{T}}\boldsymbol{y} = \boldsymbol{c}$ for $\boldsymbol{y}$. Then solve the lower triangular system $\boldsymbol{L}\boldsymbol{x} = \boldsymbol{y}$ for $\boldsymbol{x}$. We can use the `rforwardsolve` and `rbacksolve` algorithms for this. The following code tests that we find the solution to the system in this way. The code also tests that the factorization from **a)** is correct. A random positive definite matrix is generated.

code/ltlfact_solve.m

```
n=6;
C=rand(6);
B=C'*C;
L=ltlfact(B);
L'*L-B

c=rand(n,1);
y = rbacksolve(L',c,n);
x = rforwardsolve(L,y,n);
```

```
B*x-c
```

Listing 5.2: For a symmetric and positive definite matrix $\boldsymbol{B}$, solve the system $\boldsymbol{B}\boldsymbol{x} = \boldsymbol{c}$ using the $\boldsymbol{L}^{\mathrm{T}}\boldsymbol{L}$ factorization of $\boldsymbol{B}$.

We find $b_{i,i} = \sum_{k=i}^{n} l_{k,i}^2$. By summing we find $\|\boldsymbol{L}\|_F^2 = \sum_{i=1}^{n} b_{i,i}$. Thus the elements in $\boldsymbol{L}$ cannot become too large compared to the diagonal elements of $\boldsymbol{B}$, and we conclude that the algorithm is stable.

c) Show that every nonsingular matrix $\boldsymbol{A} \in \mathbb{R}^{n\times n}$ can be factored in the form $\boldsymbol{A} = \boldsymbol{Q}\boldsymbol{L}$, where $\boldsymbol{Q} \in \mathbb{R}^{n\times n}$ is orthogonal and $\boldsymbol{L} \in \mathbb{R}^{n\times n}$ is lower triangular with positive diagonal elements.

Solution. The matrix $\boldsymbol{A}^{\mathrm{T}}\boldsymbol{A}$ is symmetric and positive definite since $\boldsymbol{A}$ is nonsingular. Let $\boldsymbol{L}^{\mathrm{T}}\boldsymbol{L}$ be the corresponding factorization of $\boldsymbol{A}^{\mathrm{T}}\boldsymbol{A}$ and define $\boldsymbol{Q} := \boldsymbol{A}\boldsymbol{L}^{-1}$. Then $\boldsymbol{Q} \in \mathbb{R}^{n\times n}$ and $\boldsymbol{Q}^{\mathrm{T}}\boldsymbol{Q} = \boldsymbol{L}^{-\mathrm{T}}\boldsymbol{A}^{\mathrm{T}}\boldsymbol{A}\boldsymbol{L}^{-1} = \boldsymbol{L}^{-\mathrm{T}}\boldsymbol{L}^{\mathrm{T}}\boldsymbol{L}\boldsymbol{L}^{-1} = \boldsymbol{I}$. Thus $\boldsymbol{Q}$ is orthogonal and $\boldsymbol{A} = \boldsymbol{Q}\boldsymbol{L}$.

d) Show that the $\boldsymbol{Q}\boldsymbol{L}$ factorization in **c)** is unique.

Solution. Every $\boldsymbol{L}^{\mathrm{T}}\boldsymbol{L}$ factorization of $\boldsymbol{B}$ where $\boldsymbol{L}$ has positive diagonal elements must be given by the formulas in **a**). It follows that $\boldsymbol{L}$ in the $\boldsymbol{L}^{\mathrm{T}}\boldsymbol{L}$ factorization of $\boldsymbol{B}$ is unique. If $\boldsymbol{A} = \boldsymbol{Q}\boldsymbol{L}$ then $\boldsymbol{A}^{\mathrm{T}}\boldsymbol{A} = \boldsymbol{L}^{\mathrm{T}}\boldsymbol{Q}^{\mathrm{T}}\boldsymbol{Q}\boldsymbol{L} = \boldsymbol{L}^{\mathrm{T}}\boldsymbol{L}$. Thus $\boldsymbol{L}$ is the same as the $\boldsymbol{L}$ in the $\boldsymbol{L}^{\mathrm{T}}\boldsymbol{L}$ factorization of $\boldsymbol{A}^{\mathrm{T}}\boldsymbol{A}$ and therefore unique. Since $\boldsymbol{A} = \boldsymbol{Q}\boldsymbol{L}$ implies $\boldsymbol{Q} = \boldsymbol{A}\boldsymbol{L}^{-1}$ it follows that $\boldsymbol{Q}$ must also be unique.

Exercise 5.12: QL-factorization (Exam exercise 1982-3) ?

In this exercise we will develop an algorithm to find a QL-factorization of $\boldsymbol{A} \in \mathbb{R}^{n\times n}$ (cf. Exercise 5.11) using Householder transformations.

a) Given vectors $\boldsymbol{a} := [a_1, \ldots, a_n]^{\mathrm{T}} \in \mathbb{R}^n$ and $\boldsymbol{e}_n := [0, \ldots, 0, 1]^{\mathrm{T}}$. Find $\boldsymbol{v} \in \mathbb{R}^n$ such that the Householder transformation $\boldsymbol{H} := \boldsymbol{I} - 2\frac{\boldsymbol{v}\boldsymbol{v}^*}{\boldsymbol{v}^*\boldsymbol{v}}$ satisfies $\boldsymbol{H}\boldsymbol{a} = -s\boldsymbol{e}_n$, where $|s| = \|\boldsymbol{a}\|_2$. How should we choose the sign of s?

Solution. By Lemma 5.2, if $\boldsymbol{v} = \boldsymbol{a} + s\boldsymbol{e}_n$, where $|s| = \|\boldsymbol{a}\|_2$, then $\boldsymbol{H} := \boldsymbol{I} - 2\frac{\boldsymbol{v}\boldsymbol{v}^*}{\boldsymbol{v}^*\boldsymbol{v}}$ is a Householder transformation so that $\boldsymbol{H}\boldsymbol{a} = -s\boldsymbol{e}_n$. From the proof of Theorem 5.8 we see that we can avoid a cancellation error in $\boldsymbol{v}$ if we choose s to have the same sign as the last component of $\boldsymbol{a}$.

b) Let $1 \le r \le n$, $\boldsymbol{v}_r \in \mathbb{R}^r$, $\boldsymbol{v}_r \neq \boldsymbol{0}$, and

$$\boldsymbol{V}_r := \boldsymbol{I}_r - 2\frac{\boldsymbol{v}_r\boldsymbol{v}_r^*}{\boldsymbol{v}_r^*\boldsymbol{v}_r} = \boldsymbol{I}_r - \boldsymbol{u}_r\boldsymbol{u}_r^*, \text{ with } \boldsymbol{u}_r := \sqrt{2}\frac{\boldsymbol{v}_r}{\|\boldsymbol{v}_r\|_2}.$$

Show that $\boldsymbol{H} := \begin{bmatrix} \boldsymbol{V}_r & \boldsymbol{0} \\ \boldsymbol{0} & \boldsymbol{I}_{n-r} \end{bmatrix}$ is a Householder transformation. Show also that, if $a_{i,j} = 0$ for $i = 1, \dots, r$ and $j = r+1, \dots, n$ then the last $n-r$ columns of $\boldsymbol{A}$ and $\boldsymbol{HA}$ are the same.

Solution. Let $\boldsymbol{U} := \boldsymbol{I} - \boldsymbol{u}\boldsymbol{u}^*$, where $\boldsymbol{u} := \begin{bmatrix} \boldsymbol{u}_r \\ \boldsymbol{0} \end{bmatrix} \in \mathbb{R}^n$ and $\|\boldsymbol{u}_r\|_2 = \sqrt{2}$. Since $\|\boldsymbol{u}\|_2 = \|\boldsymbol{u}_r\|_2 = \sqrt{2}$ it follows that $\boldsymbol{U}$ is a Householder transformation. Moreover,

$$\boldsymbol{U} = \begin{bmatrix} \boldsymbol{I}_r & \boldsymbol{0} \\ \boldsymbol{0} & \boldsymbol{I}_{n-r} \end{bmatrix} - \begin{bmatrix} \boldsymbol{u}_r \\ \boldsymbol{0} \end{bmatrix} \begin{bmatrix} \boldsymbol{u}_r^* & \boldsymbol{0} \end{bmatrix} = \begin{bmatrix} \boldsymbol{I}_r & \boldsymbol{0} \\ \boldsymbol{0} & \boldsymbol{I}_{n-r} \end{bmatrix} - \begin{bmatrix} \boldsymbol{u}_r\boldsymbol{u}_r^* & \boldsymbol{0} \\ \boldsymbol{0} & \boldsymbol{0} \end{bmatrix} = \begin{bmatrix} \boldsymbol{V}_r & \boldsymbol{0} \\ \boldsymbol{0} & \boldsymbol{I}_{n-r} \end{bmatrix} =: \boldsymbol{H}.$$

If $a_{i,j} = 0$ for $i = 1, \dots, r$ and $j = r+1, \dots, n$ then $\boldsymbol{A} = \begin{bmatrix} \boldsymbol{A}_1 & \boldsymbol{0} \\ \boldsymbol{A}_2 & \boldsymbol{A}_3 \end{bmatrix}$, where $\boldsymbol{A}_3 \in \mathbb{R}^{(n-r)\times(n-r)}$. But then

$$\boldsymbol{HA} = \begin{bmatrix} \boldsymbol{V}_r & \boldsymbol{0} \\ \boldsymbol{0} & \boldsymbol{I}_{n-r} \end{bmatrix} \begin{bmatrix} \boldsymbol{A}_1 & \boldsymbol{0} \\ \boldsymbol{A}_2 & \boldsymbol{A}_3 \end{bmatrix} = \begin{bmatrix} \boldsymbol{V}_r\boldsymbol{A}_1 & \boldsymbol{0} \\ \boldsymbol{A}_2 & \boldsymbol{A}_3 \end{bmatrix}$$

has the same last $n-r$ columns as $\boldsymbol{A}$.

c) Explain, without making a detailed algorithm, how we to a given matrix $\boldsymbol{A} \in \mathbb{R}^{n\times n}$ can find Householder transformations $\boldsymbol{H}_1, \dots, \boldsymbol{H}_{n-1}$ such that $\boldsymbol{H}_{n-1}, \dots, \boldsymbol{H}_1\boldsymbol{A}$ is lower triangular. Give a $\boldsymbol{QL}$ factorization of $\boldsymbol{A}$.

Solution. We let $\boldsymbol{H}_1$ be a Householder transformation such that $\boldsymbol{H}_1\boldsymbol{a}_n = -s_n\boldsymbol{e}_n$, with $|s_n| = \|\boldsymbol{a}_n\|_2$. The transformation $\boldsymbol{H}_1$ is given in **a)** with $\boldsymbol{a} = \boldsymbol{a}_n$, the last column of $\boldsymbol{A}$. Suppose $\boldsymbol{H}_1, \dots, \boldsymbol{H}_{k-1}$ are determined such that $\boldsymbol{A}_k := \boldsymbol{H}_{k-1}\cdots\boldsymbol{H}_1\boldsymbol{A}$ is lower triangular in its $k-1$ last columns. Let $\boldsymbol{V}_k$ be a Householder transformation in $\mathbb{R}^{(n-k+1)\times(n-k+1)}$ so that the first $n-k+1$ entries in column k in $\boldsymbol{A}_k$ are sent to a scalar multiple of $\boldsymbol{e}_{n-k+1}$. Set $\boldsymbol{H}_k := \begin{bmatrix} \boldsymbol{V}_k & \boldsymbol{0} \\ \boldsymbol{0} & \boldsymbol{I}_{k-1} \end{bmatrix}$. By **b)** $\boldsymbol{H}_k$ is a Householder transformation such that $\boldsymbol{A}_{k+1} := \boldsymbol{H}_k\boldsymbol{A}_k$ is lower triangular in its last k columns. Continuing we obtain $\boldsymbol{L} := \boldsymbol{A}_n$ lower triangular. Moreover, $\boldsymbol{A} = \boldsymbol{QL}$, where

$$\boldsymbol{Q} := (\boldsymbol{H}_n\cdots\boldsymbol{H}_1)^{\mathrm{T}} = \boldsymbol{H}_1^{\mathrm{T}}\cdots\boldsymbol{H}_n^{\mathrm{T}} = \boldsymbol{H}_1\cdots\boldsymbol{H}_n$$

is orthogonal.

Exercise 5.13: QR Fact. of band matrices (Exam exercise 2006-2)

Let $\boldsymbol{A} \in \mathbb{R}^{n\times n}$ be a nonsingular symmetric band matrix with bandwidth $d \leq n-1$, so that $a_{ij} = 0$ for all i, j with $|i-j| > d$. We define $\boldsymbol{B} := \boldsymbol{A}^{\mathrm{T}}\boldsymbol{A}$ and let $\boldsymbol{A} = \boldsymbol{Q}\boldsymbol{R}$ be the QR factorization of $\boldsymbol{A}$ where $\boldsymbol{R}$ has positive diagonal entries.

a) Show that $\boldsymbol{B}$ is symmetric.

Solution. $\boldsymbol{B}^{\mathrm{T}} = (\boldsymbol{A}^{\mathrm{T}}\boldsymbol{A})^{\mathrm{T}} = \boldsymbol{A}^{\mathrm{T}}\boldsymbol{A} = \boldsymbol{B}$.

b) Show that $\boldsymbol{B}$ has bandwidth $\leq 2d$.

Solution. Let $1 \leq i \leq j \leq n$. Since $a_{ij} = 0$ for all i, j with $|i-j| > d$ we find

$$b_{ij} = \sum_{k=1}^{n} a_{ki}a_{kj} = \sum_{k=\max\{1,j-d\}}^{\min\{n,i+d\}} a_{ki}a_{kj}.$$

If $j > i + 2d$ then $j - d > i + d$ and $b_{ij} = 0$. Since $\boldsymbol{B}$ is symmetric we also have $b_{ji} = 0$. But then $b_{ij} = 0$ for all i, j with $|j - i| > 2d$.

c) Write a MATLAB `function B=ata(A,d)` which computes $\boldsymbol{B}$. You shall exploit the symmetry and the function should only use $\mathcal{O}(cn^2)$ flops, where c only depends on d.

Solution.

code/ata.m

```
function B=ata(A,d)
[m,n]=size(A); B=zeros(n);
for i=1:n
    for j=i:n
        il=max(1,j-d);
        iu=min(n,i+d);
        B(i,j)=A(il:iu,i)'*A(il:iu,j);
        B(j,i)=B(i,j);
    end
end
```

Listing 5.3: Compute the product $\boldsymbol{B} = \boldsymbol{A}^{\mathrm{T}}\boldsymbol{A}$, for a nonsingular symmetric band matrix $\boldsymbol{A}$.

d) Estimate the number of arithmetic operations in your algorithm.

Solution. Since $j \geq i$ we have $iu - il \leq i + d - j + d \leq 2d$, so that the computation of each $B(i,j)$ requires at most $2 \cdot 2d - 1 = \mathcal{O}(4d)$ arithmetic operations, yielding a total of $\mathcal{O}\big(\int_0^n \int_i^n (4d)dj\,di\big) = \mathcal{O}(2dn^2)$ operations.

e) Show that $\boldsymbol{A}^{\mathrm{T}}\boldsymbol{A} = \boldsymbol{R}^{\mathrm{T}}\boldsymbol{R}$.

Solution. Since $\boldsymbol{Q}^{\mathrm{T}}\boldsymbol{Q} = \boldsymbol{I}$, we find

$$\boldsymbol{A}^{\mathrm{T}}\boldsymbol{A} = (\boldsymbol{Q}\boldsymbol{R})^{\mathrm{T}}(\boldsymbol{Q}\boldsymbol{R}) = \boldsymbol{R}^{\mathrm{T}}(\boldsymbol{Q}^{\mathrm{T}}\boldsymbol{Q})\boldsymbol{R} = \boldsymbol{R}^{\mathrm{T}}\boldsymbol{R}.$$

f) Explain why $\boldsymbol{R}$ has upper bandwidth $2d$.

Solution. Since $\boldsymbol{R}$ is upper triangular with positive diagonal elements we see that $\boldsymbol{R}^{\mathrm{T}}\boldsymbol{R}$ is the Cholesky factorization of $\boldsymbol{A}^{\mathrm{T}}\boldsymbol{A}$. From Theorem 4.6 we know that $\boldsymbol{R}$ has the same upper bandwidth $2d$ as $\boldsymbol{A}^{\mathrm{T}}\boldsymbol{A}$.

g) We consider 3 methods for finding the QR factorization of the band matrix $\boldsymbol{A}$, where we assume that n is much bigger than d. The methods are based on

1. Gram-Schmidt orthogonalization,
2. Householder transformations,
3. Givens rotations.

Which method would you recommend for a computer program using floating point arithmetic? Give reasons for your answer.

Solution.

- *Stability:* Method 1 is not stable, while methods 2 and 3 are stable. Method 1 can produce vectors which are not orthogonal.
- *Complexity:* Since $\boldsymbol{A}$ is a band matrix with small bandwidth, method 2 requires more operations than method 1 or 3.
- *Conclusion:* Method 3 is recommended.

Exercise 5.14: Find QR factorization (Exam exercise 2008-2)

Let

$$\boldsymbol{A} := \begin{bmatrix} 2 & 1 \\ 2 & -3 \\ -2 & -1 \\ -2 & 3 \end{bmatrix}$$

a) Find the Cholesky factorization of $\boldsymbol{A}^{\mathrm{T}}\boldsymbol{A}$.

Solution. We find $\boldsymbol{A}^{\mathrm{T}}\boldsymbol{A} = \begin{bmatrix} 16 & -8 \\ -8 & 20 \end{bmatrix}$ and want

$$\boldsymbol{A}^{\mathrm{T}}\boldsymbol{A} = \boldsymbol{R}^{\mathrm{T}}\boldsymbol{R}, \qquad \boldsymbol{R} = \begin{bmatrix} r_{11} & r_{12} \\ 0 & r_{22} \end{bmatrix}, \qquad r_{11}, r_{22} > 0.$$

Since $\boldsymbol{R}^{\mathrm{T}}\boldsymbol{R} = \begin{bmatrix} r_{11}^2 & r_{11}r_{12} \\ r_{11}r_{12} & r_{12}^2 + r_{22}^2 \end{bmatrix}$ we need $r_{11}^2 = 16$, $r_{11}r_{12} = -8$ and $r_{12}^2 + r_{22}^2 = 20$. The solution is $r_{11} = 4$, $r_{12} = -2$ and $r_{22} = 4$.

b) Find the QR factorization of $\boldsymbol{A}$.

Solution. We have already found $\boldsymbol{R} = \begin{bmatrix} 4 & -2 \\ 0 & 4 \end{bmatrix}$. Then, to get $\boldsymbol{A} = \boldsymbol{Q}\boldsymbol{R}$ we need $\boldsymbol{Q} = \boldsymbol{A}\boldsymbol{R}^{-1}$, and since $\boldsymbol{R}^{-1} = \frac{1}{16}\begin{bmatrix} 4 & 2 \\ 0 & 4 \end{bmatrix}$ we find

$$\boldsymbol{Q} = \boldsymbol{A}\boldsymbol{R}^{-1} = \frac{1}{2}\begin{bmatrix} 1 & 1 \\ 1 & -1 \\ -1 & -1 \\ -1 & 1 \end{bmatrix}.$$

Exercises section 5.5

Exercise 5.15: QR using Gram-Schmidt, II

Construct $\boldsymbol{Q}_1$ and $\boldsymbol{R}_1$ in Example 5.2 using Gram-Schmidt orthogonalization.

Solution. Let

$$\boldsymbol{A} = [\boldsymbol{a}_1, \boldsymbol{a}_2, \boldsymbol{a}_3] = \begin{bmatrix} 1 & 3 & 1 \\ 1 & 3 & 7 \\ 1 & -1 & -4 \\ 1 & -1 & 2 \end{bmatrix}.$$

Applying Gram-Schmidt orthogonalization, we find

$$\boldsymbol{v}_1 = \boldsymbol{a}_1 = \begin{bmatrix} 1 \\ 1 \\ 1 \\ 1 \end{bmatrix}, \qquad \boldsymbol{q}_1 = \frac{1}{2}\begin{bmatrix} 1 \\ 1 \\ 1 \\ 1 \end{bmatrix},$$

$$\boldsymbol{v}_2 = \boldsymbol{a}_2 - \underbrace{\frac{\boldsymbol{a}_2^{\mathrm{T}}\boldsymbol{v}_1}{\boldsymbol{v}_1^{\mathrm{T}}\boldsymbol{v}_1}}_{=1}\boldsymbol{v}_1 = \begin{bmatrix} 2 \\ 2 \\ -2 \\ -2 \end{bmatrix}, \qquad \boldsymbol{q}_2 = \frac{1}{2}\begin{bmatrix} 1 \\ 1 \\ -1 \\ -1 \end{bmatrix},$$

$$v_3 = a_3 - \underbrace{\frac{a_3^T v_1}{v_1^T v_1}}_{=3/2} v_1 - \underbrace{\frac{a_3^T v_2}{v_2^T v_2}}_{=5/4} v_2 = \begin{bmatrix} -3 \\ 3 \\ -3 \\ 3 \end{bmatrix}, \qquad q_3 = \frac{1}{2} \begin{bmatrix} -1 \\ 1 \\ -1 \\ 1 \end{bmatrix}.$$

Since

$$(R_1)_{11} = \|v_1\| = 2, \qquad (R_1)_{22} = \|v_2\| = 4, \qquad (R_1)_{33} = \|v_3\| = 6,$$

and since also

$$(R_1)_{ij} = a_j^T q_i = ||v_i|| \frac{a_j^T v_i}{v_i^T v_i}, \qquad i > j,$$

we get

$$(R_1)_{12} = 2 \cdot 1 = 2, \qquad (R_1)_{13} = 2 \cdot \frac{3}{2} = 3, \qquad (R_1)_{23} = 4 \cdot \frac{5}{4} = 5,$$

so that $A = Q_1 R_1$ with

$$Q_1 = [q_1, q_2, q_3] = \frac{1}{2} \begin{bmatrix} 1 & 1 & -1 \\ 1 & 1 & 1 \\ 1 & -1 & -1 \\ 1 & -1 & 1 \end{bmatrix}, \qquad R_1 = \begin{bmatrix} 2 & 2 & 3 \\ 0 & 4 & 5 \\ 0 & 0 & 6 \end{bmatrix}.$$

Exercises section 5.6

Exercise 5.16: Plane rotation

Suppose

$$x = \begin{bmatrix} r\cos\alpha \\ r\sin\alpha \end{bmatrix}, \qquad P = \begin{bmatrix} \cos\theta & \sin\theta \\ -\sin\theta & \cos\theta \end{bmatrix}.$$

Show that

$$Px = \begin{bmatrix} r\cos(\alpha-\theta) \\ r\sin(\alpha-\theta) \end{bmatrix}.$$

Solution. Using the angle difference identities for the sine and cosine functions,

$$\cos(\theta-\alpha) = \cos\theta\cos\alpha + \sin\theta\sin\alpha,$$
$$\sin(\theta-\alpha) = \sin\theta\cos\alpha - \cos\theta\sin\alpha,$$

we find

$$Px = r \begin{bmatrix} \cos\theta\cos\alpha + \sin\theta\sin\alpha \\ -\sin\theta\cos\alpha + \cos\theta\sin\alpha \end{bmatrix} = \begin{bmatrix} r\cos(\theta-\alpha) \\ -r\sin(\theta-\alpha) \end{bmatrix}.$$

Exercise 5.17: Updating the QR decomposition

Let $\boldsymbol{H} \in \mathbb{R}^{4,4}$ be upper Hessenberg. Find Givens rotation matrices $\boldsymbol{G}_1$, $\boldsymbol{G}_2$, $\boldsymbol{G}_3$ such that

$$\boldsymbol{G}_3\boldsymbol{G}_2\boldsymbol{G}_1\boldsymbol{H} = \boldsymbol{R}$$

is upper triangular (here each $\boldsymbol{G}_k = \boldsymbol{P}_{i,j}$ for suitable i, j, c and s, and for each k you are meant to find suitable i and j).

Solution. Recall how rotations in the i, j-plane were defined, see Definition 5.6. To bring an upper Hessenberg 4×4-matrix to upper triangular form, the following three entries need to be zeroed out:

1. $h_{2,1}$: This can be zeroed out with a rotation $\boldsymbol{G}_1$ in the 12-plane.
2. $h_{3,2}$: This can be zeroed out with a rotation $\boldsymbol{G}_2$ in the 23-plane. This does not affect the zeroes in column 1.
3. $h_{4,3}$: This can be zeroed out with a rotation $\boldsymbol{G}_3$ in the 34-plane. This does not affect the zeroes in column 1 and 2.

Exercise 5.18: Solving upper Hessenberg system using rotations

Let $\boldsymbol{A} \in \mathbb{R}^{n \times n}$ be upper Hessenberg and nonsingular, and let $\boldsymbol{b} \in \mathbb{R}^n$. The following algorithm (Algorithm 5.3 in the book) solves the linear system $\boldsymbol{Ax} = \boldsymbol{b}$ using rotations $\boldsymbol{P}_{k,k+1}$ for $k = 1, \ldots, n-1$. It uses Algorithm 3.2 (`backsolve`). Determine the number of arithmetic operations of this algorithm.

code/rothesstri.m

```
function x=rothesstri(A,b)
n=length(A); A=[A b];
for k=1:n-1
   r=norm([A(k,k),A(k+1,k)]);
   if r>0
      c=A(k,k)/r; s=A(k+1,k)/r;
      A([k k+1],k+1:n+1)
            =[c s;-s c]*A([k k+1],k+1:n+1);
   end
   A(k,k)=r; A(k+1,k)=0;
end
x=rbacksolve(A(:,1:n),A(:,n+1),n);
```

Listing 5.4: Solve the upper Hessenberg system $\boldsymbol{Ax} = \boldsymbol{b}$ using rotations.

Solution. To determine the number of arithmetic operations of Algorithm 5.3, we first consider the arithmetic operations in each step. Initially the algorithm stores the length of the matrix and appends the right hand side as the $(n+1)$-th column to the matrix. Such copying and storing operations do not count as arithmetic operations.

The second big step is the loop. Let us consider the arithmetic operations at the k-th iteration of this loop. First we have to compute the norm of a two-dimensional vector, which comprises 4 arithmetic operations: two multiplications, one addition and one square root operation. Assuming $r > 0$ we compute c and s each in one division, adding 2 arithmetic operations to our count. Computing the product of the Givens rotation and $\boldsymbol{A}$ includes 2 multiplications and one addition for each entry of the result. As we have $2(n+1-k)$ entries, this amounts to $6(n+1-k)$ arithmetic operations. The last operation in the loop is just the storage of two entries of $\boldsymbol{A}$, which again does not count as an arithmetic operation.

The final step of the whole algorithm is a backward substitution, known to require $\mathcal{O}(n^2)$ arithmetic operations. We conclude that the algorithm uses

$$\begin{aligned}
&\mathcal{O}(n^2) + \sum_{k=1}^{n-1} \big(4 + 2 + 6(n+1-k)\big) \\
&= \mathcal{O}(n^2) + 6\sum_{k=1}^{n-1} (n+2-k) \\
&= \mathcal{O}(n^2) + 3n^2 + 9n - 12 \\
&= \mathcal{O}(4n^2)
\end{aligned}$$

arithmetic operations.

Exercise 5.19: A Givens transformation (Exam exercise 2013-2)

A Givens rotation of order 2 has the form $\boldsymbol{G} := \begin{bmatrix} c & s \\ -s & c \end{bmatrix} \in \mathbb{R}^{2\times 2}$, where $s^2 + c^2 = 1$.

a) Is $\boldsymbol{G}$ symmetric and unitary?

Solution. The matrix $\boldsymbol{G}$ is only symmetric for $s = 0$. In addition, $\boldsymbol{G}$ is unitary since

$$\boldsymbol{G}^*\boldsymbol{G} = \boldsymbol{G}^{\mathrm{T}}\boldsymbol{G} = \begin{bmatrix} s^2 + c^2 & 0 \\ 0 & s^2 + c^2 \end{bmatrix} = \boldsymbol{I}.$$

b) Given $x_1, x_2 \in \mathbb{R}$ and set $r := \sqrt{x_1^2 + x_2^2}$. Find $\boldsymbol{G}$ and y_1, y_2 so that $y_1 = y_2$, where $\begin{bmatrix} y_1 \\ y_2 \end{bmatrix} = \boldsymbol{G}\begin{bmatrix} x_1 \\ x_2 \end{bmatrix}$.

Solution. The goal of this exercise is to find Givens rotations in the plane that maps to the diagonal; see Figure 5.2.

We find $y_1 = y_2$ if and only if $cx_1 + sx_2 = -sx_1 + cx_2$ and $s^2 + c^2 = 1$. Thus s and c must be solutions of

$$(x_1 + x_2)s + (x_1 - x_2)c = 0$$
$$s^2 + c^2 = 1.$$

If $x_1 = x_2$ then the point is already on the diagonal and the solution is $s = 0$ and $c = \pm 1$, corresponding to the identity map and the half-turn. Suppose $x_1 \neq x_2$. Substituting $c = \frac{x_1+x_2}{x_2-x_1} s$ into $1 = s^2 + c^2$ we find

$$1 = s^2 \left(1 + \frac{(x_1 + x_2)^2}{(x_2 - x_1)^2}\right) = s^2 \frac{(x_2 - x_1)^2 + (x_1 + x_2)^2}{(x_2 - x_1)^2} = s^2 \frac{2r^2}{(x_2 - x_1)^2}.$$

There are two solutions

$$s_1 = \frac{x_2 - x_1}{r\sqrt{2}}, \; c_1 = \frac{x_2 + x_1}{r\sqrt{2}}, \qquad s_2 = -s_1, \; c_2 = -c_1.$$

We find corresponding solutions

$$y_1 = y_2 = c_1 x_1 + s_1 x_2 = \frac{(x_1 + x_2)x_1 + (x_2 - x_1)x_2}{r\sqrt{2}} = r/\sqrt{2}$$

and

$$y_1 = y_2 = c_2 x_1 + s_2 x_2 = -(c_1 x_1 + s_1 x_2) = -r/\sqrt{2}.$$

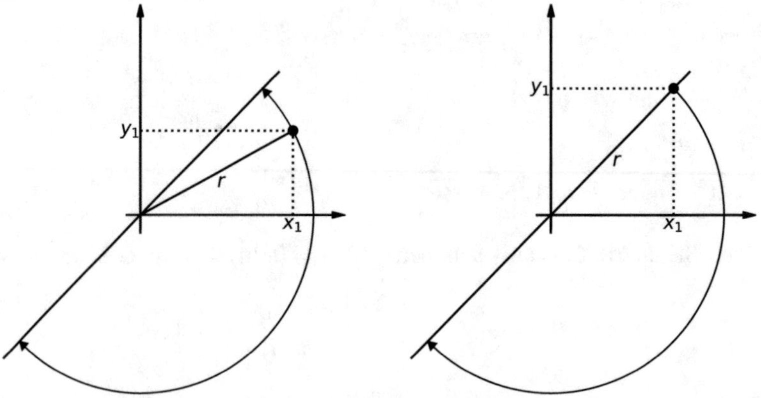

Figure 5.2: The Givens transformations in Exercise 5.19 mapping generic points (left) to the diagonal, and each point on the diagonal (right) to itself and its antipodal.

Exercise 5.20: Givens transformations (Exam exercise 2016-3)

Recall that a rotation in the ij-plane is an $m \times m$-matrix, denoted $\boldsymbol{P}_{i,j}$, which differs from the identity matrix only in the entries ii, ij, ji, jj, which equal

$$\begin{bmatrix} p_{ii} & p_{ij} \\ p_{ji} & p_{jj} \end{bmatrix} = \begin{bmatrix} \cos\theta & \sin\theta \\ -\sin\theta & \cos\theta \end{bmatrix},$$

i.e., these four entries are those of a Givens rotation.

a) For $\theta \in \mathbb{R}$, let $\boldsymbol{P}$ be a Givens rotation of the form

$$\boldsymbol{P} = \begin{bmatrix} \cos\theta & \sin\theta \\ -\sin\theta & \cos\theta \end{bmatrix}$$

and let $\boldsymbol{x}$ be a fixed vector in $\mathbb{R}^2$. Show that there exists a unique $\theta \in (-\pi/2, \pi/2]$ so that $\boldsymbol{P}\boldsymbol{x} = \pm\|\boldsymbol{x}\|_2\boldsymbol{e}_1$, where $\boldsymbol{e}_1 = (1,0)^{\mathrm{T}}$.

Solution. The matrix $\boldsymbol{P}$ represents a clockwise rotation with angle θ, and this is unitary, i.e., it preserves length. Write $\boldsymbol{x} = \|\boldsymbol{x}\|_2[\cos\phi, \sin\phi]$ in polar coordinates, with $-\pi \leq \phi < \pi$. We can always find an angle $\theta \in (-\pi/2, \pi/2]$ so that $\boldsymbol{x}$ is rotated onto the positive or negative x-axis, i.e., onto $\pm\|\boldsymbol{x}\|_2\boldsymbol{e}_1$ and so that $\boldsymbol{P}\boldsymbol{x} = \pm\|\boldsymbol{x}\|_2\boldsymbol{e}_1$:

- If $\phi \in (-\pi/2, \pi/2]$, a rotation with $\theta = \phi$ rotates $\boldsymbol{x}$ to the positive x-axis.
- If $\phi \in [-\pi, -\pi/2]$, a rotation with $\theta = \pi + \phi$ rotates $\boldsymbol{x}$ to the negative x-axis.
- If $\phi \in (\pi/2, \pi)$, a rotation with $\theta = \phi - \pi$ rotates $\boldsymbol{x}$ to the negative x-axis.

b) Show that, for any vector $\boldsymbol{w} \in \mathbb{R}^m$, one can find rotations in the 12-plane, 23-plane, …, $(m-1)m$-plane, so that

$$\boldsymbol{P}_{1,2}\boldsymbol{P}_{2,3}\cdots\boldsymbol{P}_{m-2,m-1}\boldsymbol{P}_{m-1,m}\boldsymbol{w} = \begin{bmatrix} \alpha \\ 0 \\ \vdots \\ 0 \end{bmatrix},$$

where $\alpha = \pm\|\boldsymbol{w}\|_2$.

Solution. A rotation $\boldsymbol{P}_{m-1,m}$ only changes the last two entries, and by **a)** we can find such a rotation that zeros out component m. We now proceed by induction: Assume that we have found rotations in the corresponding planes so that $\boldsymbol{P}_{k-1,k}\cdots\boldsymbol{P}_{m-1,m}\boldsymbol{w}$ is zero in components $k, \ldots, m$. A rotation $\boldsymbol{P}_{k-2,k-1}$ only changes components $k-2$ and $k-1$, and clearly we can find such a rotation that zeros out component $k-1$ as well. At the final step we find a plane rotation $\boldsymbol{P}_{1,2}$ which zeroes out the second component, and we now have a vector of the form $k\boldsymbol{e}_1$. We must have that $k = \pm\|\boldsymbol{w}\|_2$, since all plane rotations preserve length.

c) Assume that $m \geq n$. Recall that an $m \times n$-matrix $\boldsymbol{A}$ with entries $a_{i,j}$ is called upper trapezoidal if there are no nonzero entries below the main diagonal

$$(a_{1,1}, a_{2,2}, \ldots, a_{n,n})$$

(for $m = n$, upper trapezoidal is the same as upper triangular). Recall also that an $m \times n$-matrix is said to be in upper Hessenberg form if there are no nonzero entries below the subdiagonal

$$(a_{2,1}, a_{3,2}, \ldots, a_{n,n-1}).$$

Explain that, if an $m \times n$-matrix $\boldsymbol{H}$ is in upper Hessenberg form, one can find plane rotations so that

$$\boldsymbol{P}_{m-1,m}\boldsymbol{P}_{m-2,m-1} \cdots \boldsymbol{P}_{2,3}\boldsymbol{P}_{1,2}\boldsymbol{H}$$

is upper trapezoidal.

Solution. The plane rotation $\boldsymbol{P}_{1,2}$ only changes the two first components in any column of $\boldsymbol{H}$. Using **a)** and **b)** we can find such a plane rotation which zeros out the entry $(2, 1)$ in $\boldsymbol{H}$. Such a plane rotation will keep $\boldsymbol{H}$ in upper Hessenberg form. We then find a plane rotation $\boldsymbol{P}_{2,3}$ which zeros out entry $(3, 2)$. After this the first two columns will have zeros below the diagonal, and $\boldsymbol{H}$ will still be in upper Hessenberg form. We can continue in this way to find plane rotations so that

$$\boldsymbol{P}_{m-1,m}\boldsymbol{P}_{m-2,m-1} \cdots \boldsymbol{P}_{2,3}\boldsymbol{P}_{1,2}\boldsymbol{H}$$

is upper trapezoidal.

d) Let again $\boldsymbol{A}$ be an $m \times n$-matrix with $m \geq n$, and let $\boldsymbol{A}_-$ be the matrix obtained by removing column k in $\boldsymbol{A}$. Explain how you can find a QR decomposition of $\boldsymbol{A}_-$, when we already have a QR decomposition $\boldsymbol{A} = \boldsymbol{Q}\boldsymbol{R}$ of $\boldsymbol{A}$.

Hint.
Consider the matrix $\boldsymbol{Q}^{\mathrm{T}}\boldsymbol{A}_-$.

Solution. Let $\boldsymbol{a}_i$ be the columns of $\boldsymbol{A}$, and let $\boldsymbol{r}_i$ be the columns of $\boldsymbol{R}$. Since $\boldsymbol{Q}^{\mathrm{T}}\boldsymbol{A} = \boldsymbol{R}$ we have that

$$\begin{aligned}\boldsymbol{Q}^{\mathrm{T}}\boldsymbol{A}_- &= \left[\boldsymbol{Q}^{\mathrm{T}}\boldsymbol{a}_1 \cdots \boldsymbol{Q}^{\mathrm{T}}\boldsymbol{a}_{k-1}\ \boldsymbol{Q}^{\mathrm{T}}\boldsymbol{a}_{k+1} \cdots \boldsymbol{Q}^{\mathrm{T}}\boldsymbol{a}_n\right] \\ &= \left[\boldsymbol{r}_1 \cdots \boldsymbol{r}_{k-1}\ \boldsymbol{r}_{k+1} \cdots \boldsymbol{r}_n\right],\end{aligned}$$

which clearly is in upper Hessenberg form. Due to **c)** we can find plane rotations so that

$$\boldsymbol{P}_{m-1,m}\boldsymbol{P}_{m-2,m-1} \cdots \boldsymbol{P}_{2,3,n-1}\boldsymbol{P}_{1,2}\boldsymbol{Q}^{\mathrm{T}}\boldsymbol{A}_- = \boldsymbol{R}_1,$$

where $\boldsymbol{R}_1$ is upper trapezoidal. We thus have that

$$\boldsymbol{A}_- = \boldsymbol{Q}\boldsymbol{P}_{1,2}^{\mathrm{T}}\boldsymbol{P}_{2,3}^{\mathrm{T}} \cdots \boldsymbol{P}_{m-2,m-1}^{\mathrm{T}}\boldsymbol{P}_{m-1,m}^{\mathrm{T}}\boldsymbol{R}_1,$$

which gives a QR decomposition of $\boldsymbol{A}_-$.

Exercise 5.21: Cholesky and Givens (Exam exercise 2018-2)

Assume that $\boldsymbol{A}$ is $n \times n$ symmetric positive definite, and with Cholesky factorization $\boldsymbol{A} = \boldsymbol{L}\boldsymbol{L}^*$. Assume also that $\boldsymbol{z}$ is a given column vector of length n.

a) Explain why $\boldsymbol{A} + \boldsymbol{z}\boldsymbol{z}^*$ has a unique Cholesky factorization.

Solution. Since $\boldsymbol{A}$ is symmetric positive definite, and $\boldsymbol{z}\boldsymbol{z}^*$ is symmetric positive semidefinite, it follows that

$$\boldsymbol{x} \neq 0 \qquad \Longrightarrow \qquad \boldsymbol{x}^{\mathrm{T}}(\boldsymbol{A} + \boldsymbol{z}\boldsymbol{z}^*)\boldsymbol{x} = \boldsymbol{x}^{\mathrm{T}}\boldsymbol{A}\boldsymbol{x} + \boldsymbol{x}^{\mathrm{T}}\boldsymbol{z}\boldsymbol{z}^*\boldsymbol{x} > 0,$$

and hence that $\boldsymbol{A} + \boldsymbol{z}\boldsymbol{z}^*$ is symmetric positive definite. Any symmetric matrix is positive definite if and only if it has a (unique) Cholesky factorization (see Theorem 4.2).

b) Assume that we are given a QR decomposition

$$\begin{bmatrix} \boldsymbol{L}^* \\ \boldsymbol{z}^* \end{bmatrix} = \boldsymbol{Q} \begin{bmatrix} \boldsymbol{R} \\ 0 \end{bmatrix},$$

with $\boldsymbol{R}$ square and upper triangular. Explain why $\boldsymbol{R}$ is nonsingular. Explain also that, if $\boldsymbol{R}$ also has nonnegative diagonal entries, then $\boldsymbol{A} + \boldsymbol{z}\boldsymbol{z}^*$ has the Cholesky factorization $\boldsymbol{R}^*\boldsymbol{R}$.

Solution. We obtain

$$\begin{aligned} \boldsymbol{A} + \boldsymbol{z}\boldsymbol{z}^* &= \boldsymbol{L}\boldsymbol{L}^* + \boldsymbol{z}\boldsymbol{z}^* = \begin{bmatrix} \boldsymbol{L} & \boldsymbol{z} \end{bmatrix} \begin{bmatrix} \boldsymbol{L}^* \\ \boldsymbol{z}^* \end{bmatrix} \\ &= \begin{bmatrix} \boldsymbol{R}^* & 0 \end{bmatrix} \boldsymbol{Q}^{\mathrm{T}}\boldsymbol{Q} \begin{bmatrix} \boldsymbol{R} \\ 0 \end{bmatrix} \\ &= \begin{bmatrix} \boldsymbol{R}^* & 0 \end{bmatrix} \begin{bmatrix} \boldsymbol{R} \\ 0 \end{bmatrix} = \boldsymbol{R}^*\boldsymbol{R}. \end{aligned}$$

Here $\begin{bmatrix} \boldsymbol{L}^* \\ \boldsymbol{z}^* \end{bmatrix}$ has rank n, and hence the same holds for $\begin{bmatrix} \boldsymbol{R} \\ 0 \end{bmatrix}$. But this is the case if and only if $\boldsymbol{R}$ is nonsingular. If the diagonal elements of $\boldsymbol{R}$ are nonnegative, they must be positive, since $\boldsymbol{R}$ is nonsingular. Thus $\boldsymbol{R}^*$ is lower triangular with positive diagonal elements, so that $\boldsymbol{R}^*\boldsymbol{R}$ is the Cholesky factorization.

Recall that a plane rotation in the (i, j)-plane, denoted $P_{i,j}$, is an $n \times n$-matrix which differs from the identity matrix only in the entries (i,i), (i,j), (j,i), (j,j), which equal those of a Givens rotation, i.e., they are

$$\begin{bmatrix} p_{ii} & p_{ij} \\ p_{ji} & p_{jj} \end{bmatrix} = \begin{bmatrix} \cos\theta & \sin\theta \\ -\sin\theta & \cos\theta \end{bmatrix}.$$

c) Explain how one can find plane rotations $P_{i_1,n+1}$, $P_{i_2,n+1}$,...,$P_{i_n,n+1}$ so that

$$P_{i_1,n+1} P_{i_2,n+1} \cdots P_{i_n,n+1} \begin{bmatrix} \boldsymbol{L}^* \\ \boldsymbol{z}^* \end{bmatrix} = \begin{bmatrix} \boldsymbol{R}' \\ 0 \end{bmatrix}. \tag{5.23}$$

with $\boldsymbol{R}'$ upper triangular, and explain how to obtain a QR decomposition of $\begin{bmatrix} \boldsymbol{L}^* \\ \boldsymbol{z}^* \end{bmatrix}$ from this. In particular you should write down the numbers $i_1, \ldots, i_n$. Is it possible to choose the plane rotations so that $\boldsymbol{R}'$ in (5.23) also has positive diagonal entries?

Solution. In the book, rectangular matrices which are zero below the main diagonal were called upper trapezoidal; hence the matrix

$$\boldsymbol{B}_0 := \begin{bmatrix} \boldsymbol{L}^* \\ \boldsymbol{z}^* \end{bmatrix}$$

is upper trapezoidal except for the last row. We can clearly find a Givens rotation $\boldsymbol{P}_{1,n+1}$ in the $(1, n+1)$-plane so that $\boldsymbol{B}_1 := \boldsymbol{P}_{1,n+1}\boldsymbol{B}_0$ has a zero in entry $(n+1, 1)$, and a nonzero in entry $(1, 1)$. This is because a Givens rotation with angle θ maps

$$\begin{bmatrix} r\cos\alpha \\ r\sin\alpha \end{bmatrix} \longmapsto \begin{bmatrix} r\cos(\alpha-\theta) \\ r\sin(\alpha-\theta) \end{bmatrix},$$

so choosing $\theta = \alpha$ maps to the positive x-axis, while choosing $\theta = \alpha + \pi$ maps to the negative x-axis. The resulting matrix will still be upper trapezoidal except for the last row, since $\boldsymbol{P}_{1,n+1}$ changes only rows 1 and $n + 1$.

Assume now that we have found Givens rotations that have mapped $\boldsymbol{B}_0$ to a matrix $\boldsymbol{B}_k$ with zeroes in the first k entries of row $n + 1$, and upper trapezoidal with nonzero diagonal elements, except for the last row. We find a Givens rotation $\boldsymbol{P}_{k+1,n+1}$ so that $\boldsymbol{B}_{k+1} := \boldsymbol{P}_{k+1,n+1}\boldsymbol{B}_k$ has a zero in entry $(n + 1, k + 1)$. This rotation will affect only rows $n+1$ and $k+1$, and since the first k elements in both these rows in $\boldsymbol{B}_k$ are zero, the same will be the case for $\boldsymbol{B}_{k+1}$. This proves that the final matrix we obtain after n Givens rotations will be upper trapezoidal, so that

$$\boldsymbol{P}_{n,n+1}\boldsymbol{P}_{n-1,n+1} \cdots \boldsymbol{P}_{1,n+1} \begin{bmatrix} \boldsymbol{L}^* \\ \boldsymbol{z}^* \end{bmatrix} = \begin{bmatrix} \boldsymbol{R}' \\ 0 \end{bmatrix},$$

with $\boldsymbol{R}'$ upper triangular. In particular we can set $i_k := n+1-k$ for all k. We now obtain

$$\begin{bmatrix} \boldsymbol{L}^* \\ \boldsymbol{z}^* \end{bmatrix} = \boldsymbol{Q} \begin{bmatrix} \boldsymbol{R}' \\ 0 \end{bmatrix}, \qquad \boldsymbol{Q} := \boldsymbol{P}_{1,n+1}^{\mathrm{T}} \boldsymbol{P}_{2,n+1}^{\mathrm{T}} \cdots \boldsymbol{P}_{n,n+1}^{\mathrm{T}}.$$

Since all the Givens rotations are unitary, their product is also unitary, so that we have factored the matrix as a product of a unitary matrix $\boldsymbol{Q}$ and an upper trapezoidal matrix, i.e., we have a QR decomposition.

If we choose the angles in the Givens rotations so that all vectors are mapped to the positive x-axis, the diagonal elements of $\boldsymbol{R}'$ will also be positive.

Chapter 6
Eigenpairs and Similarity Transformations

Exercises section 6.1

Exercise 6.1: Eigenvalues of a block triangular matrix

What are the eigenvalues of the matrix

$$\begin{bmatrix} 2&1&0&0&0&0&0&0\\ 0&2&1&0&0&0&0&0\\ 0&0&2&0&0&0&0&0\\ 0&0&0&2&0&0&0&0\\ 0&0&0&1&2&0&0&0\\ 0&0&0&0&0&2&0&0\\ 0&0&0&0&0&0&3&0\\ 0&0&0&0&0&0&1&3 \end{bmatrix} \in \mathbb{R}^{8,8}?$$

Solution. Successive cofactor expansion along the 1st column, 2nd column, 3rd column, 4th row, 5th column, 6th row, and 7th row, yields

$$\det(\boldsymbol{A}-\lambda\boldsymbol{I}) = (2-\lambda)\cdot(2-\lambda)\cdot(2-\lambda)\cdot(2-\lambda)\cdot(2-\lambda)\cdot(2-\lambda)\cdot(3-\lambda)\cdot(3-\lambda).$$

It follows that $\boldsymbol{A}$ has eigenvalue 2 (with algebraic multiplicity 6) and eigenvalue 3 (with algebraic multiplicity 2).

Exercise 6.2: Characteristic polynomial of transpose

We have $\det(\boldsymbol{B}^{\mathrm{T}}) = \det(\boldsymbol{B})$ and $\det(\overline{\boldsymbol{B}}) = \overline{\det(\boldsymbol{B})}$ for any square matrix $\boldsymbol{B}$. Use this to show that

a) $\pi_{\boldsymbol{A}^{\mathrm{T}}} = \pi_{\boldsymbol{A}}$,

Solution. One obtains

T. Lyche et al., *Exercises in Numerical Linear Algebra and Matrix Factorizations*, Texts in Computational Science and Engineering 23, https://doi.org/10.1007/978-3-030-59789-4_6

$$\pi_{\boldsymbol{A}^{\mathrm{T}}}(\lambda) = \det(\boldsymbol{A}^{\mathrm{T}} - \lambda \boldsymbol{I}) = \det((\boldsymbol{A} - \lambda \boldsymbol{I})^{\mathrm{T}}) = \det(\boldsymbol{A} - \lambda \boldsymbol{I}) = \pi_{\boldsymbol{A}}(\lambda).$$

b) $\pi_{\boldsymbol{A}^*}(\bar{\lambda}) = \overline{\pi_{\boldsymbol{A}}(\lambda)}$.

Solution. One obtains

$$\pi_{\boldsymbol{A}^*}(\bar{\lambda}) = \det(\boldsymbol{A}^* - \bar{\lambda}\boldsymbol{I}) = \det(\overline{\boldsymbol{A}^{\mathrm{T}} - \lambda \boldsymbol{I}}) = \overline{\det(\boldsymbol{A}^{\mathrm{T}} - \lambda \boldsymbol{I})} = \overline{\pi_{\boldsymbol{A}^{\mathrm{T}}}(\lambda)} = \overline{\pi_{\boldsymbol{A}}(\lambda)}.$$

Exercise 6.3: Characteristic polynomial of inverse

Suppose $(\lambda, \boldsymbol{x})$ is an eigenpair for $\boldsymbol{A} \in \mathbb{C}^{n \times n}$. Show that

a) if $\boldsymbol{A}$ is nonsingular then $(\lambda^{-1}, \boldsymbol{x})$ is an eigenpair for $\boldsymbol{A}^{-1}$.

Solution. If $\boldsymbol{A}\boldsymbol{x} = \lambda \boldsymbol{x}$ we have that $\boldsymbol{A}^{-1}(\lambda \boldsymbol{x}) = \boldsymbol{x}$, hence $\boldsymbol{A}^{-1}\boldsymbol{x} = \lambda^{-1}\boldsymbol{x}$, so that $(\lambda^{-1}, \boldsymbol{x})$ is an eigenpair for $\boldsymbol{A}^{-1}$.

b) $(\lambda^k, \boldsymbol{x})$ is an eigenpair for $\boldsymbol{A}^k$ for $k \in \mathbb{Z}$.

Solution. Since $\boldsymbol{I}\boldsymbol{x} = 1 \cdot \boldsymbol{x}$, the statement is immediate for $k = 0$. Repeatedly multiplying by $\boldsymbol{A}$, and using that $\boldsymbol{A}\boldsymbol{x} = \lambda \boldsymbol{x}$, yields

$$\boldsymbol{A}^k \boldsymbol{x} = \boldsymbol{A}^{k-1}\boldsymbol{A}\boldsymbol{x} = \boldsymbol{A}^{k-2}\boldsymbol{A}\lambda \boldsymbol{x} = \boldsymbol{A}^{k-3}\boldsymbol{A}\lambda^2 \boldsymbol{x} = \cdots = \lambda^k \boldsymbol{x}.$$

Applying this result to the matrix $\boldsymbol{A}^{-1}$, which has eigenpair $(\lambda^{-1}, \boldsymbol{x})$ by 1., it follows that 2. holds for any $k \in \mathbb{Z}$.

Exercise 6.4: The power of the eigenvector expansion

Show that if $\boldsymbol{A} \in \mathbb{C}^{n \times n}$ is nondefective with eigenpairs $(\lambda_j, \boldsymbol{x}_j)$, $j = 1, \ldots, n$ then for any $\boldsymbol{x} \in \mathbb{C}^n$ and $k \in \mathbb{N}$

$$\boldsymbol{A}^k \boldsymbol{x} = \sum_{j=1}^{n} c_j \lambda_j^k \boldsymbol{x}_j, \qquad \text{for some scalars } c_1, \ldots, c_n. \tag{6.19}$$

Show that if $\boldsymbol{A}$ is nonsingular then (6.19) holds for all $k \in \mathbb{Z}$.

Solution. If the eigenvectors form a basis we can write

$$\boldsymbol{x} = \sum_{j=1}^{n} c_j \boldsymbol{x}_j$$

for some scalars $c_1, \ldots, c_n$. But then

$$\boldsymbol{A}\boldsymbol{x} = \sum_{j=1}^{n} c_j \boldsymbol{A}\boldsymbol{x}_j = \sum_{j=1}^{n} c_j \lambda_j \boldsymbol{x}_j.$$

Iterating this we obtain

$$\boldsymbol{A}^k\boldsymbol{x} = \sum_{j=1}^{n} c_j \boldsymbol{A}^k\boldsymbol{x}_j = \sum_{j=1}^{n} c_j \lambda_j^k \boldsymbol{x}_j.$$

Exercise 6.5: Eigenvalues of an idempotent matrix

Let $\lambda \in \sigma(\boldsymbol{A})$ where $\boldsymbol{A}^2 = \boldsymbol{A} \in \mathbb{C}^{n\times n}$. Show that $\lambda = 0$ or $\lambda = 1$. (A matrix is called *idempotent* if $\boldsymbol{A}^2 = \boldsymbol{A}$.)

Solution. Suppose that $(\lambda, \boldsymbol{x})$ is an eigenpair of a matrix $\boldsymbol{A}$ satisfying $\boldsymbol{A}^2 = \boldsymbol{A}$. Then

$$\lambda\boldsymbol{x} = \boldsymbol{A}\boldsymbol{x} = \boldsymbol{A}^2\boldsymbol{x} = \lambda\boldsymbol{A}\boldsymbol{x} = \lambda^2\boldsymbol{x}.$$

Since any eigenvector is nonzero, one has $\lambda = \lambda^2$, from which it follows that either $\lambda = 0$ or $\lambda = 1$. We conclude that the eigenvalues of any idempotent matrix can only be zero or one.

Exercise 6.6: Eigenvalues of a nilpotent matrix

Let $\lambda \in \sigma(\boldsymbol{A})$ where $\boldsymbol{A}^k = 0$ for some $k \in \mathbb{N}$. Show that $\lambda = 0$. (A matrix $\boldsymbol{A} \in \mathbb{C}^{n\times n}$ such that $\boldsymbol{A}^k = 0$ for some $k \in \mathbb{N}$ is called *nilpotent*.)

Solution. Suppose that $(\lambda, \boldsymbol{x})$ is an eigenpair of a matrix $\boldsymbol{A}$ satisfying $\boldsymbol{A}^k = 0$ for some natural number k. Then

$$0 = \boldsymbol{A}^k\boldsymbol{x} = \lambda\boldsymbol{A}^{k-1}\boldsymbol{x} = \lambda^2\boldsymbol{A}^{k-2}\boldsymbol{x} = \cdots = \lambda^k\boldsymbol{x}.$$

Since any eigenvector is nonzero, one has $\lambda^k = 0$, from which it follows that $\lambda = 0$. We conclude that any eigenvalue of a nilpotent matrix is zero.

Exercise 6.7: Eigenvalues of a unitary matrix

Let $\lambda \in \sigma(\boldsymbol{A})$, where $\boldsymbol{A}^*\boldsymbol{A} = \boldsymbol{I}$. Show that $|\lambda| = 1$.

Solution. Let $\boldsymbol{x}$ be an eigenvector corresponding to λ. Then $\boldsymbol{A}\boldsymbol{x} = \lambda\boldsymbol{x}$ and, as a consequence, $\boldsymbol{x}^*\boldsymbol{A}^* = \boldsymbol{x}^*\overline{\lambda}$. To use that $\boldsymbol{A}^*\boldsymbol{A} = \boldsymbol{I}$, it is tempting to multiply the left hand sides of these equations, yielding

$$|\lambda|^2\|\boldsymbol{x}\|^2 = \boldsymbol{x}^*\overline{\lambda}\lambda\boldsymbol{x} = \boldsymbol{x}^*\boldsymbol{A}^*\boldsymbol{A}\boldsymbol{x} = \boldsymbol{x}^*\boldsymbol{I}\boldsymbol{x} = \|\boldsymbol{x}\|^2.$$

Since $\boldsymbol{x}$ is an eigenvector, it must be nonzero. Nonzero vectors have nonzero norms, and we can therefore divide the above equation by $\|\boldsymbol{x}\|^2$, which results in $|\lambda|^2 =$

1. Taking square roots we find that $|\lambda| = 1$, which is what needed to be shown. Apparently the eigenvalues of any unitary matrix reside on the unit circle in the complex plane.

Exercise 6.8: Nonsingular approximation of a singular matrix

Suppose $\boldsymbol{A} \in \mathbb{C}^{n \times n}$ is singular. Then we can find $\epsilon_0 > 0$ such that $\boldsymbol{A} + \epsilon \boldsymbol{I}$ is nonsingular for all $\epsilon \in \mathbb{C}$ with $|\epsilon| < \epsilon_0$.

Hint.
Use that $\det(\boldsymbol{A}) = \lambda_1 \lambda_2 \cdots \lambda_n$, where λ_i are the eigenvalues of $\boldsymbol{A}$.

Solution. Let $\lambda_1, \ldots, \lambda_n$ be the eigenvalues of the matrix $\boldsymbol{A}$. As the matrix $\boldsymbol{A}$ is singular, its determinant $\det(\boldsymbol{A}) = \lambda_1 \cdots \lambda_n$ is zero, implying that one of its eigenvalues is zero. If all the eigenvalues of $\boldsymbol{A}$ are zero let $\varepsilon_0 := 1$. Otherwise, let $\varepsilon_0 := \min_{\lambda_i \neq 0} |\lambda_i|$ be the absolute value of the eigenvalue closest to zero. By definition of the eigenvalues, $\det(\boldsymbol{A} - \lambda \boldsymbol{I})$ is zero for $\lambda = \lambda_1, \ldots, \lambda_n$, and nonzero otherwise. In particular $\det(\boldsymbol{A} - \varepsilon \boldsymbol{I})$ is nonzero for any $\varepsilon \in (0, \varepsilon_0)$, and $\boldsymbol{A} - \varepsilon \boldsymbol{I}$ will be nonsingular in this interval.

Exercise 6.9: Companion matrix

For $q_0, \ldots, q_{n-1} \in \mathbb{C}$ let $p(\lambda) = \lambda^n + q_{n-1}\lambda^{n-1} + \cdots + q_0$ be a polynomial of degree n in λ. We derive two matrices that have $(-1)^n p$ as its characteristic polynomial.

a) Show that $p = (-1)^n \pi_{\boldsymbol{A}}$ where

$$\boldsymbol{A} = \begin{bmatrix} -q_{n-1} & -q_{n-2} & \cdots & -q_1 & -q_0 \\ 1 & 0 & \cdots & 0 & 0 \\ 0 & 1 & \cdots & 0 & 0 \\ \vdots & \vdots & \ddots & \vdots & \vdots \\ 0 & 0 & \cdots & 1 & 0 \end{bmatrix}.$$

$\boldsymbol{A}$ is called a *companion matrix* of p.

Solution. To show that $(-1)^n f$ is the characteristic polynomial $\pi_{\boldsymbol{A}}$ of the matrix $\boldsymbol{A}$, we need to compute

$$\pi_{\boldsymbol{A}}(\lambda) = \det(\boldsymbol{A} - \lambda \boldsymbol{I}) = \det \begin{bmatrix} -q_{n-1} - \lambda & -q_{n-2} & \cdots & -q_1 & -q_0 \\ 1 & -\lambda & \cdots & 0 & 0 \\ 0 & 1 & \cdots & 0 & 0 \\ \vdots & \vdots & \ddots & \vdots & \vdots \\ 0 & 0 & \cdots & 1 & -\lambda \end{bmatrix}.$$

By the rules of determinant evaluation, we can subtract from any column a linear combination of the other columns without changing the value of the determinant. Multiply columns $1, 2, \ldots, n-1$ by $\lambda^{n-1}, \lambda^{n-2}, \ldots, \lambda$ and adding the corresponding linear combination to the final column, we find

$$\pi_{\boldsymbol{A}}(\lambda) = \det \begin{bmatrix} -q_{n-1}-\lambda & -q_{n-2} & \cdots & -q_1 & -f(\lambda) \\ 1 & -\lambda & \cdots & 0 & 0 \\ 0 & 1 & \cdots & 0 & 0 \\ \vdots & \vdots & \ddots & \vdots & \vdots \\ 0 & 0 & \cdots & 1 & 0 \end{bmatrix} = (-1)^n f(\lambda),$$

where the second equality follows from cofactor expansion along the final column. Multiplying this equation by $(-1)^n$ yields the statement of the Exercise.

b) Show that $p = (-1)^n \pi_{\boldsymbol{B}}$ where

$$\boldsymbol{B} = \begin{bmatrix} 0 & 0 & \cdots & 0 & -q_0 \\ 1 & 0 & \cdots & 0 & -q_1 \\ 0 & 1 & \cdots & 0 & -q_2 \\ \vdots & \vdots & \ddots & \vdots & \vdots \\ 0 & 0 & \cdots & 1 & -q_{n-1} \end{bmatrix}.$$

Thus $\boldsymbol{B}$ can also be regarded as a companion matrix for p.

Solution. Similar to **a)**, by multiplying rows $2, 3, \ldots, n$ by $\lambda, \lambda^2, \ldots, \lambda^{n-1}$ and adding the corresponding linear combination to the first row.

Exercise 6.10: Find eigenpair example

Find the eigenvalues and eigenvectors of $\boldsymbol{A} = \begin{bmatrix} 1 & 2 & 3 \\ 0 & 2 & 3 \\ 0 & 0 & 2 \end{bmatrix}$. Is $\boldsymbol{A}$ defective?

Solution. As $\boldsymbol{A}$ is a triangular matrix, its eigenvalues are the diagonal entries. One finds two eigenvalues $\lambda_1 = 1$ and $\lambda_2 = 2$, the latter with algebraic multiplicity two. Solving $\boldsymbol{A}\boldsymbol{x}_1 = \lambda_1 \boldsymbol{x}_1$ and $\boldsymbol{A}\boldsymbol{x}_2 = \lambda_2 \boldsymbol{x}_2$, one finds (valid choices of) eigenpairs, for instance

$$(\lambda_1, \boldsymbol{x}_1) = (1, \begin{bmatrix} 1 \\ 0 \\ 0 \end{bmatrix}), \qquad (\lambda_2, \boldsymbol{x}_2) = (2, \begin{bmatrix} 2 \\ 1 \\ 0 \end{bmatrix}).$$

It follows that the eigenvectors span a space of dimension 2, and this means that $\boldsymbol{A}$ is defective.

Exercise 6.11: Right or wrong? (Exam exercise 2005-1)

Decide if the following statements are right or wrong. Give supporting arguments for your decisions.

a) The matrix

$$A = \frac{1}{6}\begin{bmatrix} 3 & 4 \\ 4 & -3 \end{bmatrix}$$

is orthogonal?

Solution. Wrong! Since $[3, 4]\,[4, -3]^{\mathrm{T}} = 0$ it follows that $\boldsymbol{A}$ has orthogonal columns, but since $\frac{1}{36}(3^2 + 4^2) \neq 1$ the columns are not orthonormal. If we change $\frac{1}{6}$ to $\frac{1}{5}$, then $\boldsymbol{A}$ becomes orthogonal.

b) Let

$$A = \begin{bmatrix} a & 1 \\ 0 & a \end{bmatrix}$$

where $a \in \mathbb{R}$. There is a nonsingular matrix $\boldsymbol{Y} \in \mathbb{R}^{2\times 2}$ and a diagonal matrix $\boldsymbol{D} \in \mathbb{R}^{2\times 2}$ such that $\boldsymbol{A} = \boldsymbol{YDY}^{-1}$?

Solution. Wrong! Let $\boldsymbol{Y} = [\boldsymbol{y}_1, \boldsymbol{y}_2]$ and $\boldsymbol{D} = \mathrm{diag}(\lambda_1, \lambda_2)$. Since $\boldsymbol{AY} = \boldsymbol{YD}$ we have $\boldsymbol{Ay}_j = \lambda_j \boldsymbol{y}_j$ for $j = 1, 2$, i.e., $\boldsymbol{y}_1$ and $\boldsymbol{y}_2$ are eigenvectors of $\boldsymbol{A}$ with eigenvalues λ_1 and λ_2. Since $\boldsymbol{A}$ is upper triangular the eigenvalues of $\boldsymbol{A}$ are $\lambda_1 = \lambda_2 = a$. Furthermore $\boldsymbol{y}_1$ and $\boldsymbol{y}_2$ are linearly independent since $\boldsymbol{Y}$ is nonsingular. Let $\boldsymbol{x} := [x_1, x_2]^{\mathrm{T}}$ be an eigenvector to $\boldsymbol{A}$ with eigenvalue λ so that $\boldsymbol{Ax} = \lambda \boldsymbol{x}$ or $ax_1 + x_2 = ax_1$ and $ax_2 = ax_2$. The solution is x_1 arbitrary and $x_2 = 0$. But this means that $\boldsymbol{A}$ does not have linearly independent eigenvectors and therefore cannot be diagonalized.

Exercise 6.12: Eigenvalues of tridiagonal matrix (Exam exercise 2009-3)

Let $\boldsymbol{A} \in \mathbb{R}^{n,n}$ be tridiagonal (i.e., $a_{ij} = 0$ when $|i - j| > 1$) and suppose also that $a_{i+1,i}a_{i,i+1} > 0$ for $i = 1, \ldots, n-1$. Show that the eigenvalues of $\boldsymbol{A}$ are real.

Hint.
Show that there is a diagonal matrix $\boldsymbol{D}$ such that $\boldsymbol{D}^{-1}\boldsymbol{AD}$ is symmetric.

Solution. $\boldsymbol{B} := \boldsymbol{D}^{-1}\boldsymbol{AD}$ will be symmetric if we choose

$$d_1 = 1, \quad \text{and} \quad d_{i+1} = d_i\sqrt{\frac{a_{i+1,i}}{a_{i,i+1}}}, \qquad i = 1, \ldots, n-1.$$

Since a real symmetric matrix has real eigenvalues and $\boldsymbol{A}$ is similar to $\boldsymbol{B}$ it follows that $\boldsymbol{A}$ has real eigenvalues.

Exercises section 6.2

Exercise 6.13: Jordan example

Find $\boldsymbol{S}$ in the Jordan factorization

$$\boldsymbol{AS} = \boldsymbol{SJ}, \qquad \boldsymbol{A} = \begin{bmatrix} 3 & 0 & 1 \\ -4 & 1 & -2 \\ -4 & 0 & -1 \end{bmatrix}, \qquad \boldsymbol{J} = \begin{bmatrix} 1 & 1 & 0 \\ 0 & 1 & 0 \\ 0 & 0 & 1 \end{bmatrix}.$$

Solution. This exercise shows that it matters in which order we solve for the columns of $\boldsymbol{S}$. One would here need to find the second column first before solving for the other two. We are asked to find $\boldsymbol{S} = [\boldsymbol{s}_1, \boldsymbol{s}_2, \boldsymbol{s}_3]$ satisfying

$$[\boldsymbol{As}_1, \boldsymbol{As}_2, \boldsymbol{As}_3] = \boldsymbol{AS} = \boldsymbol{SJ} = [\boldsymbol{s}_1, \boldsymbol{s}_2, \boldsymbol{s}_3]\boldsymbol{J} = \left[\boldsymbol{s}_1, \boldsymbol{s}_1 + \boldsymbol{s}_2, \boldsymbol{s}_3\right].$$

The equations for the first and third columns say that $\boldsymbol{s}_1$ and $\boldsymbol{s}_3$ are eigenvectors for $\lambda = 1$, so that they can be found by row reducing $\boldsymbol{A} - \boldsymbol{I}$:

$$\boldsymbol{A} - \boldsymbol{I} = \begin{bmatrix} 2 & 0 & 1 \\ -4 & 0 & -2 \\ -4 & 0 & -2 \end{bmatrix} \sim \begin{bmatrix} 2 & 0 & 1 \\ 0 & 0 & 0 \\ 0 & 0 & 0 \end{bmatrix}.$$

Hence $[1, 0, -2]^{\mathrm{T}}$ and $[0, 1, 0]^{\mathrm{T}}$ span the space $\ker(\boldsymbol{A} - \boldsymbol{I})$ of eigenvectors for $\lambda = 1$.

The vector $\boldsymbol{s}_2$ can be found by solving $\boldsymbol{As}_2 = \boldsymbol{s}_1 + \boldsymbol{s}_2$, so that $(\boldsymbol{A} - \boldsymbol{I})\boldsymbol{s}_2 = \boldsymbol{s}_1$. This means that $(\boldsymbol{A} - \boldsymbol{I})^2\boldsymbol{s}_2 = (\boldsymbol{A} - \boldsymbol{I})\boldsymbol{s}_1 = \boldsymbol{0}$, so that $\boldsymbol{s}_2 \in \ker(\boldsymbol{A} - \boldsymbol{I})^2$. A simple computation shows that $(\boldsymbol{A} - \boldsymbol{I})^2 = 0$ so that any $\boldsymbol{s}_2$ will do, but we must also choose $\boldsymbol{s}_2$ so that $(\boldsymbol{A} - \boldsymbol{I})\boldsymbol{s}_2 = \boldsymbol{s}_1$ is an eigenvector of $\boldsymbol{A}$. Since $\boldsymbol{A} - \boldsymbol{I}$ has rank one, we may choose any $\boldsymbol{s}_2$ so that $(\boldsymbol{A} - \boldsymbol{I})\boldsymbol{s}_2$ is nonzero. In particular we can choose $\boldsymbol{s}_2 = [1, 0, 0]^{\mathrm{T}}$, and then $\boldsymbol{s}_1 = (\boldsymbol{A} - \boldsymbol{I})\boldsymbol{s}_2 = [2, -4, -4]^{\mathrm{T}}$.

We can also choose $\boldsymbol{s}_3 = [0, 1, 0]^{\mathrm{T}}$, since it is an eigenvector not spanned by the $\boldsymbol{s}_1$ and $\boldsymbol{s}_2$ which we just defined. All this means that we can choose

$$\boldsymbol{S} = \begin{bmatrix} 2 & 1 & 0 \\ -4 & 0 & 1 \\ -4 & 0 & 0 \end{bmatrix}.$$

Exercise 6.14: A nilpotent matrix

Show that $(\boldsymbol{J}_m(\lambda) - \lambda \boldsymbol{I})^r = \begin{bmatrix} 0 & \boldsymbol{I}_{m-r} \\ 0 & 0 \end{bmatrix}$ for $1 \le r \le m-1$ and conclude that $\left(\boldsymbol{J}_m(\lambda) - \lambda \boldsymbol{I}\right)^m = 0$.

Solution. We show this by induction. For $r = 1$ the statement is obvious. Define $\boldsymbol{E}_r = \begin{bmatrix} 0 & \boldsymbol{I}_{m-r} \\ 0 & 0 \end{bmatrix}$. We have that

$$(\boldsymbol{E}_1 \boldsymbol{E}_r)_{i,j} = \sum_k (\boldsymbol{E}_1)_{i,k} (\boldsymbol{E}_r)_{k,j}.$$

In the sum on the right hand side only one term can contribute (since any row/column in $\boldsymbol{E}_1$ and $\boldsymbol{E}_r$ contains only one nonzero entry, being a one). This occurs when there is a k so that $k = i+1$, $k+r = j$, i.e., when $j = i+r+1$. $\boldsymbol{E}_{r+1}$ has all nonzero entries when $j = i+r+1$, and this proves that $\boldsymbol{E}_{r+1} = \boldsymbol{E}_1 \boldsymbol{E}_r$. It now follows that

$$\left(\boldsymbol{J}_m(\lambda) - \boldsymbol{I}\right)^{r+1} = \left(\boldsymbol{J}_m(\lambda) - \boldsymbol{I}\right)\left(\boldsymbol{J}_m(\lambda) - \boldsymbol{I}\right)^r = \boldsymbol{E}_1 \boldsymbol{E}_r = \boldsymbol{E}_{r+1},$$

and the result follows.

Exercise 6.15: Properties of the Jordan factorization

Let $\boldsymbol{J}$ be the Jordan factorization of a matrix $\boldsymbol{A} \in \mathbb{C}^{n \times n}$ as given in Theorem 6.4. Then for $r = 0, 1, 2, \ldots, m = 2, 3, \ldots$, and any $\lambda \in \mathbb{C}$,

a) $\boldsymbol{A}^r = \boldsymbol{S}\boldsymbol{J}^r\boldsymbol{S}^{-1}$,

b) $\boldsymbol{J}^r = \operatorname{diag}(\boldsymbol{U}_1^r, \ldots, \boldsymbol{U}_k^r)$,

c) $\boldsymbol{U}_i^r = \operatorname{diag}(\boldsymbol{J}_{m_{i,1}}(\lambda_i)^r, \ldots, \boldsymbol{J}_{m_{i,g_i}}(\lambda_i)^r)$,

d) $\boldsymbol{J}_m(\lambda)^r = (\boldsymbol{E}_m + \lambda \boldsymbol{I}_m)^r = \sum_{k=0}^{\min\{r,m-1\}} \binom{r}{k} \lambda^{r-k} \boldsymbol{E}_m^k$.

Solution. Let $\boldsymbol{J} = \boldsymbol{S}^{-1}\boldsymbol{A}\boldsymbol{S}$ be the Jordan form of the matrix $\boldsymbol{A}$ as in Theorem 6.4. Items **a)**–**c)** are easily shown by induction, making use of the rules of block multiplication in **b)** and **c)**. For **d)**, write $\boldsymbol{E}_m := \boldsymbol{J}_m(\lambda) - \lambda \boldsymbol{I}_m$, with $\boldsymbol{J}_m(\lambda)$ the Jordan block of order m. By the binomial theorem,

$$\boldsymbol{J}_m(\lambda)^r = (\boldsymbol{E}_m + \lambda \boldsymbol{I}_m)^r = \sum_{k=0}^{r} \binom{r}{k} \boldsymbol{E}_m^k (\lambda \boldsymbol{I}_m)^{r-k} = \sum_{k=0}^{r} \binom{r}{k} \lambda^{r-k} \boldsymbol{E}_m^k.$$

Since $\boldsymbol{E}_m^k = \boldsymbol{0}$ for any $k \ge m$, we obtain

$$J_m(\lambda)^r = \sum_{k=0}^{\min\{r,m-1\}} \binom{r}{k} \lambda^{r-k} E_m^k.$$

Exercise 6.16: Powers of a Jordan block

Find J^{100} and A^{100} for the matrix in Exercise 6.13.

Solution. Let S be as in Exercise 6.13. J is block-diagonal so that we can write

$$J^n = \begin{bmatrix} 1 & 1 & 0 \\ 0 & 1 & 0 \\ 0 & 0 & 1 \end{bmatrix}^n = \begin{bmatrix} \begin{bmatrix} 1 & 1 \\ 0 & 1 \end{bmatrix}^n & 0 \\ 0 & 1^n \end{bmatrix} = \begin{bmatrix} 1 & n & 0 \\ 0 & 1 & 0 \\ 0 & 0 & 1 \end{bmatrix}, \tag{6.i}$$

where we used property **d)** in Exercise 6.15 on the upper left block. It follows that

$$A^{100} = (SJS^{-1})^{100} = SJ^{100}S^{-1} = \begin{bmatrix} 2 & 1 & 0 \\ -4 & 0 & 1 \\ -4 & 0 & 0 \end{bmatrix} \begin{bmatrix} 1 & 100 & 0 \\ 0 & 1 & 0 \\ 0 & 0 & 1 \end{bmatrix} \begin{bmatrix} 2 & 1 & 0 \\ -4 & 0 & 1 \\ -4 & 0 & 0 \end{bmatrix}^{-1}$$

$$= \begin{bmatrix} 2 & 1 & 0 \\ -4 & 0 & 1 \\ -4 & 0 & 0 \end{bmatrix} \begin{bmatrix} 1 & 100 & 0 \\ 0 & 1 & 0 \\ 0 & 0 & 1 \end{bmatrix} \begin{bmatrix} 0 & 0 & -\frac{1}{4} \\ 1 & 0 & \frac{1}{2} \\ 0 & 1 & -1 \end{bmatrix} = \begin{bmatrix} 201 & 0 & 100 \\ -400 & 1 & -200 \\ -400 & 0 & -199 \end{bmatrix}.$$

Exercise 6.17: The minimal polynomial

Let J be the Jordan factorization of a matrix $A \in \mathbb{C}^{n \times n}$ as given in Theorem 6.4. The polynomial

$$\mu_A(\lambda) := \prod_{i=1}^{k} (\lambda_i - \lambda)^{m_i}, \qquad \text{where } m_i := \max_{1 \le j \le g_i} m_{i,j},$$

is called the *minimal polynomial* of A. We define the matrix polynomial $\mu_A(A)$ by replacing the factors $\lambda_i - \lambda$ by $\lambda_i I - A$.

a) We have $\pi_A(\lambda) = \prod_{i=1}^{k} \prod_{j=1}^{g_i} (\lambda_i - \lambda)^{m_{i,j}}$. Use this to show that the minimal polynomial divides the characteristic polynomial, i.e., $\pi_A = \mu_A \nu_A$ for some polynomial ν_A.

Solution. For each i, $(\lambda_i - \lambda)^{a_i} = (\lambda_i - \lambda)^{\sum_{j=1}^{g_i} m_{i,j}}$ divides $\pi_A(\lambda)$. Since $\sum_{j=1}^{g_i} m_{i,j} \ge \max_{1 \le j \le g_i} m_{i,j} = m_i$, also $(\lambda_i - \lambda)^{m_i}$ divides $\pi_A(\lambda)$. From this it follows that also $\mu_A(\lambda)$ divides $\pi_A(\lambda)$.

b) Show that $\mu_A(A) = 0 \iff \mu_A(J) = 0$.

Solution. We have that

$$\mu_{\boldsymbol{A}}(\boldsymbol{A}) = \prod_{i=1}^{k}(\lambda_i \boldsymbol{I} - \boldsymbol{A})^{m_i} = \prod_{i=1}^{k}(\lambda_i \boldsymbol{I} - \boldsymbol{S}\boldsymbol{J}\boldsymbol{S}^{-1})^{m_i}$$
$$= \boldsymbol{S}\left(\prod_{i=1}^{k}(\lambda_i \boldsymbol{I} - \boldsymbol{J})^{m_i}\right)\boldsymbol{S}^{-1} = \boldsymbol{S}\mu_{\boldsymbol{A}}(\boldsymbol{J})\boldsymbol{S}^{-1}.$$

It follows that $\mu_{\boldsymbol{A}}(\boldsymbol{A}) = \boldsymbol{0}$ if and only if $\mu_{\boldsymbol{A}}(\boldsymbol{J}) = \boldsymbol{0}$.

c) (can be difficult) Use Exercises 6.14, 6.15 and the maximality of m_i to show that $\mu_{\boldsymbol{A}}(\boldsymbol{A}) = 0$. Thus a matrix satisfies its minimal equation. Finally show that the degree of any polynomial p such that $p(\boldsymbol{A}) = \boldsymbol{0}$ is at least as large as the degree of the minimal polynomial.

Solution. We have that

$$\mu_{\boldsymbol{A}}(\boldsymbol{J}) = \prod_{i=1}^{k}(\lambda_i \boldsymbol{I} - \boldsymbol{J})^{m_i} = \prod_{i=1}^{k}\left(\operatorname{diag}(\lambda_i \boldsymbol{I} - \boldsymbol{U}_1, \ldots, \lambda_i \boldsymbol{I} - \boldsymbol{U}_k)\right)^{m_i}$$
$$= \prod_{i=1}^{k}\operatorname{diag}\left((\lambda_i \boldsymbol{I} - \boldsymbol{U}_1)^{m_i}, \ldots, (\lambda_i \boldsymbol{I} - \boldsymbol{U}_k)^{m_i}\right)$$
$$= \operatorname{diag}\left(\prod_{i=1}^{k}(\lambda_i \boldsymbol{I} - \boldsymbol{U}_1)^{m_i}, \ldots, \prod_{i=1}^{k}(\lambda_i \boldsymbol{I} - \boldsymbol{U}_k)^{m_i}\right).$$

Now we also have

$$(\lambda_i \boldsymbol{I} - \boldsymbol{U}_i)^{m_i} = \left(\lambda_i \boldsymbol{I} - \operatorname{diag}\left(\boldsymbol{J}_{m_{i,1}}(\lambda_i), \ldots, \boldsymbol{J}_{m_{i,g_i}}(\lambda_i)\right)\right)^{m_i}$$
$$= \operatorname{diag}\left(\lambda_i \boldsymbol{I} - \boldsymbol{J}_{m_{i,1}}(\lambda_i), \ldots, \lambda_i \boldsymbol{I} - \boldsymbol{J}_{m_{i,g_i}}(\lambda_i)\right)^{m_i}$$
$$= \operatorname{diag}\left(\left(\lambda_i \boldsymbol{I} - \boldsymbol{J}_{m_{i,1}}(\lambda_i)\right)^{m_i}, \ldots, \left(\lambda_i \boldsymbol{I} - \boldsymbol{J}_{m_{i,g_i}}(\lambda_i)\right)^{m_i}\right) = \boldsymbol{0},$$

since

$$\left(\lambda_i \boldsymbol{I} - \boldsymbol{J}_{m_{i,j}}(\lambda_i)\right)^{m_i} = \left(\lambda_i \boldsymbol{I} - \boldsymbol{J}_{m_{i,j}}(\lambda_i)\right)^{m_{i,j}}\left(\lambda_i \boldsymbol{I} - \boldsymbol{J}_{m_{i,j}}(\lambda_i)\right)^{m_i - m_{i,j}}$$
$$= \boldsymbol{0}(\lambda_i \boldsymbol{I} - \boldsymbol{J}_{m_{i,j}}(\lambda_i))^{m_i - m_{i,j}} = \boldsymbol{0}.$$

We now get that

$$\prod_{i=1}^{k}(\lambda_i \boldsymbol{I} - \boldsymbol{U}_j)^{m_i} = (\lambda_j \boldsymbol{I} - \boldsymbol{U}_j)^{m_j} \prod_{i=1, i\neq j}^{k}(\lambda_i \boldsymbol{I} - \boldsymbol{U}_j)^{m_i} = \boldsymbol{0},$$

so that

$$\mu_{\boldsymbol{A}}(\boldsymbol{J}) = \operatorname{diag}\left(\prod_{i=1}^{k}(\lambda_i \boldsymbol{I} - \boldsymbol{U}_1)^{m_i}, \ldots, \prod_{i=1}^{k}(\lambda_i \boldsymbol{I} - \boldsymbol{U}_k)^{m_i}\right) = \boldsymbol{0}.$$

It follows that $\mu_{\boldsymbol{A}}(\boldsymbol{A}) = \boldsymbol{0}$.

Suppose now that $p(\boldsymbol{A}) = \boldsymbol{0}$. We can write $p(\boldsymbol{A}) = C\prod_{i=1}^{r}(k_i \boldsymbol{I} - \boldsymbol{A})^{s_i}$, where k_i are the zeros of p, with multiplicity s_i. As above it follows that $p(\boldsymbol{A}) = \boldsymbol{0}$ if and only if $p(\boldsymbol{J}) = \boldsymbol{0}$. Factor $p(\boldsymbol{J})$ as above to obtain

$$p(\boldsymbol{J}) = \operatorname{diag}\left(\prod_{i=1}^{r}(k_i \boldsymbol{I} - \boldsymbol{U}_1)^{s_i}, \ldots, \prod_{i=1}^{k}(k_i \boldsymbol{I} - \boldsymbol{U}_k)^{s_i}\right).$$

Note that

$$k_i \boldsymbol{I} - \boldsymbol{U}_j = \operatorname{diag}\left(k_i \boldsymbol{I} - \boldsymbol{J}_{m_{j,1}}(\lambda_j), \ldots, k_i \boldsymbol{I} - \boldsymbol{J}_{m_{j,g_j}}(\lambda_j)\right)$$

is upper triangular with $k_i - \lambda_j$ on the diagonal. If $k_i \neq \lambda_j$, then $k_i \boldsymbol{I} - \boldsymbol{U}_j$ must be invertible, but then $(k_i \boldsymbol{I} - \boldsymbol{U}_j)^{s_i}$ is invertible as well. In order for $p(\boldsymbol{J}) = \boldsymbol{0}$, we must then have that for each j there exists a t so that $k_t = \lambda_j$.

The qth diagonal block entry in $(k_i \boldsymbol{I} - \boldsymbol{U}_j)^{s_i}$ is

$$\begin{aligned}\prod_{i=1}^{r}\left(k_i \boldsymbol{I} - \boldsymbol{J}_{m_{j,q}}(\lambda_j)\right)^{s_i} &= \left(k_t \boldsymbol{I} - \boldsymbol{J}_{m_{j,q}}(\lambda_j)\right)^{s_t} \prod_{i=1, i\neq t}^{r}\left(k_i \boldsymbol{I} - \boldsymbol{J}_{m_{j,q}}(\lambda_j)\right)^{s_i} \\ &= \left(\lambda_j \boldsymbol{I} - \boldsymbol{J}_{m_{j,q}}(\lambda_j)\right)^{s_t} \prod_{i=1, i\neq t}^{r}\left(k_i \boldsymbol{I} - \boldsymbol{J}_{m_{j,q}}(\lambda_j)\right)^{s_i}.\end{aligned}$$

The last matrix here is invertible (all $k_i \neq \lambda_j$ when $i \neq t$), so that we must have that $\left(\lambda_j \boldsymbol{I} - \boldsymbol{J}_{m_{j,q}}(\lambda_j)\right)^{s_t} = \boldsymbol{0}$ in order for $p(\boldsymbol{J}) = \boldsymbol{0}$. We know from the exercises that this happens only when $s_t \geq m_{j,q}$. Since q was arbitrary we obtain that $s_t \geq m_j$, i.e., that λ_j is a zero in p of multiplicity $\geq m_j$. Since this applied for any j, it follows that the minimal polynomial divides p, and the result follows.

d) Use **c)** to show the *Cayley-Hamilton Theorem*, which says that a matrix satisfies its characteristic equation $\pi_{\boldsymbol{A}}(\boldsymbol{A}) = \boldsymbol{0}$.

Solution. We have that $\pi_{\boldsymbol{A}}(\boldsymbol{A}) = \mu_{\boldsymbol{A}}(\boldsymbol{A})\nu_{\boldsymbol{A}}(\boldsymbol{A}) = \boldsymbol{0}$.

Exercise 6.18: Cayley Hamilton Theorem (Exam exercise 1996-3)

Suppose p is a polynomial given by $p(t) := \sum_{j=0}^{r} b_j t^j$, where $b_j \in \mathbb{C}$ and $\boldsymbol{A} \in \mathbb{C}^{n\times n}$. We define the matrix $p(\boldsymbol{A}) \in \mathbb{C}^{n\times n}$ by

$$p(\boldsymbol{A}) := \sum_{j=0}^{r} b_j \boldsymbol{A}^j,$$

where $\boldsymbol{A}^0 := \boldsymbol{I}$. From this it follows that if $p(t) := (t-\alpha_1)\cdots(t-\alpha_r)$ for some $\alpha_0,\dots,\alpha_r \in \mathbb{C}$ then $p(\boldsymbol{A}) = (\boldsymbol{A}-\alpha_1)\cdots(\boldsymbol{A}-\alpha_r)$. We accept this without proof.

Let $\boldsymbol{U}^*\boldsymbol{A}\boldsymbol{U} = \boldsymbol{T}$, where $\boldsymbol{U}$ is unitary and $\boldsymbol{T}$ upper triangular with the eigenvalues of $\boldsymbol{A}$ on the diagonal.

a) Find the characteristic polynomial $\pi_{\boldsymbol{A}}$ to $\begin{bmatrix} 2 & 1 \\ -1 & 4 \end{bmatrix}$. Show that $\pi(\boldsymbol{A}) = \boldsymbol{0}$.

Solution. We find

$$\pi_{\boldsymbol{A}} = \det(\boldsymbol{A}-\lambda\boldsymbol{I}) = \begin{vmatrix} 2-\lambda & 1 \\ -1 & 4-\lambda \end{vmatrix} = (2-\lambda)(4-\lambda)+1 = \lambda^2 - 6\lambda + 9$$

and hence

$$\begin{aligned} \pi_{\boldsymbol{A}}(\boldsymbol{A}) &= \boldsymbol{A}^2 - 6\boldsymbol{A} + 9\boldsymbol{I} \\ &= \begin{bmatrix} 2 & 1 \\ -1 & 4 \end{bmatrix}\begin{bmatrix} 2 & 1 \\ -1 & 4 \end{bmatrix} - 6\begin{bmatrix} 2 & 1 \\ -1 & 4 \end{bmatrix} + 9\begin{bmatrix} 1 & 0 \\ 0 & 1 \end{bmatrix} \\ &= \begin{bmatrix} 3 & 6 \\ -6 & 15 \end{bmatrix} + \begin{bmatrix} -12 & -6 \\ 6 & -24 \end{bmatrix} + \begin{bmatrix} 9 & 0 \\ 0 & 9 \end{bmatrix} = \begin{bmatrix} 0 & 0 \\ 0 & 0 \end{bmatrix}. \end{aligned}$$

b) Let now $\boldsymbol{A} \in \mathbb{C}^{n\times n}$ be arbitrary. For any polynomial p show that $p(\boldsymbol{A}) = \boldsymbol{U}p(\boldsymbol{T})\boldsymbol{U}^*$.

Solution. We first show by induction that

$$\boldsymbol{A}^j = \boldsymbol{U}\boldsymbol{T}^j\boldsymbol{U}^*, \quad j \in \mathbb{N}. \tag{6.ii}$$

For $j = 1$ this follows form the Schur decomposition. If $\boldsymbol{A}^j = \boldsymbol{U}\boldsymbol{T}^j\boldsymbol{U}^*$ for some $j \geq 1$, then

$$\boldsymbol{A}^{j+1} = \boldsymbol{A}^j\boldsymbol{A} = (\boldsymbol{U}\boldsymbol{T}^j\boldsymbol{U}^*)(\boldsymbol{U}\boldsymbol{T}\boldsymbol{U}^*) = \boldsymbol{U}\boldsymbol{T}^j(\boldsymbol{U}^*\boldsymbol{U})\boldsymbol{T}\boldsymbol{U}^* = \boldsymbol{U}\boldsymbol{T}^{j+1}\boldsymbol{U}^*,$$

since $\boldsymbol{U}^*\boldsymbol{U} = \boldsymbol{I}$. Now (6.ii) follows.

If $p(t) := \sum_{j=0}^{r} b_j t^j$ then

$$p(\boldsymbol{A}) := \sum_{j=0}^{r} b_j \boldsymbol{A}^j \overset{(6.\text{ii})}{=} \sum_{j=0}^{r} b_j \boldsymbol{U}\boldsymbol{T}^j\boldsymbol{U}^* = \boldsymbol{U}\Big(\sum_{j=0}^{r} b_j \boldsymbol{T}^j\Big)\boldsymbol{U}^* = \boldsymbol{U}p(\boldsymbol{T})\boldsymbol{U}^*.$$

c) Let $n, k \in \mathbb{N}$ with $1 \le k < n$. Let $\boldsymbol{C}, \boldsymbol{D} \in \mathbb{C}^{n\times n}$ be upper triangular. Moreover, $c_{i,j} = 0$ for $i, j \le k$ and $d_{k+1,k+1} = 0$. Define $\boldsymbol{E} := \boldsymbol{C}\boldsymbol{D}$, and show that $e_{i,j} = 0$ for $i, j \le k+1$.

Solution. Let $\boldsymbol{M}_{[k]} = (m_{ij})_{i,j=1}^{k}$ be the leading principal $k \times k$-submatrix of a matrix $\boldsymbol{M}$. Since $\boldsymbol{C}$ and $\boldsymbol{D}$ are upper triangular we find

$$\boldsymbol{E} := \boldsymbol{C}\boldsymbol{D} = \begin{bmatrix} \boldsymbol{C}_{[k+1]} & \boldsymbol{C}_{1,2} \\ \boldsymbol{0} & \boldsymbol{C}_{2,2} \end{bmatrix} \begin{bmatrix} \boldsymbol{D}_{[k+1]} & \boldsymbol{D}_{1,2} \\ \boldsymbol{0} & \boldsymbol{D}_{2,2} \end{bmatrix} = \begin{bmatrix} \boldsymbol{C}_{[k+1]}\boldsymbol{D}_{[k+1]} & \boldsymbol{E}_{1,2} \\ \boldsymbol{0} & \boldsymbol{E}_{2,2} \end{bmatrix},$$

and it follows that $\boldsymbol{E}_{[k+1]} = \boldsymbol{C}_{[k+1]}\boldsymbol{D}_{[k+1]}$. But since $\boldsymbol{C}_{[k]} = \boldsymbol{0}$ and $d_{k+1,k+1} = 0$ we find

$$\begin{bmatrix} \boldsymbol{E}_{[k]} & \boldsymbol{e} \\ \boldsymbol{0} & \alpha \end{bmatrix} = \boldsymbol{E}_{[k+1]} = \boldsymbol{C}_{[k+1]}\boldsymbol{D}_{[k+1]} = \begin{bmatrix} \boldsymbol{0} & \boldsymbol{c} \\ \boldsymbol{0} & \beta \end{bmatrix} \begin{bmatrix} \boldsymbol{D}_{[k]} & \boldsymbol{d} \\ \boldsymbol{0} & 0 \end{bmatrix} = \begin{bmatrix} \boldsymbol{0} & \boldsymbol{0} \\ \boldsymbol{0} & \boldsymbol{0} \end{bmatrix}$$

for some $\boldsymbol{e}, \boldsymbol{c}, \boldsymbol{d} \in \mathbb{R}^k$ and $\alpha, \beta \in \mathbb{R}$. We conclude that $e_{i,j} = 0$ for $i, j \le k+1$.

d) Now let $p := \pi_{\boldsymbol{A}}$ be the characteristic polynomial of $\boldsymbol{A}$. Show that $p(\boldsymbol{T}) = \boldsymbol{0}$. Then show that $p(\boldsymbol{A}) = \boldsymbol{0}$. (Cayley Hamilton Theorem)

Hint.
Use a suitable factorization of p and use **c)**.

Solution. Since $p(t) = (t - \lambda_1) \cdots (t - \lambda_n)$ we have

$$p(\boldsymbol{T}) = (\boldsymbol{T} - \lambda_1 \boldsymbol{I}) \cdots (\boldsymbol{T} - \lambda_n \boldsymbol{I}),$$

where each $\boldsymbol{T} - \lambda_i \boldsymbol{I}$ is upper triangular with $t_{ii} = 0$. Define

$$\boldsymbol{W}_s := (\boldsymbol{T} - \lambda_1 \boldsymbol{I}) \cdots (\boldsymbol{T} - \lambda_s \boldsymbol{I}), \quad s = 1, \ldots, n.$$

We show by induction that $w_s(i,j) = 0$ for $i, j \le s$. This holds for $s = 1$. Suppose it holds for some $s \ge 1$. Since $\boldsymbol{W}_s$ and $\boldsymbol{T} - \lambda_{s+1}\boldsymbol{I}$ are upper triangular and $(\boldsymbol{T} - \lambda_{s+1}\boldsymbol{I})(s+1, s+1) = 0$ we can apply **c)** and obtain

$$\boldsymbol{W}_{s+1}(i,j) = \big(\boldsymbol{W}_s(\boldsymbol{T} - \lambda_{s+1}\boldsymbol{I})\big)(i,j) = 0, \quad i, j \le s+1.$$

Taking $s = n$ we obtain $\boldsymbol{W}_n = \boldsymbol{0}$ which means that $p(\boldsymbol{T}) = \boldsymbol{0}$. But then also $p(\boldsymbol{A}) = \boldsymbol{U}p(\boldsymbol{T})\boldsymbol{U}^* = \boldsymbol{0}$.

Exercises section 6.3

Exercise 6.19: Schur factorization example

Show that a Schur factorization of $\boldsymbol{A} = \left[\begin{smallmatrix} 1 & 2 \\ 3 & 2 \end{smallmatrix}\right]$ is $\boldsymbol{U}^{\mathrm{T}}\boldsymbol{A}\boldsymbol{U} = \left[\begin{smallmatrix} -1 & -1 \\ 0 & 4 \end{smallmatrix}\right]$, where $\boldsymbol{U} = \frac{1}{\sqrt{2}}\left[\begin{smallmatrix} 1 & 1 \\ -1 & 1 \end{smallmatrix}\right]$.

Solution. The matrix $\boldsymbol{U}$ is unitary, as $\boldsymbol{U}^*\boldsymbol{U} = \boldsymbol{U}^{\mathrm{T}}\boldsymbol{U} = \boldsymbol{I}$. One directly verifies that

$$\boldsymbol{R} := \boldsymbol{U}^{\mathrm{T}}\boldsymbol{A}\boldsymbol{U} = \begin{bmatrix} -1 & -1 \\ 0 & 4 \end{bmatrix}.$$

Since this matrix is upper triangular, $\boldsymbol{A} = \boldsymbol{U}\boldsymbol{R}\boldsymbol{U}^{\mathrm{T}}$ is a Schur decomposition of $\boldsymbol{A}$.

Exercise 6.20: Skew-Hermitian matrix

Suppose $\boldsymbol{C} = \boldsymbol{A} + i\boldsymbol{B}$, where $\boldsymbol{A}, \boldsymbol{B} \in \mathbb{R}^{n\times n}$. Show that $\boldsymbol{C}$ is skew-Hermitian if and only if $\boldsymbol{A}^{\mathrm{T}} = -\boldsymbol{A}$ and $\boldsymbol{B}^{\mathrm{T}} = \boldsymbol{B}$.

Solution. By definition, a matrix $\boldsymbol{C}$ is *skew-Hermitian* if $\boldsymbol{C}^* = -\boldsymbol{C}$.

"$\Longrightarrow$": Suppose that $\boldsymbol{C} = \boldsymbol{A} + i\boldsymbol{B}$, with $\boldsymbol{A}, \boldsymbol{B} \in \mathbb{R}^{m,m}$, is skew-Hermitian. Then

$$-\boldsymbol{A} - i\boldsymbol{B} = -\boldsymbol{C} = \boldsymbol{C}^* = (\boldsymbol{A} + i\boldsymbol{B})^* = \boldsymbol{A}^{\mathrm{T}} - i\boldsymbol{B}^{\mathrm{T}},$$

which implies that $\boldsymbol{A}^{\mathrm{T}} = -\boldsymbol{A}$ and $\boldsymbol{B}^{\mathrm{T}} = \boldsymbol{B}$ (use that two complex numbers coincide if and only if their real parts coincide and their imaginary parts coincide). In other words, $\boldsymbol{A}$ is skew-Hermitian and $\boldsymbol{B}$ is real symmetric.

"$\Longleftarrow$": Suppose that we are given matrices $\boldsymbol{A}, \boldsymbol{B} \in \mathbb{R}^{m,m}$ such that $\boldsymbol{A}$ is skew-Hermitian and $\boldsymbol{B}$ is real symmetric. Let $\boldsymbol{C} = \boldsymbol{A} + i\boldsymbol{B}$. Then

$$\boldsymbol{C}^* = (\boldsymbol{A} + i\boldsymbol{B})^* = \boldsymbol{A}^{\mathrm{T}} - i\boldsymbol{B}^{\mathrm{T}} = -\boldsymbol{A} - i\boldsymbol{B} = -(\boldsymbol{A} + i\boldsymbol{B}) = -\boldsymbol{C},$$

meaning that $\boldsymbol{C}$ is skew-Hermitian.

Exercise 6.21: Eigenvalues of a skew-Hermitian matrix

Show that any eigenvalue of a skew-Hermitian matrix is purely imaginary.

Solution. Let $\boldsymbol{A}$ be a skew-Hermitian matrix and consider a Schur triangularization $\boldsymbol{A} = \boldsymbol{U}\boldsymbol{R}\boldsymbol{U}^*$ of $\boldsymbol{A}$. Then

$$\boldsymbol{R} = \boldsymbol{U}^*\boldsymbol{A}\boldsymbol{U} = \boldsymbol{U}^*(-\boldsymbol{A}^*)\boldsymbol{U} = -\boldsymbol{U}^*\boldsymbol{A}^*\boldsymbol{U} = -(\boldsymbol{U}^*\boldsymbol{A}\boldsymbol{U})^* = -\boldsymbol{R}^*.$$

Since $\boldsymbol{R}$ differs from $\boldsymbol{A}$ by a similarity transform, their eigenvalues coincide (use the multiplicative property of the determinant to show that

$$\det(\boldsymbol{A} - \lambda \boldsymbol{I}) = \det(\boldsymbol{U}^*) \det(\boldsymbol{U}\boldsymbol{R}\boldsymbol{U}^* - \lambda \boldsymbol{I})) \det(\boldsymbol{U}) = \det(\boldsymbol{R} - \lambda \boldsymbol{I}).)$$

As $\boldsymbol{R}$ is a triangular matrix, its eigenvalues λ_i appear on its diagonal. From the equation $\boldsymbol{R} = -\boldsymbol{R}^*$ it then follows that $\lambda_i = -\overline{\lambda_i}$, implying that each λ_i is purely imaginary.

Exercise 6.22: Eigenvector expansion using orthogonal eigenvectors

Show that if the eigenpairs $(\lambda_1, \boldsymbol{u}_1), \ldots, (\lambda_n, \boldsymbol{u}_n)$ of $\boldsymbol{A} \in \mathbb{C}^{n\times n}$ are orthogonal, i.e., $\boldsymbol{u}_j^* \boldsymbol{u}_k = 0$ for $j \neq k$, then the eigenvector expansions of $\boldsymbol{x}, \boldsymbol{A}\boldsymbol{x} \in \mathbb{C}^n$ take the form

$$\boldsymbol{x} = \sum_{j=1}^{n} c_j \boldsymbol{u}_j, \quad \boldsymbol{A}\boldsymbol{x} = \sum_{j=1}^{n} c_j \lambda_j \boldsymbol{u}_j, \text{ where } c_j = \frac{\boldsymbol{u}_j^* \boldsymbol{x}}{\boldsymbol{u}_j^* \boldsymbol{u}_j}.$$

Solution. If $\boldsymbol{x} = \sum_{j=1}^{n} c_j \boldsymbol{u}_j$ and the eigenvectors are orthogonal we get that

$$\boldsymbol{u}_i^* \boldsymbol{x} = \sum_{j=1}^{n} c_j \boldsymbol{u}_i^* \boldsymbol{u}_j = c_i \boldsymbol{u}_i^* \boldsymbol{u}_i,$$

so that $c_i = \boldsymbol{u}_i^* \boldsymbol{x} / (\boldsymbol{u}_i^* \boldsymbol{u}_i)$.

Exercise 6.23: Rayleigh quotient (Exam exercise 2015-3)

a) Let $\boldsymbol{A} \in \mathbb{R}^{n\times n}$ be a symmetric matrix. Explain how we can use the spectral theorem for symmetric matrices to show that

$$\lambda_{\min} = \min_{\boldsymbol{x} \neq 0} R(\boldsymbol{x}) = \min_{\|\boldsymbol{x}\|_2 = 1} R(\boldsymbol{x}),$$

where $\lambda_{\min}$ is the smallest eigenvalue of $\boldsymbol{A}$, and $R(\boldsymbol{x})$ is the Rayleigh quotient given by

$$R(\boldsymbol{x}) := \frac{\boldsymbol{x}^{\mathrm{T}} \boldsymbol{A} \boldsymbol{x}}{\boldsymbol{x}^{\mathrm{T}} \boldsymbol{x}}.$$

Solution. The spectral theorem says that we can write any real symmetric matrix as $\boldsymbol{A} = \boldsymbol{U}\boldsymbol{D}\boldsymbol{U}^{\mathrm{T}}$, where $\boldsymbol{U}$ is orthogonal and $\boldsymbol{D} = \mathrm{diag}(\lambda_1, \ldots, \lambda_n)$ is diagonal. We now get that

$$\begin{aligned} R(\boldsymbol{x}) &= \frac{\boldsymbol{x}^{\mathrm{T}} \boldsymbol{A} \boldsymbol{x}}{\boldsymbol{x}^{\mathrm{T}} \boldsymbol{x}} = \frac{\boldsymbol{x}^{\mathrm{T}} \boldsymbol{U}\boldsymbol{D}\boldsymbol{U}^{\mathrm{T}} \boldsymbol{x}}{\boldsymbol{x}^{\mathrm{T}} \boldsymbol{x}} = \frac{(\boldsymbol{U}^{\mathrm{T}}\boldsymbol{x})^{\mathrm{T}} \boldsymbol{D} (\boldsymbol{U}^{\mathrm{T}}\boldsymbol{x})}{\|\boldsymbol{x}\|^2} \\ &= \frac{(\boldsymbol{U}^{\mathrm{T}}\boldsymbol{x})^{\mathrm{T}} \boldsymbol{D} (\boldsymbol{U}^{\mathrm{T}}\boldsymbol{x})}{\|\boldsymbol{U}^{\mathrm{T}}\boldsymbol{x}\|^2} = R_{\boldsymbol{D}}(\boldsymbol{U}^{\mathrm{T}}\boldsymbol{x}) \end{aligned}$$

since $\boldsymbol{U}$ is orthogonal ($R_{\boldsymbol{D}}$ is the Rayleigh quotient using $\boldsymbol{D}$ instead of $\boldsymbol{A}$). We thus have that

$$\min_{\boldsymbol{x}\neq 0} R(\boldsymbol{x}) = \min_{\boldsymbol{x}\neq 0} R_{\boldsymbol{D}}(\boldsymbol{U}^{\mathrm{T}}\boldsymbol{x}) = \min_{\boldsymbol{x}\neq 0} R_{\boldsymbol{D}}(\boldsymbol{x}) = \min_{\boldsymbol{x}\neq 0} \sum_{i=1}^{n} \lambda_i \frac{x_i^2}{\|\boldsymbol{x}\|^2} = \lambda_{\min},$$

where the minimum is attained for $\boldsymbol{x} = \boldsymbol{e}_i$ with $\lambda_i = \lambda_{\min}$.

b) Let $\boldsymbol{x}, \boldsymbol{y} \in \mathbb{R}^n$ such that $\|\boldsymbol{x}\|_2 = 1$ and $\boldsymbol{y} \neq 0$. Show that

$$R(\boldsymbol{x} - t\boldsymbol{y}) = R(\boldsymbol{x}) - 2t\big(\boldsymbol{A}\boldsymbol{x} - R(\boldsymbol{x})\boldsymbol{x}\big)^{\mathrm{T}}\boldsymbol{y} + \mathcal{O}(t^2),$$

where $t > 0$ is small.

Hint.
Use Taylor's theorem for the function $f(t) = R(\boldsymbol{x} - t\boldsymbol{y})$.

Solution. Using the hint we have that $f(0) = R(\boldsymbol{x})$ and

$$f(t) = \frac{(\boldsymbol{x} - t\boldsymbol{y})^{\mathrm{T}}\boldsymbol{A}(\boldsymbol{x} - t\boldsymbol{y})}{(\boldsymbol{x} - t\boldsymbol{y})^{\mathrm{T}}(\boldsymbol{x} - t\boldsymbol{y})} = \frac{\boldsymbol{x}^{\mathrm{T}}\boldsymbol{A}\boldsymbol{x} - 2t\boldsymbol{x}^{\mathrm{T}}\boldsymbol{A}\boldsymbol{y} + t^2\boldsymbol{y}^{\mathrm{T}}\boldsymbol{A}\boldsymbol{y}}{\|\boldsymbol{x}\|^2 - 2t\boldsymbol{x}^{\mathrm{T}}\boldsymbol{y} + t^2\|\boldsymbol{y}\|^2} = \frac{g(t)}{h(t)}.$$

Here

$$\begin{aligned} g(0) &= \boldsymbol{x}^{\mathrm{T}}\boldsymbol{A}\boldsymbol{x}, & g'(t) &= -2\boldsymbol{x}^{\mathrm{T}}\boldsymbol{A}\boldsymbol{y} + 2t\boldsymbol{y}^{\mathrm{T}}\boldsymbol{A}\boldsymbol{y}, & g'(0) &= -2\boldsymbol{x}^{\mathrm{T}}\boldsymbol{A}\boldsymbol{y}, \\ h(0) &= \|\boldsymbol{x}\|^2 = 1, & h'(t) &= -2\boldsymbol{x}^{\mathrm{T}}\boldsymbol{y} + 2t\|\boldsymbol{y}\|^2, & h'(0) &= -2\boldsymbol{x}^{\mathrm{T}}\boldsymbol{y}. \end{aligned}$$

We now get that

$$\begin{aligned} f'(0) &= \frac{g'(0)h(0) - g(0)h'(0)}{h(0)^2} = -2\boldsymbol{x}^{\mathrm{T}}\boldsymbol{A}\boldsymbol{y} + 2\boldsymbol{x}^{\mathrm{T}}\boldsymbol{y}\boldsymbol{x}^{\mathrm{T}}\boldsymbol{A}\boldsymbol{x} \\ &= -2\big((\boldsymbol{A}\boldsymbol{x})^{\mathrm{T}}\boldsymbol{y} - R(\boldsymbol{x})\boldsymbol{x}^{\mathrm{T}}\boldsymbol{y}\big) = -2\big(\boldsymbol{A}\boldsymbol{x} - R(\boldsymbol{x})\boldsymbol{x}\big)^{\mathrm{T}}\boldsymbol{y}. \end{aligned}$$

Clearly the second derivative of f is bounded close to 0, so that $f(t) = f(0) + tf'(0) + \mathcal{O}(t^2)$. Inserting $f(0) = R(\boldsymbol{x})$ and $f'(0) = -2(\boldsymbol{A}\boldsymbol{x} - R(\boldsymbol{x})\boldsymbol{x})^{\mathrm{T}}\boldsymbol{y}$ gives the desired result.

c) Based on the characterisation given in **a)** above it is tempting to develop an algorithm for computing $\lambda_{\min}$ by approximating the minimum of $R(\boldsymbol{x})$ over the unit ball

$$B_1 := \{\boldsymbol{x} \in \mathbb{R}^n \mid \|\boldsymbol{x}\|_2 = 1\}.$$

Assume that $\boldsymbol{x}^0 \in B_1$ satisfies $\boldsymbol{A}\boldsymbol{x}^0 - R(\boldsymbol{x}^0)\boldsymbol{x}^0 \neq 0$, i.e., $(R(\boldsymbol{x}^0), \boldsymbol{x}^0)$ is not an eigenpair for $\boldsymbol{A}$. Explain how we can find a vector $\boldsymbol{x}^1 \in B_1$ such that $R(\boldsymbol{x}^1) < R(\boldsymbol{x}^0)$.

Solution. If $\boldsymbol{A}\boldsymbol{x}^0 - R(\boldsymbol{x}^0)\boldsymbol{x}^0 \neq 0$ we can choose a vector $\boldsymbol{y}$ so that $(\boldsymbol{A}\boldsymbol{x}^0 - R(\boldsymbol{x}^0)\boldsymbol{x}^0)^{\mathrm{T}}\boldsymbol{y} > 0$ ($\boldsymbol{y}$ can for instance be a vector pointing in the same direction as $\boldsymbol{A}\boldsymbol{x}^0 - R(\boldsymbol{x}^0)\boldsymbol{x}^0$). But then $-2t(\boldsymbol{A}\boldsymbol{x}^0 - R(\boldsymbol{x}^0)\boldsymbol{x}^0)^{\mathrm{T}}\boldsymbol{y} < 0$ (t is assumed to be positive here) and since this term dominates $\mathcal{O}(t^2)$ for small t, we see that $R(\boldsymbol{x}^0 - t\boldsymbol{y}) < R(\boldsymbol{x}^0)$. In other words, we can reduce the Rayleigh quotient by taking a small step from $\boldsymbol{x}^0$ in the direction of $\boldsymbol{A}\boldsymbol{x}^0 - R(\boldsymbol{x}^0)\boldsymbol{x}^0$.

Exercises section 6.4

Exercise 6.24: Eigenvalue perturbation for Hermitian matrices

Show that in Theorem 6.13, if $\boldsymbol{E}$ is symmetric positive semidefinite then $\beta_i \geq \alpha_i$.

Solution. Let $\varepsilon_1 \geq \varepsilon_2 \geq \cdots \geq \varepsilon_n$ be the eigenvalues of $\boldsymbol{E} := \boldsymbol{B} - \boldsymbol{A}$. Since a positive semidefinite matrix has no negative eigenvalues, one has $\varepsilon_n \geq 0$. It immediately follows from $\alpha_i + \varepsilon_n \leq \beta_i$ that in this case $\beta_i \geq \alpha_i$.

Exercise 6.25: Hoffman-Wielandt

Show that Equation (6.15) does not hold for the matrices $\boldsymbol{A} := \left[\begin{smallmatrix} 0 & 0 \\ 0 & 4 \end{smallmatrix}\right]$ and $\boldsymbol{B} := \left[\begin{smallmatrix} -1 & -1 \\ 1 & 1 \end{smallmatrix}\right]$. Why does this not contradict the Hoffman-Wielandt theorem (Theorem 6.14)?

Solution. The matrix $\boldsymbol{A}$ has eigenvalues 0 and 4, and the matrix $\boldsymbol{B}$ has eigenvalue 0 with algebraic multiplicity two. Independently of the choice of the permutation $i_1, \ldots, i_n$, the Hoffman-Wielandt Theorem would yield

$$16 = \sum_{j=1}^{n} |\mu_{i_j} - \lambda_j|^2 \leq \sum_{i=1}^{n}\sum_{j=1}^{n} |a_{ij} - b_{ij}|^2 = 12,$$

which clearly cannot be valid. However, the Hoffman-Wielandt Theorem cannot be applied to these matrices, because $\boldsymbol{B}$ is not normal,

$$\boldsymbol{B}^*\boldsymbol{B} = \begin{bmatrix} 2 & 2 \\ 2 & 2 \end{bmatrix} \neq \begin{bmatrix} 2 & -2 \\ -2 & 2 \end{bmatrix} = \boldsymbol{B}\boldsymbol{B}^*.$$

Exercise 6.26: Biorthogonal expansion

Determine right and left eigenpairs for the matrix $\boldsymbol{A} := \left[\begin{smallmatrix} 3 & 1 \\ 2 & 2 \end{smallmatrix}\right]$ and the two expansions in Equation (6.16) for any $\boldsymbol{v} \in \mathbb{R}^2$.

Solution. The matrix $\boldsymbol{A}$ has characteristic polynomial $\det(\boldsymbol{A} - \lambda\boldsymbol{I}) = (\lambda - 4)(\lambda - 1)$ and right eigenpairs $(\lambda_1, \boldsymbol{x}_1) = (4, [1, 1]^{\mathrm{T}})$ and $(\lambda_2, \boldsymbol{x}_2) = (1, [1, -2]^{\mathrm{T}})$. Since the right eigenvectors $\boldsymbol{x}_1, \boldsymbol{x}_2$ are linearly independent, there exists vectors $\boldsymbol{y}_1, \boldsymbol{y}_2$ satisfying $\langle \boldsymbol{y}_i, \boldsymbol{x}_j \rangle = \delta_{ij}$.

A vector orthogonal to $\boldsymbol{x}_1 = [1,1]^{\mathrm{T}}$ must be of the form $\boldsymbol{y}_2 = \alpha[1,-1]^{\mathrm{T}}$, and a vector orthogonal to $\boldsymbol{x}_2 = [1,-2]^{\mathrm{T}}$ must be of the form $\boldsymbol{y}_1 = \beta[2,1]^{\mathrm{T}}$. These choices secure that $\langle \boldsymbol{y}_i, \boldsymbol{x}_j \rangle = 0$ when $i \neq j$. We must also have that

$$1 = \langle \boldsymbol{y}_1, \boldsymbol{x}_1 \rangle = 2\beta + \beta = 3\beta, \qquad 1 = \langle \boldsymbol{y}_2, \boldsymbol{x}_2 \rangle = \alpha + 2\alpha = 3\alpha,$$

so that $\alpha = \beta = 1/3$, and we can choose the dual basis as $\boldsymbol{y}_1 = \frac{1}{3}[2,1]^{\mathrm{T}}$ and $\boldsymbol{y}_2 = \frac{1}{3}[1,-1]^{\mathrm{T}}$. Equation (6.16) then gives us the biorthogonal expansions

$$\begin{aligned}\boldsymbol{v} &= \langle \boldsymbol{v}, \boldsymbol{y}_1 \rangle \boldsymbol{x}_1 + \langle \boldsymbol{v}, \boldsymbol{y}_2 \rangle \boldsymbol{x}_2 = \frac{1}{3}(2v_1 + v_2)\boldsymbol{x}_1 + \frac{1}{3}(v_1 - v_2)\boldsymbol{x}_2 \\ &= \langle \boldsymbol{v}, \boldsymbol{x}_1 \rangle \boldsymbol{y}_1 + \langle \boldsymbol{v}, \boldsymbol{x}_2 \rangle \boldsymbol{y}_2 = (v_1 + v_2)\boldsymbol{y}_1 + (v_1 - 2v_2)\boldsymbol{y}_2.\end{aligned}$$

Exercise 6.27: Generalized Rayleigh quotient

For $\boldsymbol{A} \in \mathbb{C}^{n \times n}$ and any $\boldsymbol{y}, \boldsymbol{x} \in \mathbb{C}^n$ with $\boldsymbol{y}^*\boldsymbol{x} \neq 0$ the quantity $R(\boldsymbol{y}, \boldsymbol{x}) = R_{\boldsymbol{A}}(\boldsymbol{y}, \boldsymbol{x}) := \frac{\boldsymbol{y}^*\boldsymbol{A}\boldsymbol{x}}{\boldsymbol{y}^*\boldsymbol{x}}$ is called a *generalized Rayleigh quotient* for $\boldsymbol{A}$. Show that if $(\lambda, \boldsymbol{x})$ is a right eigenpair for $\boldsymbol{A}$ then $R(\boldsymbol{y}, \boldsymbol{x}) = \lambda$ for any $\boldsymbol{y}$ with $\boldsymbol{y}^*\boldsymbol{x} \neq 0$. Also show that if $(\lambda, \boldsymbol{y})$ is a left eigenpair for $\boldsymbol{A}$ then $R(\boldsymbol{y}, \boldsymbol{x}) = \lambda$ for any $\boldsymbol{x}$ with $\boldsymbol{y}^*\boldsymbol{x} \neq 0$.

Solution. Suppose $(\lambda, \boldsymbol{x})$ is a right eigenpair for $\boldsymbol{A}$, so that $\boldsymbol{A}\boldsymbol{x} = \lambda\boldsymbol{x}$. Then the generalized Rayleigh quotient for $\boldsymbol{A}$ is

$$R(\boldsymbol{y}, \boldsymbol{x}) := \frac{\boldsymbol{y}^*\boldsymbol{A}\boldsymbol{x}}{\boldsymbol{y}^*\boldsymbol{x}} = \frac{\boldsymbol{y}^*\lambda\boldsymbol{x}}{\boldsymbol{y}^*\boldsymbol{x}} = \lambda,$$

which is well defined whenever $\boldsymbol{y}^*\boldsymbol{x} \neq 0$. On the other hand, if $(\lambda, \boldsymbol{y})$ is a left eigenpair for $\boldsymbol{A}$, then $\boldsymbol{y}^*\boldsymbol{A} = \lambda\boldsymbol{y}^*$ and it follows that

$$R(\boldsymbol{y}, \boldsymbol{x}) := \frac{\boldsymbol{y}^*\boldsymbol{A}\boldsymbol{x}}{\boldsymbol{y}^*\boldsymbol{x}} = \frac{\lambda\boldsymbol{y}^*\boldsymbol{x}}{\boldsymbol{y}^*\boldsymbol{x}} = \lambda.$$

Chapter 7
The Singular Value Decomposition

Exercises section 7.1

Exercise 7.1: SVD1

Show that the decomposition

$$A := \begin{bmatrix} 1 & 1 \\ 1 & 1 \end{bmatrix} = \frac{1}{\sqrt{2}} \begin{bmatrix} 1 & 1 \\ 1 & -1 \end{bmatrix} \begin{bmatrix} 2 & 0 \\ 0 & 0 \end{bmatrix} \frac{1}{\sqrt{2}} \begin{bmatrix} 1 & 1 \\ 1 & -1 \end{bmatrix} = UDU^{\mathrm{T}}$$

is both a spectral decomposition and a singular value decomposition.

Solution. The factorization is easily verified. Here

$$U = V = \frac{1}{\sqrt{2}} \begin{bmatrix} 1 & 1 \\ 1 & -1 \end{bmatrix},$$

which is clearly unitary. Since in addition the matrix in the middle is diagonal, it follows that this is both a spectral and a singular value decomposition.

Exercise 7.2: SVD2

Show that the decomposition

$$A := \begin{bmatrix} 1 & -1 \\ 1 & -1 \end{bmatrix} = \frac{1}{\sqrt{2}} \begin{bmatrix} 1 & 1 \\ 1 & -1 \end{bmatrix} \begin{bmatrix} 2 & 0 \\ 0 & 0 \end{bmatrix} \frac{1}{\sqrt{2}} \begin{bmatrix} 1 & -1 \\ 1 & 1 \end{bmatrix} =: U\Sigma V^{\mathrm{T}}$$

is a singular value decomposition. Show that A is defective, so it cannot be diagonalized by any similarity transformation.

Solution. Again the factorization is easily verified. Here

$$U = \frac{1}{\sqrt{2}} \begin{bmatrix} 1 & 1 \\ 1 & -1 \end{bmatrix}, \qquad V = \frac{1}{\sqrt{2}} \begin{bmatrix} 1 & 1 \\ -1 & 1 \end{bmatrix},$$

T. Lyche et al., *Exercises in Numerical Linear Algebra and Matrix Factorizations*, Texts in Computational Science and Engineering 23, https://doi.org/10.1007/978-3-030-59789-4_7

which are both unitary (but different). Since in addition the matrix in the middle is diagonal and nonnegative, we have a singular value decomposition. The characteristic equation of the matrix $\boldsymbol{A}$ is

$$0 = (1 - \lambda)(-1 - \lambda) + 1 = \lambda^2,$$

so that 0 is an eigenvalue with algebraic multiplicity 2. The corresponding eigenvectors are found by row reducing

$$\begin{bmatrix} 1 & -1 \\ 1 & -1 \end{bmatrix} \sim \begin{bmatrix} 1 & -1 \\ 0 & 0 \end{bmatrix},$$

so that $[1, 1]^{\mathrm{T}}$ spans this eigenspace, which is thus one-dimensional. It follows that $\boldsymbol{A}$ is defective.

Exercise 7.3: SVD examples

Find the singular value decomposition of the following matrices

a) $\boldsymbol{A} = \begin{bmatrix} 3 \\ 4 \end{bmatrix}$.

Solution. For $\boldsymbol{A} = [3, 4]^{\mathrm{T}}$ we find a 1×1 matrix $\boldsymbol{A}^{\mathrm{T}}\boldsymbol{A} = 25$, which has the eigenvalue $\lambda_1 = 25$. This provides us with the singular value $\sigma_1 = \sqrt{\lambda_1} = 5$ for $\boldsymbol{A}$. Hence the matrix $\boldsymbol{A}$ has rank 1 and an SVD of the form

$$\boldsymbol{A} = [\boldsymbol{U}_1\, \boldsymbol{U}_2] \begin{bmatrix} 5 \\ 0 \end{bmatrix} [\boldsymbol{V}_1], \qquad \text{with } \boldsymbol{U}_1, \boldsymbol{U}_2 \in \mathbb{R}^{2,1},\ \boldsymbol{V} = \boldsymbol{V}_1 \in \mathbb{R}.$$

The eigenvector of $\boldsymbol{A}^{\mathrm{T}}\boldsymbol{A}$ that corresponds to the eigenvalue $\lambda_1 = 25$ is given by $\boldsymbol{v}_1 = 1$, providing us with $\boldsymbol{V} = [1]$. Using part 3 of Theorem 7.2, one finds $\boldsymbol{u}_1 = \frac{1}{5}[3, 4]^{\mathrm{T}}$. Extending $\boldsymbol{u}_1$ to an orthonormal basis for $\mathbb{R}^2$ gives $\boldsymbol{u}_2 = \frac{1}{5}[-4, 3]^{\mathrm{T}}$. An SVD of $\boldsymbol{A}$ is therefore

$$\boldsymbol{A} = \frac{1}{5}\begin{bmatrix} 3 & -4 \\ 4 & 3 \end{bmatrix} \begin{bmatrix} 5 \\ 0 \end{bmatrix} [1].$$

b) $\boldsymbol{A} = \begin{bmatrix} 1 & 1 \\ 2 & 2 \\ 2 & 2 \end{bmatrix}$.

Solution. One has

$$\boldsymbol{A} = \begin{bmatrix} 1 & 1 \\ 2 & 2 \\ 2 & 2 \end{bmatrix}, \qquad \boldsymbol{A}^{\mathrm{T}} = \begin{bmatrix} 1 & 2 & 2 \\ 1 & 2 & 2 \end{bmatrix}, \qquad \boldsymbol{A}^{\mathrm{T}}\boldsymbol{A} = \begin{bmatrix} 9 & 9 \\ 9 & 9 \end{bmatrix}.$$

The eigenvalues of $\boldsymbol{A}^{\mathrm{T}}\boldsymbol{A}$ are the zeros of $\det(\boldsymbol{A}^{\mathrm{T}}\boldsymbol{A}-\lambda\boldsymbol{I}) = (9-\lambda)^2-81$, yielding $\lambda_1 = 18$ and $\lambda_2 = 0$, and therefore $\sigma_1 = \sqrt{18}$ and $\sigma_2 = 0$. Note that since there is only one nonzero singular value, the rank of $\boldsymbol{A}$ is one. Following the dimensions of $\boldsymbol{A}$, one finds

$$\boldsymbol{\Sigma} = \begin{bmatrix} \sqrt{18} & 0 \\ 0 & 0 \\ 0 & 0 \end{bmatrix}.$$

The normalized eigenvectors $\boldsymbol{v}_1, \boldsymbol{v}_2$ of $\boldsymbol{A}^{\mathrm{T}}\boldsymbol{A}$ corresponding to the eigenvalues λ_1, λ_2 are the columns of the matrix

$$\boldsymbol{V} = [\boldsymbol{v}_1\ \boldsymbol{v}_2] = \frac{1}{\sqrt{2}}\begin{bmatrix} 1 & -1 \\ 1 & 1 \end{bmatrix}.$$

Using part 3 of Theorem 7.2 one finds $\boldsymbol{u}_1$, which can be extended to an orthonormal basis $\{\boldsymbol{u}_1, \boldsymbol{u}_2, \boldsymbol{u}_3\}$ using Gram-Schmidt orthogonalization (see Theorem 5.4). The vectors $\boldsymbol{u}_1, \boldsymbol{u}_2, \boldsymbol{u}_3$ constitute a matrix

$$\boldsymbol{U} = [\boldsymbol{u}_1\ \boldsymbol{u}_2\ \boldsymbol{u}_3] = \frac{1}{3}\begin{bmatrix} 1 & -2 & -2 \\ 2 & 2 & -1 \\ 2 & -1 & 2 \end{bmatrix}.$$

An SVD of $\boldsymbol{A}$ is therefore given by

$$\boldsymbol{A} = \frac{1}{3}\begin{bmatrix} 1 & -2 & -2 \\ 2 & 2 & -1 \\ 2 & -1 & 2 \end{bmatrix} \begin{bmatrix} \sqrt{18} & 0 \\ 0 & 0 \\ 0 & 0 \end{bmatrix} \frac{1}{\sqrt{2}}\begin{bmatrix} 1 & 1 \\ -1 & 1 \end{bmatrix}.$$

Exercise 7.4: More SVD examples

Find the singular value decomposition of the following matrices

a) $\boldsymbol{A} = \boldsymbol{e}_1$ the first unit vector in $\mathbb{R}^m$.

Solution. We have $\boldsymbol{A} = \boldsymbol{e}_1$ and $\boldsymbol{A}^{\mathrm{T}}\boldsymbol{A} = \boldsymbol{e}_1^{\mathrm{T}}\boldsymbol{e}_1 = \begin{bmatrix}1\end{bmatrix}$. This gives the eigenpair $(\lambda_1, \boldsymbol{v}_1) = (1, [1])$ of $\boldsymbol{A}^{\mathrm{T}}\boldsymbol{A}$. Hence $\sigma_1 = 1$ and $\boldsymbol{\Sigma} = \boldsymbol{e}_1 = \boldsymbol{A}$. As $\boldsymbol{\Sigma} = \boldsymbol{A}$ and $\boldsymbol{V} = \boldsymbol{I}_1$ we must have $\boldsymbol{U} = \boldsymbol{I}_m$ yielding a singular value decomposition

$$\boldsymbol{A} = \boldsymbol{I}_m \boldsymbol{e}_1 \boldsymbol{I}_1.$$

b) $\boldsymbol{A} = \boldsymbol{e}_n^{\mathrm{T}}$ the last unit vector in $\mathbb{R}^n$.

Solution. For $\boldsymbol{A} = \boldsymbol{e}_n^{\mathrm{T}}$, the matrix

$$\boldsymbol{A}^{\mathrm{T}}\boldsymbol{A} = \begin{bmatrix} 0 & \cdots & 0 & 0 \\ \vdots & \ddots & \vdots & \vdots \\ 0 & \cdots & 0 & 0 \\ 0 & \cdots & 0 & 1 \end{bmatrix}.$$

has eigenpairs $(0, \boldsymbol{e}_j)$ for $j = 1, \ldots, n-1$ and $(1, \boldsymbol{e}_n)$. Then $\boldsymbol{\Sigma} = \boldsymbol{e}_1^{\mathrm{T}} \in \mathbb{R}^{1,n}$ and $\boldsymbol{V} = \left[\boldsymbol{e}_n, \boldsymbol{e}_{n-1}, \ldots, \boldsymbol{e}_1\right] \in \mathbb{R}^{n,n}$. Using part 3 of Theorem 7.2 we get $\boldsymbol{u}_1 = 1$, yielding $\boldsymbol{U} = \left[1\right]$. An SVD for $\boldsymbol{A}$ is therefore given by

$$\boldsymbol{A} = \boldsymbol{e}_n^{\mathrm{T}} = \left[1\right] \boldsymbol{e}_1^{\mathrm{T}} \left[\boldsymbol{e}_n, \boldsymbol{e}_{n-1}, \ldots, \boldsymbol{e}_1\right].$$

c) $\boldsymbol{A} = \left[\begin{smallmatrix} -1 & 0 \\ 0 & 3 \end{smallmatrix}\right]$.

Solution. In this exercise

$$\boldsymbol{A} = \begin{bmatrix} -1 & 0 \\ 0 & 3 \end{bmatrix}, \qquad \boldsymbol{A}^{\mathrm{T}} = \boldsymbol{A}, \qquad \boldsymbol{A}^{\mathrm{T}}\boldsymbol{A} = \begin{bmatrix} 1 & 0 \\ 0 & 9 \end{bmatrix}.$$

The eigenpairs of $\boldsymbol{A}^{\mathrm{T}}\boldsymbol{A}$ are given by $(\lambda_1, \boldsymbol{v}_1) = (9, \boldsymbol{e}_2)$ and $(\lambda_2, \boldsymbol{v}_2) = (1, \boldsymbol{e}_1)$, from which we find

$$\boldsymbol{\Sigma} = \begin{bmatrix} 3 & 0 \\ 0 & 1 \end{bmatrix}, \qquad \boldsymbol{V} = \begin{bmatrix} 0 & 1 \\ 1 & 0 \end{bmatrix}.$$

Using part 3 of Theorem 7.2 one finds $\boldsymbol{u}_1 = \boldsymbol{e}_2$ and $\boldsymbol{u}_2 = -\boldsymbol{e}_1$, which constitute the matrix

$$\boldsymbol{U} = \begin{bmatrix} 0 & -1 \\ 1 & 0 \end{bmatrix}.$$

An SVD of $\boldsymbol{A}$ is therefore given by

$$\boldsymbol{A} = \begin{bmatrix} 0 & -1 \\ 1 & 0 \end{bmatrix} \begin{bmatrix} 3 & 0 \\ 0 & 1 \end{bmatrix} \begin{bmatrix} 0 & 1 \\ 1 & 0 \end{bmatrix}.$$

Exercise 7.5: Singular values of a normal matrix

Show that

a) the singular values of a normal matrix are the absolute values of its eigenvalues,

Solution. A normal matrix can be written in the form $\boldsymbol{A} = \boldsymbol{U}\boldsymbol{\Sigma}\boldsymbol{U}^*$, with $\boldsymbol{U}$ unitary, $\boldsymbol{\Sigma}$ diagonal. We now obtain that $\boldsymbol{A}^*\boldsymbol{A} = \boldsymbol{U}\boldsymbol{\Sigma}^*\boldsymbol{\Sigma}\boldsymbol{U}^*$, so that the eigenvalues of $\boldsymbol{A}^*\boldsymbol{A}$ are $|\lambda_i|^2$, where λ_i are the eigenvalues of $\boldsymbol{A}$. It follows that $|\lambda_i|$ are the singular values of $\boldsymbol{A}$.

b) the singular values of a symmetric positive semidefinite matrix are its eigenvalues.

Solution. A symmetric positive definite matrix is also normal, so that the singular values are $|\lambda_i|$. Since the eigenvalues are also positive, the result follows.

Exercise 7.6: The matrices $\boldsymbol{A}^*\boldsymbol{A}$, $\boldsymbol{A}\boldsymbol{A}^*$ and SVD

Show the following:
If $\boldsymbol{A} = \boldsymbol{U}\boldsymbol{\Sigma}\boldsymbol{V}^*$ is a singular value decomposition of $\boldsymbol{A} \in \mathbb{C}^{m\times n}$ then

a) $\boldsymbol{A}^*\boldsymbol{A} = \boldsymbol{V}\operatorname{diag}(\sigma_1^2,\ldots,\sigma_n^2)\boldsymbol{V}^*$ is a spectral decomposition of $\boldsymbol{A}^*\boldsymbol{A}$.

b) $\boldsymbol{A}\boldsymbol{A}^* = \boldsymbol{U}\operatorname{diag}(\sigma_1^2,\ldots,\sigma_m^2)\boldsymbol{U}^*$ is a spectral decomposition of $\boldsymbol{A}\boldsymbol{A}^*$.

c) The columns of $\boldsymbol{U}$ are orthonormal eigenvectors of $\boldsymbol{A}\boldsymbol{A}^*$.

d) The columns of $\boldsymbol{V}$ are orthonormal eigenvectors of $\boldsymbol{A}^*\boldsymbol{A}$.

Solution.

a) and **d)**: We have that $\boldsymbol{A}^*\boldsymbol{A} = \boldsymbol{V}\boldsymbol{\Sigma}^*\boldsymbol{U}^*\boldsymbol{U}\boldsymbol{\Sigma}\boldsymbol{V}^* = \boldsymbol{V}\boldsymbol{\Sigma}^*\boldsymbol{\Sigma}\boldsymbol{V}^*$. This is a spectral decomposition, and the diagonal entries of $\boldsymbol{\Sigma}^*\boldsymbol{\Sigma}$ are σ_i^2. In particular it follows that the columns of $\boldsymbol{V}$ are the corresponding eigenvectors.

b) and **c)**: We have that $\boldsymbol{A}\boldsymbol{A}^* = \boldsymbol{U}\boldsymbol{\Sigma}\boldsymbol{V}^*\boldsymbol{V}\boldsymbol{\Sigma}^*\boldsymbol{U}^* = \boldsymbol{U}\boldsymbol{\Sigma}^*\boldsymbol{\Sigma}\boldsymbol{U}^*$. This is a spectral decomposition, and the diagonal entries of $\boldsymbol{\Sigma}^*\boldsymbol{\Sigma}$ are again σ_i^2. In particular it follows that the columns of $\boldsymbol{U}$ are the corresponding eigenvectors.

Exercise 7.7: Singular values (Exam exercise 2005-2)

Given the statement: "If $\boldsymbol{A} \in \mathbb{R}^{n\times n}$ has singular values $(\sigma_1,\ldots,\sigma_n)$ then $\boldsymbol{A}^2$ has singular values $(\sigma_1^2,\ldots,\sigma_n^2)$". Find a class of matrices for which the statement is true. Show that the statement is not true in general.

Solution. The singular values are the nonnegative square roots of the eigenvalues of $\boldsymbol{A}^{\mathrm{T}}\boldsymbol{A}$ so the statement is true if and only if $\boldsymbol{A}^{\mathrm{T}}\boldsymbol{A}$ and $\boldsymbol{A}^2$ have the same eigenvalues. The claim is correct if $\boldsymbol{A}$ is symmetric. An example where the statement is not true is given by $\boldsymbol{A} = \begin{bmatrix} 1 & 1 \\ 0 & 1 \end{bmatrix}$. Note that $\boldsymbol{A}^2$ is upper triangular with eigenvalue 1 with algebraic multiplicity 2, while $\boldsymbol{A}^{\mathrm{T}}\boldsymbol{A} = \begin{bmatrix} 1 & 1 \\ 1 & 2 \end{bmatrix}$ has eigenvalues $(3 \pm \sqrt{5})/2$.

Suppose $\boldsymbol{A}$ has singular value decomposition $\boldsymbol{A} = \boldsymbol{U}\boldsymbol{\Sigma}\boldsymbol{V}^{\mathrm{T}}$. The statement is correct if $\boldsymbol{V} = \boldsymbol{U}$ for then $\boldsymbol{A}^2 = \boldsymbol{U}\boldsymbol{\Sigma}\boldsymbol{V}^{\mathrm{T}}\boldsymbol{U}\boldsymbol{\Sigma}\boldsymbol{V}^{\mathrm{T}} = \boldsymbol{U}\boldsymbol{\Sigma}^2\boldsymbol{V}^{\mathrm{T}}$ and this is the singular value decomposition of $\boldsymbol{A}^2$. The identity $\boldsymbol{V} = \boldsymbol{U}$ holds if $\boldsymbol{A}$ is normal.

Exercises section 7.2

Exercise 7.8: Nonsingular matrix

Derive the SVF and SVD of the matrix $\boldsymbol{A} = \frac{1}{25}\begin{bmatrix} 11 & 48 \\ 48 & 39 \end{bmatrix}$. Also, possibly using a computer, find its spectral decomposition $\boldsymbol{U}\boldsymbol{D}\boldsymbol{U}^{\mathrm{T}}$. The matrix $\boldsymbol{A}$ is normal, but the spectral decomposition is not an SVD. Why?

Hint.

Answer: $\boldsymbol{A} = \frac{1}{5}\begin{bmatrix} 3 & -4 \\ 4 & 3 \end{bmatrix}\begin{bmatrix} 3 & 0 \\ 0 & 1 \end{bmatrix}\frac{1}{5}\begin{bmatrix} 3 & 4 \\ 4 & -3 \end{bmatrix}$.

Solution. We have that

$$\boldsymbol{A}^{\mathrm{T}}\boldsymbol{A} = \frac{1}{625}\begin{bmatrix} 2425 & 2400 \\ 2400 & 3825 \end{bmatrix} = \frac{1}{25}\begin{bmatrix} 97 & 96 \\ 96 & 153 \end{bmatrix}.$$

The characteristic equation of $\begin{bmatrix} 97 & 96 \\ 96 & 153 \end{bmatrix}$ is

$$0 = (97 - \lambda)(153 - \lambda) - 96^2 = \lambda^2 - 250\lambda + 5625.$$

This gives $\lambda = \frac{250 \pm \sqrt{40000}}{2} = 125 \pm 100$, i.e., 25 and 225. Scaling gives the eigenvalues 1 and 9 of $\boldsymbol{A}^{\mathrm{T}}\boldsymbol{A}$, so that the singular values of $\boldsymbol{A}$ are 1 and 3.

A corresponding eigenvector for $\lambda = 9$ for $\boldsymbol{A}^{\mathrm{T}}\boldsymbol{A}$ is found by row reducing

$$9\boldsymbol{I} - \boldsymbol{A}^{\mathrm{T}}\boldsymbol{A} = \begin{bmatrix} 5.12 & -3.84 \\ -3.84 & 2.88 \end{bmatrix} \sim \begin{bmatrix} 4 & -3 \\ 0 & 0 \end{bmatrix},$$

which gives $\boldsymbol{v}_1 = \frac{1}{5}[3, 4]^{\mathrm{T}}$. For $\lambda = 1$ we get

$$\boldsymbol{I} - \boldsymbol{A}^{\mathrm{T}}\boldsymbol{A} = \begin{bmatrix} -2.88 & -3.84 \\ -3.84 & -5.12 \end{bmatrix} \sim \begin{bmatrix} 3 & 4 \\ 0 & 0 \end{bmatrix},$$

which gives $\boldsymbol{v}_2 = \frac{1}{5}[4, -3]^{\mathrm{T}}$. This gives

$$\boldsymbol{V}^{\mathrm{T}} = \begin{bmatrix} \boldsymbol{v}_1 & \boldsymbol{v}_2 \end{bmatrix}^{\mathrm{T}} = \frac{1}{5}\begin{bmatrix} 3 & 4 \\ 4 & -3 \end{bmatrix}.$$

We also obtain

$$\boldsymbol{u}_1 = \frac{1}{\sigma_1}\boldsymbol{A}\boldsymbol{v}_1 = \frac{1}{75}\begin{bmatrix} 11 & 48 \\ 48 & 39 \end{bmatrix}\frac{1}{5}\begin{bmatrix} 3 \\ 4 \end{bmatrix} = \frac{1}{5}\begin{bmatrix} 3 \\ 4 \end{bmatrix}$$

$$\boldsymbol{u}_2 = \frac{1}{\sigma_2}\boldsymbol{A}\boldsymbol{v}_2 = \frac{1}{25}\begin{bmatrix}11 & 48\\ 48 & 39\end{bmatrix}\frac{1}{5}\begin{bmatrix}4\\ -3\end{bmatrix} = \frac{1}{5}\begin{bmatrix}-4\\ 3\end{bmatrix}.$$

This gives $\boldsymbol{U} = \begin{bmatrix}\boldsymbol{u}_1 & \boldsymbol{u}_2\end{bmatrix} = \frac{1}{5}\begin{bmatrix}3 & -4\\ 4 & 3\end{bmatrix}$. Finally we obtain

$$\boldsymbol{A} = \boldsymbol{U}\boldsymbol{\Sigma}\boldsymbol{V}^{\mathrm{T}} = \frac{1}{5}\begin{bmatrix}3 & -4\\ 4 & 3\end{bmatrix}\begin{bmatrix}3 & 0\\ 0 & 1\end{bmatrix}\frac{1}{5}\begin{bmatrix}3 & 4\\ 4 & -3\end{bmatrix}.$$

The spectral decomposition is easily found to be

$$\boldsymbol{A} = \frac{1}{5}\begin{bmatrix}3 & -4\\ 4 & 3\end{bmatrix}\begin{bmatrix}3 & 0\\ 0 & -1\end{bmatrix}\frac{1}{5}\begin{bmatrix}3 & 4\\ -4 & 3\end{bmatrix}.$$

In particular the matrix is unitarily diagonalizable, so that it is normal. The spectral decomposition is not an SVD since one of the eigenvalues was negative: In the SVD the singular values are positive, so this is something different.

Exercise 7.9: Full row rank

Find the SVF and SVD of

$$\boldsymbol{A} := \frac{1}{15}\begin{bmatrix}14 & 4 & 16\\ 2 & 22 & 13\end{bmatrix} \in \mathbb{R}^{2\times 3}.$$

Hint.

Take the transpose of the matrix in (7.2).

Solution. Transposing the matrix in (7.2) we obtain

$$\boldsymbol{A} = \frac{1}{5}\begin{bmatrix}3 & 4\\ 4 & -3\end{bmatrix}\begin{bmatrix}2 & 0 & 0\\ 0 & 1 & 0\end{bmatrix}\frac{1}{3}\begin{bmatrix}1 & 2 & 2\\ 2 & -2 & 1\\ 2 & 1 & -2\end{bmatrix},$$

which is an SVD. An SVF is

$$\boldsymbol{A} = \frac{1}{5}\begin{bmatrix}3 & 4\\ 4 & -3\end{bmatrix}\begin{bmatrix}2 & 0\\ 0 & 1\end{bmatrix}\frac{1}{3}\begin{bmatrix}1 & 2 & 2\\ 2 & -2 & 1\end{bmatrix}.$$

Exercise 7.10: Counting dimensions of fundamental subspaces

Suppose $\boldsymbol{A} \in \mathbb{C}^{m\times n}$. Show using SVD that

a) $\operatorname{rank}(\boldsymbol{A}) = \operatorname{rank}(\boldsymbol{A}^*)$.

Solution. Let $\boldsymbol{A}$ have singular value decomposition $\boldsymbol{U}\boldsymbol{\Sigma}\boldsymbol{V}^*$.

By parts 1. and 3. of Theorem 7.3, $\operatorname{span}(\boldsymbol{A})$ and $\operatorname{span}(\boldsymbol{A}^*)$ are vector spaces of the same dimension r, implying that $\operatorname{rank}(\boldsymbol{A}) = \operatorname{rank}(\boldsymbol{A}^*)$.

b) $\operatorname{rank}(\boldsymbol{A}) + \operatorname{null}(\boldsymbol{A}) = n$,

Solution. This statement is known as the *rank-nullity theorem*, and it follows immediately from combining parts 1. and 4. in Theorem 7.3.

c) $\operatorname{rank}(\boldsymbol{A}) + \operatorname{null}(\boldsymbol{A}^*) = m$, where $\operatorname{null}(\boldsymbol{A})$ is defined as the dimension of $\mathcal{N}(\boldsymbol{A})$.

Solution. As $\operatorname{rank}(\boldsymbol{A}^*) = \operatorname{rank}(\boldsymbol{A})$ by 1., this follows by replacing $\boldsymbol{A}$ by $\boldsymbol{A}^*$ in 2.

Exercise 7.11: Rank and nullity relations

Use Theorem 7.1 to show that for any $\boldsymbol{A} \in \mathbb{C}^{m\times n}$

a) $\operatorname{rank} \boldsymbol{A} = \operatorname{rank}(\boldsymbol{A}^*\boldsymbol{A}) = \operatorname{rank}(\boldsymbol{A}\boldsymbol{A}^*)$,

Solution. Let $\boldsymbol{A} = \boldsymbol{U}\boldsymbol{\Sigma}\boldsymbol{V}^*$ be a singular value decomposition of a matrix $\boldsymbol{A} \in \mathbb{C}^{m\times n}$.

By part 5 of Theorem 7.1, $\operatorname{rank}(\boldsymbol{A})$ is the number of positive eigenvalues of

$$\boldsymbol{A}\boldsymbol{A}^* = \boldsymbol{U}\boldsymbol{\Sigma}\boldsymbol{V}^*\boldsymbol{V}\boldsymbol{\Sigma}^*\boldsymbol{U}^* = \boldsymbol{U}\boldsymbol{D}\boldsymbol{U}^*,$$

where $\boldsymbol{D} := \boldsymbol{\Sigma}\boldsymbol{\Sigma}^*$ is a diagonal matrix with real nonnegative elements. Since $\boldsymbol{U}\boldsymbol{D}\boldsymbol{U}^*$ is an orthogonal diagonalization of $\boldsymbol{A}\boldsymbol{A}^*$, the number of positive eigenvalues of $\boldsymbol{A}\boldsymbol{A}^*$ is the number of nonzero diagonal elements in $\boldsymbol{D}$. Moreover, $\operatorname{rank}(\boldsymbol{A}\boldsymbol{A}^*)$ is the number of positive eigenvalues of

$$\boldsymbol{A}\boldsymbol{A}^*(\boldsymbol{A}\boldsymbol{A}^*)^* = \boldsymbol{A}\boldsymbol{A}^*\boldsymbol{A}\boldsymbol{A}^* = \boldsymbol{U}\boldsymbol{\Sigma}\boldsymbol{\Sigma}^*\boldsymbol{\Sigma}\boldsymbol{\Sigma}^*\boldsymbol{U}^* = \boldsymbol{U}\boldsymbol{D}^2\boldsymbol{U}^*,$$

which is the number of nonzero diagonal elements in $\boldsymbol{D}^2$, so that $\operatorname{rank}(\boldsymbol{A}) = \operatorname{rank}(\boldsymbol{A}\boldsymbol{A}^*)$. From a similar argument for $\operatorname{rank}(\boldsymbol{A}^*\boldsymbol{A})$, we conclude that

$$\operatorname{rank}(\boldsymbol{A}) = \operatorname{rank}(\boldsymbol{A}\boldsymbol{A}^*) = \operatorname{rank}(\boldsymbol{A}^*\boldsymbol{A}).$$

b) $\operatorname{null}(\boldsymbol{A}^*\boldsymbol{A}) = \operatorname{null} \boldsymbol{A}$, and $\operatorname{null}(\boldsymbol{A}\boldsymbol{A}^*) = \operatorname{null}(\boldsymbol{A}^*)$.

Solution. Let $r := \operatorname{rank}(\boldsymbol{A}) = \operatorname{rank}(\boldsymbol{A}^*) = \operatorname{rank}(\boldsymbol{A}\boldsymbol{A}^*) = \operatorname{rank}(\boldsymbol{A}^*\boldsymbol{A})$. Applying Theorem 7.1, parts 3 and 4, to the singular value decompositions

$$\boldsymbol{A} = \boldsymbol{U}\boldsymbol{\Sigma}\boldsymbol{V}^*,\ \boldsymbol{A}^* = \boldsymbol{V}\boldsymbol{\Sigma}\boldsymbol{U}^*,\ \boldsymbol{A}\boldsymbol{A}^* = \boldsymbol{U}\boldsymbol{\Sigma}\boldsymbol{\Sigma}^*\boldsymbol{U}^*,\ \boldsymbol{A}^*\boldsymbol{A} = \boldsymbol{V}\boldsymbol{\Sigma}^*\boldsymbol{\Sigma}\boldsymbol{V}^*,$$

one finds that $\{\boldsymbol{v}_{r+1},\dots,\boldsymbol{v}_n\}$ is a basis for both $\ker(\boldsymbol{A})$ and $\ker(\boldsymbol{A}^*\boldsymbol{A})$, while $\{\boldsymbol{u}_{r+1},\dots \boldsymbol{u}_m\}$ is a basis for both $\ker(\boldsymbol{A}^*)$ and $\ker(\boldsymbol{A}\boldsymbol{A}^*)$. In particular it follows that

$$\operatorname{null}(\boldsymbol{A}) = \operatorname{null}(\boldsymbol{A}^*\boldsymbol{A}), \qquad \operatorname{null}(\boldsymbol{A}^*) = \operatorname{null}(\boldsymbol{A}\boldsymbol{A}^*),$$

which is what needed to be shown.

Exercise 7.12: Orthonormal bases example

Let $\boldsymbol{A}$ and $\boldsymbol{B}$ be as in Example 7.2. Give orthonormal bases for $\mathcal{R}(\boldsymbol{B})$ and $\mathcal{N}(\boldsymbol{B})$.

Solution. We recall the SVD

$$\boldsymbol{A} = \frac{1}{15}\begin{bmatrix} 14 & 2 \\ 4 & 22 \\ 16 & 13 \end{bmatrix} = \frac{1}{3}\begin{bmatrix} 1 & 2 & 2 \\ 2 & -2 & 1 \\ 2 & 1 & -2 \end{bmatrix}\begin{bmatrix} 2 & 0 \\ 0 & 1 \\ 0 & 0 \end{bmatrix}\frac{1}{5}\begin{bmatrix} 3 & 4 \\ 4 & -3 \end{bmatrix} = \boldsymbol{U}\boldsymbol{\Sigma}\boldsymbol{V}^*.$$

We have that $\boldsymbol{B} = \boldsymbol{A}^*\boldsymbol{A}$, and that $\boldsymbol{B} = \boldsymbol{V}\boldsymbol{\Sigma}^2\boldsymbol{V}^*$ is an SVD for $\boldsymbol{B}$, with $\operatorname{rank}(\boldsymbol{B}) = 2$. Since $\boldsymbol{B}$ has full rank $\mathcal{N}(\boldsymbol{B}) = \{\boldsymbol{0}\}$, and we can choose any orthonormal basis (for instance the standard basis) for $\mathbb{R}^2$ as basis for $\mathcal{R}(\boldsymbol{B})$.

Exercise 7.13: Some spanning sets

Show for any $\boldsymbol{A} \in \mathbb{C}^{m\times n}$ that $\mathcal{R}(\boldsymbol{A}^*\boldsymbol{A}) = \mathcal{R}(\boldsymbol{V}_1) = \mathcal{R}(\boldsymbol{A}^*)$

Solution. The matrices $\boldsymbol{A} \in \mathbb{C}^{m\times n}$ and $\boldsymbol{A}^*\boldsymbol{A}$ have the same rank r since they have the same number of singular values, so that the vector spaces $\operatorname{span}(\boldsymbol{A}^*\boldsymbol{A})$ and $\operatorname{span}(\boldsymbol{A}^*)$ have the same dimension. It is immediate from the definition that $\operatorname{span}(\boldsymbol{A}^*\boldsymbol{A}) \subset \operatorname{span}(\boldsymbol{A}^*)$, and therefore $\operatorname{span}(\boldsymbol{A}^*\boldsymbol{A}) = \operatorname{span}(\boldsymbol{A}^*)$.

From Theorem 7.3 we know that the columns of $\boldsymbol{V}_1$ form an orthonormal basis for $\operatorname{span}(\boldsymbol{A}^*)$, and the result follows.

Exercise 7.14: Singular values and eigenpairs of composite matrix

Let $\boldsymbol{A} \in \mathbb{C}^{m\times n}$ with $m \geq n$ have singular values $\sigma_1, \ldots, \sigma_n$, left singular vectors $\boldsymbol{u}_1, \ldots, \boldsymbol{u}_m \in \mathbb{C}^m$, and right singular vectors $\boldsymbol{v}_1, \ldots, \boldsymbol{v}_n \in \mathbb{C}^n$. Show that the matrix

$$\boldsymbol{C} := \begin{bmatrix} \boldsymbol{0} & \boldsymbol{A} \\ \boldsymbol{A}^* & \boldsymbol{0} \end{bmatrix} \in \mathbb{R}^{(m+n)\times(m+n)}$$

has the $n+m$ eigenpairs

$$\{(\sigma_1, \boldsymbol{p}_1), \ldots, (\sigma_n, \boldsymbol{p}_n), (-\sigma_1, \boldsymbol{q}_1), \ldots, (-\sigma_n, \boldsymbol{q}_n), (0, \boldsymbol{r}_{n+1}), \ldots, (0, \boldsymbol{r}_m)\},$$

where, for $i = 1, \ldots, n$ and $j = n+1, \ldots, m$,

$$\boldsymbol{p}_i = \begin{bmatrix} \boldsymbol{u}_i \\ \boldsymbol{v}_i \end{bmatrix}, \qquad \boldsymbol{q}_i = \begin{bmatrix} \boldsymbol{u}_i \\ -\boldsymbol{v}_i \end{bmatrix}, \qquad \boldsymbol{r}_j = \begin{bmatrix} \boldsymbol{u}_j \\ \boldsymbol{0} \end{bmatrix}.$$

Solution. Given is a singular value decomposition $\boldsymbol{A} = \boldsymbol{U\Sigma V}^*$. Let $r = \text{rank}(\boldsymbol{A})$, so that $\sigma_1 \geq \cdots \geq \sigma_r > 0$ and $\sigma_{r+1} = \cdots = \sigma_n = 0$. Let $\boldsymbol{U} = [\boldsymbol{U}_1, \boldsymbol{U}_2]$ and $\boldsymbol{V} = [\boldsymbol{V}_1, \boldsymbol{V}_2]$ be partitioned accordingly and $\boldsymbol{\Sigma}_1 = \text{diag}(\sigma_1, \ldots, \sigma_r)$ as in Equation (7.4), so that $\boldsymbol{A} = \boldsymbol{U}_1\boldsymbol{\Sigma}_1\boldsymbol{V}_1^*$ forms a singular value factorization of $\boldsymbol{A}$.

By Theorem 7.3,

$$\boldsymbol{C}\boldsymbol{p}_i = \begin{bmatrix} \boldsymbol{0} & \boldsymbol{A} \\ \boldsymbol{A}^* & \boldsymbol{0} \end{bmatrix} \begin{bmatrix} \boldsymbol{u}_i \\ \boldsymbol{v}_i \end{bmatrix} = \begin{bmatrix} \boldsymbol{A}\boldsymbol{v}_i \\ \boldsymbol{A}^*\boldsymbol{u}_i \end{bmatrix} = \begin{cases} \sigma_i \boldsymbol{p}_i & \text{for } i = 1, \ldots, r \\ 0 \cdot \boldsymbol{p}_i & \text{for } i = r+1, \ldots, n \end{cases}$$

$$\boldsymbol{C}\boldsymbol{q}_i = \begin{bmatrix} \boldsymbol{0} & \boldsymbol{A} \\ \boldsymbol{A}^* & \boldsymbol{0} \end{bmatrix} \begin{bmatrix} \boldsymbol{u}_i \\ -\boldsymbol{v}_i \end{bmatrix} = \begin{bmatrix} -\boldsymbol{A}\boldsymbol{v}_i \\ \boldsymbol{A}^*\boldsymbol{u}_i \end{bmatrix} = \begin{cases} -\sigma_i \boldsymbol{q}_i & \text{for } i = 1, \ldots, r \\ 0 \cdot \boldsymbol{q}_i & \text{for } i = r+1, \ldots, n \end{cases}$$

$$\boldsymbol{C}\boldsymbol{r}_j = \begin{bmatrix} \boldsymbol{0} & \boldsymbol{A} \\ \boldsymbol{A}^* & \boldsymbol{0} \end{bmatrix} \begin{bmatrix} \boldsymbol{u}_j \\ \boldsymbol{0} \end{bmatrix} = \begin{bmatrix} \boldsymbol{0} \\ \boldsymbol{A}^*\boldsymbol{u}_j \end{bmatrix} = \begin{bmatrix} \boldsymbol{0} \\ \boldsymbol{0} \end{bmatrix} = 0 \cdot \boldsymbol{r}_j, \text{ for } j = n+1, \ldots, m.$$

This gives a total of $n + n + (m - n) = m + n$ eigenpairs.

Exercise 7.15: Polar decomposition (Exam exercise 2011-2)

Given $n \in \mathbb{N}$ and a singular value decomposition $\boldsymbol{A} = \boldsymbol{U\Sigma V}^{\mathrm{T}}$ of a square matrix $\boldsymbol{A} \in \mathbb{R}^{n,n}$, consider the matrices

$$\boldsymbol{Q} := \boldsymbol{U}\boldsymbol{V}^{\mathrm{T}}, \quad \boldsymbol{P} := \boldsymbol{V\Sigma V}^{\mathrm{T}}$$

of order n.

a) Show that

$$\boldsymbol{A} = \boldsymbol{QP}, \tag{7.13}$$

and show that $\boldsymbol{Q}$ is orthonormal.

Solution. Since $\boldsymbol{V}^{\mathrm{T}}\boldsymbol{V} = \boldsymbol{I}$ we find $\boldsymbol{QP} = \boldsymbol{U}\boldsymbol{V}^{\mathrm{T}}\boldsymbol{V\Sigma V}^{\mathrm{T}} = \boldsymbol{U\Sigma V}^{\mathrm{T}} = \boldsymbol{A}$. $\boldsymbol{Q}$ is orthonormal since $\boldsymbol{U}$ and $\boldsymbol{V}^{\mathrm{T}}$ are orthonormal and a product of orthonormal matrices is orthonormal.

b) Show that $\boldsymbol{P}$ is symmetric positive semidefinite and positive definite if $\boldsymbol{A}$ is nonsingular. The factorization in (7.13) is called a *polar factorization*.

Solution. Write $\boldsymbol{\Sigma} = \text{diag}(\sigma_1, \ldots, \sigma_n)$. For any $\boldsymbol{x} \in \mathbb{R}^n$

$$\boldsymbol{x}^{\mathrm{T}}\boldsymbol{P}\boldsymbol{x} = \boldsymbol{x}^{\mathrm{T}}\boldsymbol{V\Sigma V}^{\mathrm{T}}\boldsymbol{x} = \boldsymbol{y}^{\mathrm{T}}\boldsymbol{\Sigma}\boldsymbol{y} = \sum_{j=1}^{n} \sigma_j |y_j|^2,$$

where $\boldsymbol{y} = \boldsymbol{V}^{\mathrm{T}}\boldsymbol{x}$. Since the singular values σ_j are nonnegative it follows that $\boldsymbol{x}^{\mathrm{T}}\boldsymbol{P}\boldsymbol{x} \geq 0$.

If $\boldsymbol{A}$ is nonsingular then $\boldsymbol{\Sigma}$ is nonsingular and then $\sigma_j > 0$ for all j. But then $\sum_{j=1}^{n} \sigma_j |y_j|^2 > 0$ for all nonzero $\boldsymbol{y} \in \mathbb{R}^n$ so that $\boldsymbol{x}^{\mathrm{T}}\boldsymbol{P}\boldsymbol{x} > 0$ for all nonzero $\boldsymbol{x} \in \mathbb{R}^n$.

c) Use the singular value decomposition of $\boldsymbol{A}$ to give a suitable definition of $\boldsymbol{B} := \sqrt{\boldsymbol{A}^{\mathrm{T}}\boldsymbol{A}}$ so that $\boldsymbol{P} = \boldsymbol{B}$.

Solution. We have $\boldsymbol{A}^{\mathrm{T}}\boldsymbol{A} = \boldsymbol{V}\boldsymbol{\Sigma}\boldsymbol{U}^{\mathrm{T}}\boldsymbol{U}\boldsymbol{\Sigma}\boldsymbol{V}^{\mathrm{T}} = \boldsymbol{V}\boldsymbol{\Sigma}^2\boldsymbol{V}^{\mathrm{T}} = \boldsymbol{P}^2$, so we can define $\boldsymbol{B} = \sqrt{\boldsymbol{A}^{\mathrm{T}}\boldsymbol{A}} := \boldsymbol{P} = \boldsymbol{V}\boldsymbol{\Sigma}\boldsymbol{V}^{\mathrm{T}}$.

For the rest of this exercise assume that $\boldsymbol{A}$ is nonsingular. Consider the iterative method

$$\boldsymbol{X}_{k+1} = \frac{1}{2}(\boldsymbol{X}_k + \boldsymbol{X}_k^{-\mathrm{T}}), \qquad k = 0, 1, 2, \ldots, \qquad \boldsymbol{X}_0 := \boldsymbol{A}, \tag{7.14}$$

for finding $\boldsymbol{Q}$.

d) Show that the iteration (7.14) is well defined by showing that $\boldsymbol{X}_k = \boldsymbol{U}\boldsymbol{\Sigma}_k\boldsymbol{V}^{\mathrm{T}}$, for a diagonal matrix $\boldsymbol{\Sigma}_k$ with positive diagonal elements, $k = 0, 1, 2, \ldots$.

Solution. We have $\boldsymbol{X}_0 := \boldsymbol{A} = \boldsymbol{U}\boldsymbol{\Sigma}_0\boldsymbol{V}^{\mathrm{T}}$, where $\boldsymbol{\Sigma}_0 = \boldsymbol{\Sigma}$ is a diagonal matrix with positive diagonal elements. Suppose by induction that $\boldsymbol{X}_k = \boldsymbol{U}\boldsymbol{\Sigma}_k\boldsymbol{V}^{\mathrm{T}}$ for a diagonal matrix $\boldsymbol{\Sigma}_k$ with positive diagonal elements. Then

$$\boldsymbol{X}_{k+1} := \frac{1}{2}(\boldsymbol{X}_k + \boldsymbol{X}_k^{-\mathrm{T}}) = \boldsymbol{U}\boldsymbol{\Sigma}_{k+1}\boldsymbol{V}^{\mathrm{T}},$$

where $\boldsymbol{\Sigma}_{k+1} := \frac{1}{2}(\boldsymbol{\Sigma}_k + \boldsymbol{\Sigma}_k^{-1})$ is a diagonal matrix, and each diagonal element in $\boldsymbol{\Sigma}_{k+1}$ is a sum of two positive numbers and therefore positive. In particular, if $\boldsymbol{X}_k$ is nonsingular then $\boldsymbol{X}_{k+1}$ will also be nonsingular. It follows that the iteration is well defined.

e) Show that

$$\boldsymbol{X}_{k+1} - \boldsymbol{Q} = \frac{1}{2}\boldsymbol{X}_k^{-\mathrm{T}}(\boldsymbol{X}_k^{\mathrm{T}} - \boldsymbol{Q}^{\mathrm{T}})(\boldsymbol{X}_k - \boldsymbol{Q}) \tag{7.15}$$

and use (7.15) and the Frobenius norm to show (quadratic convergence to $\boldsymbol{Q}$)

$$\|\boldsymbol{X}_{k+1} - \boldsymbol{Q}\|_F \le \frac{1}{2}\|\boldsymbol{X}_k^{-1}\|_F \|\boldsymbol{X}_k - \boldsymbol{Q}\|_F^2. \tag{7.16}$$

Solution. Since $\boldsymbol{X}_k = \boldsymbol{U}\boldsymbol{\Sigma}_k\boldsymbol{V}^{\mathrm{T}}$ we have $\boldsymbol{X}_k^{-1} = \boldsymbol{V}\boldsymbol{\Sigma}_k^{-1}\boldsymbol{U}^{\mathrm{T}}$ and $\boldsymbol{X}_k^{-\mathrm{T}} = \boldsymbol{U}\boldsymbol{\Sigma}_k^{-1}\boldsymbol{V}^{\mathrm{T}}$. But then

$$\boldsymbol{X}_k^{-\mathrm{T}}\boldsymbol{Q}^{\mathrm{T}}\boldsymbol{X}_k = \boldsymbol{U}\boldsymbol{\Sigma}_k^{-1}\boldsymbol{V}^{\mathrm{T}}\boldsymbol{V}\boldsymbol{U}^{\mathrm{T}}\boldsymbol{U}\boldsymbol{\Sigma}_k\boldsymbol{V}^{\mathrm{T}} = \boldsymbol{Q}.$$

Expanding the right hand side of (7.15) and using this we find

$$\frac{1}{2}\boldsymbol{X}_k^{-\mathrm{T}}(\boldsymbol{X}_k^{\mathrm{T}} - \boldsymbol{Q}^{\mathrm{T}})(\boldsymbol{X}_k - \boldsymbol{Q})$$

$$
\begin{aligned}
&= \frac{1}{2}(\boldsymbol{I} - \boldsymbol{X}_k^{-\mathrm{T}}\boldsymbol{Q}^{\mathrm{T}})(\boldsymbol{X}_k - \boldsymbol{Q}) \\
&= \frac{1}{2}(\boldsymbol{X}_k - \boldsymbol{X}_k^{-\mathrm{T}}\boldsymbol{Q}^{\mathrm{T}}\boldsymbol{X}_k - \boldsymbol{Q} + \boldsymbol{X}_k^{-\mathrm{T}}\boldsymbol{Q}^{\mathrm{T}}\boldsymbol{Q}) \\
&= \frac{1}{2}(\boldsymbol{X}_k + \boldsymbol{X}_k^{-\mathrm{T}}) - \boldsymbol{Q} \\
&= \boldsymbol{X}_{k+1} - \boldsymbol{Q}.
\end{aligned}
$$

This shows (7.15).

We have $\|\boldsymbol{B}^{\mathrm{T}}\|_F = \|\boldsymbol{B}\|_F$ for any matrix $\boldsymbol{B}$. Moreover, the Frobenius norm is consistent on $\mathbb{R}^{n,n}$. Therefore

$$
\begin{aligned}
\|\boldsymbol{X}_{k+1} - \boldsymbol{Q}\|_F &= \left\| \frac{1}{2}\boldsymbol{X}_k^{-\mathrm{T}}(\boldsymbol{X}_k^{\mathrm{T}} - \boldsymbol{Q}^{\mathrm{T}})(\boldsymbol{X}_k - \boldsymbol{Q}) \right\|_F \\
&\le \frac{1}{2}\|\boldsymbol{X}_k^{-\mathrm{T}}\|_F \|\boldsymbol{X}_k^{\mathrm{T}} - \boldsymbol{Q}^{\mathrm{T}}\|_F \|\boldsymbol{X}_k - \boldsymbol{Q}\|_F \\
&= \frac{1}{2}\|\boldsymbol{X}_k^{-1}\|_F \|\boldsymbol{X}_k - \boldsymbol{Q}\|_F \|\boldsymbol{X}_k - \boldsymbol{Q}\|_F
\end{aligned}
$$

and (7.16) follows.

f) Write a MATLAB program `function [Q,P,k] = polardecomp(A,tol,K)` to carry out the iteration in (7.14). The output is approximations $\boldsymbol{Q}$ and $\boldsymbol{P} = \boldsymbol{Q}^{\mathrm{T}}\boldsymbol{A}$ to the polar decomposition $\boldsymbol{A} = \boldsymbol{Q}\boldsymbol{P}$ of $\boldsymbol{A}$ and the number of iterations k such that $\|\boldsymbol{X}_{k+1} - \boldsymbol{X}_k\|_F < tol \cdot \|\boldsymbol{X}_{k+1}\|_F$. Set $k = K + 1$ if convergence is not achieved in K iterations. The Frobenius norm in MATLAB is written `norm(A,'fro')`.

Solution.

code/polardecomp.m

```
function [Q,P,k] = polardecomp(A,tol,K)
X=A;
for k=1:K
  Q=(X+inv(X'))/2;
  if norm(Q-X,'fro')<tol*norm(Q,'fro')
    P=Q'*A; return
  end
  X=Q;
end
P=[];k=K+1;
```

Listing 7.1: Compute the kth iterative approximation of the polar factorization $\boldsymbol{A} = \boldsymbol{Q}\boldsymbol{P}$ of a nonsingular matrix $\boldsymbol{A}$.

Exercise 7.16: Underdetermined system (Exam exercise 2015-1)

a) Let $\boldsymbol{A}$ be the matrix

$$\boldsymbol{A} = \begin{bmatrix} 1 & 2 \\ 0 & 1 \\ -1 & 3 \end{bmatrix}.$$

Compute $\|\boldsymbol{A}\|_1$ and $\|\boldsymbol{A}\|_\infty$.

Solution. Using that $\|\boldsymbol{A}\|_1$ is the maximum absolute column sum in $\boldsymbol{A}$, we have that $\|\boldsymbol{A}\|_1 = \max\{2, 6\} = 6$. Using that $\|\boldsymbol{A}\|_\infty$ is the maximum absolute row sum in $\boldsymbol{A}$, we have that $\|\boldsymbol{A}\|_\infty = \max\{3, 1, 4\} = 4$.

b) Let $\boldsymbol{B}$ be the matrix

$$\boldsymbol{B} = \begin{bmatrix} 1 & 0 & -1 \\ 1 & 1 & 1 \end{bmatrix}.$$

Find the spaces $\operatorname{span}(\boldsymbol{B}^{\mathrm{T}})$ and $\ker(\boldsymbol{B})$.

Solution. We have that

$$\boldsymbol{B}^{\mathrm{T}} = \begin{bmatrix} 1 & 1 \\ 0 & 1 \\ -1 & 1 \end{bmatrix}.$$

Clearly the two columns here are linearly independent, so that they form a basis for $\operatorname{span}(\boldsymbol{B}^{\mathrm{T}})$. To find $\ker(\boldsymbol{B})$ first perform row reduction:

$$\begin{bmatrix} 1 & 0 & -1 \\ 1 & 1 & 1 \end{bmatrix} \sim \begin{bmatrix} 1 & 0 & -1 \\ 0 & 1 & 2 \end{bmatrix}.$$

Hence x_3 can be considered to be a free variable, and we must have that $x_1 = x_3$, and $x_2 = -2x_3$ for any $\boldsymbol{x} \in \ker \boldsymbol{B}$. Therefore

$$\boldsymbol{x} = \begin{bmatrix} x_1 \\ x_2 \\ x_3 \end{bmatrix} = x_3 \begin{bmatrix} 1 \\ -2 \\ 1 \end{bmatrix},$$

so that $\{[1, -2, 1]^{\mathrm{T}}\}$ is a basis for $\ker \boldsymbol{B}$.

c) Consider the underdetermined linear system

$$\begin{aligned} x_1 \qquad - x_3 &= 4, \\ x_1 + x_2 + x_3 &= 12. \end{aligned}$$

Find the solution $\boldsymbol{x} \in \mathbb{R}^3$ with $\|\boldsymbol{x}\|_2$ as small as possible.

Solution. There are several ways one can solve this task.

We can use that $\mathrm{span}(\boldsymbol{B}^{\mathrm{T}})$ and $\ker(\boldsymbol{B})$ are orthogonal and together form a basis for $\mathbb{R}^3$. Any solution $\boldsymbol{x}$ to the above equation can thus be written as $\boldsymbol{x} = \boldsymbol{B}^{\mathrm{T}}\boldsymbol{y} + \boldsymbol{z}$, where $\boldsymbol{y} \in \mathbb{R}^2$, $\boldsymbol{z} \in \ker \boldsymbol{B}$. The solution with minimum Euclidean norm is then $\boldsymbol{x} = \boldsymbol{B}^{\mathrm{T}}\boldsymbol{y}$, since $\mathrm{span}(\boldsymbol{B}^{\mathrm{T}})$ and $\ker(\boldsymbol{B})$ are orthogonal, so that one can solve for $\boldsymbol{y}$ first in $\boldsymbol{B}\boldsymbol{B}^{\mathrm{T}}\boldsymbol{y} = \begin{bmatrix} 4 \\ 12 \end{bmatrix}$. Since $\boldsymbol{B}\boldsymbol{B}^{\mathrm{T}} = \begin{bmatrix} 2 & 0 \\ 0 & 3 \end{bmatrix}$, we get that $\boldsymbol{y} = \begin{bmatrix} 2 \\ 4 \end{bmatrix}$, and finally $\boldsymbol{x} = \boldsymbol{B}^{\mathrm{T}}\boldsymbol{y} = [6, 4, 2]^{\mathrm{T}}$.

Alternatively, applying row reduction to the augmented matrix gives

$$\left[\begin{array}{ccc|c} 1 & 0 & -1 & 4 \\ 1 & 1 & 1 & 12 \end{array}\right] \sim \left[\begin{array}{ccc|c} 1 & 0 & -1 & 4 \\ 0 & 1 & 2 & 8 \end{array}\right],$$

so that the general solution is

$$\boldsymbol{x} = \begin{bmatrix} x_3 + 4 \\ -2x_3 + 8 \\ x_3 \end{bmatrix}.$$

We have that $\|\boldsymbol{x}\|_2^2 = (x_3 + 4)^2 + (-2x_3 + 8)^2 + x_3^2 = 6x_3^2 - 24x_2 + 80$. This is minimized when $12x_3 - 24 = 0$, i.e., when $x_3 = 2$, which gives $\boldsymbol{x} = [6, 4, 2]^{\mathrm{T}}$.

It is also rather straightforward to solve this exercise using the pseudoinverse. Consider $\boldsymbol{B}\boldsymbol{B}^{\mathrm{T}} = \begin{bmatrix} 2 & 0 \\ 0 & 3 \end{bmatrix}$ (rather than $\boldsymbol{B}^{\mathrm{T}}\boldsymbol{B}$, which is a 3×3 matrix). We see that the singular values of $\boldsymbol{B}$ are $\sigma_1 = \sqrt{3}$ and $\sigma_2 = \sqrt{2}$. The corresponding eigenvectors for $\boldsymbol{B}\boldsymbol{B}^{\mathrm{T}}$ are $\boldsymbol{e}_2$ and $\boldsymbol{e}_1$, respectively. Since

$$\frac{1}{\sigma_1}\boldsymbol{B}^{\mathrm{T}}\boldsymbol{e}_2 = [1/\sqrt{3}, 1/\sqrt{3}, 1/\sqrt{3}]^{\mathrm{T}}$$
$$\frac{1}{\sigma_2}\boldsymbol{B}^{\mathrm{T}}\boldsymbol{e}_1 = [1/\sqrt{2}, 0, -1/\sqrt{2}]^{\mathrm{T}},$$

a singular value factorization of $\boldsymbol{B}^{\mathrm{T}}$ is

$$\boldsymbol{B}^{\mathrm{T}} = \begin{bmatrix} 1/\sqrt{3} & 1/\sqrt{2} \\ 1/\sqrt{3} & 0 \\ 1/\sqrt{3} & -1/\sqrt{2} \end{bmatrix} \begin{bmatrix} \sqrt{3} & 0 \\ 0 & \sqrt{2} \end{bmatrix} \begin{bmatrix} 0 & 1 \\ 1 & 0 \end{bmatrix}.$$

Transposing this we get a singular value factorization for $\boldsymbol{B}$, and we then easily get the following expression for the pseudoinverse:

$$\boldsymbol{B}^{\dagger} = \begin{bmatrix} 1/\sqrt{3} & 1/\sqrt{2} \\ 1/\sqrt{3} & 0 \\ 1/\sqrt{3} & -1/\sqrt{2} \end{bmatrix} \begin{bmatrix} 1/\sqrt{3} & 0 \\ 0 & 1/\sqrt{2} \end{bmatrix} \begin{bmatrix} 0 & 1 \\ 1 & 0 \end{bmatrix} = \begin{bmatrix} 1/2 & 1/3 \\ 0 & 1/3 \\ -1/2 & 1/3 \end{bmatrix}.$$

The least squares solution to the system $\boldsymbol{A}\boldsymbol{x} = \boldsymbol{b}$ with minimum Euclidean norm can now be obtained by computing

$$B^\dagger b = \begin{bmatrix} 1/2 & 1/3 \\ 0 & 1/3 \\ -1/2 & 1/3 \end{bmatrix} \begin{bmatrix} 4 \\ 12 \end{bmatrix} = \begin{bmatrix} 6 \\ 4 \\ 2 \end{bmatrix}.$$

d) Let $\boldsymbol{A} \in \mathbb{R}^{m \times n}$ be a matrix with linearly independent columns, and $\boldsymbol{b} \in \mathbb{R}^m$ a vector. Assume that we use the Gauss-Seidel method to solve the normal equations $\boldsymbol{A}^{\mathrm{T}}\boldsymbol{A}\boldsymbol{x} = \boldsymbol{A}^{\mathrm{T}}\boldsymbol{b}$. Will the method converge? Justify your answer.

Solution. If $\boldsymbol{A}$ has linearly independent columns, $\boldsymbol{A}^{\mathrm{T}}\boldsymbol{A}$ is invertible (by the characterization of least squares solutions in terms of the normal equations), so that it is also positive definite. But from Theorem 12.7 we know that the Gauss-Seidel method converges for any positive definite matrix.

Exercises section 7.4

Exercise 7.17: Rank example

Consider the singular value decomposition

$$\boldsymbol{A} := \begin{bmatrix} 0 & 3 & 3 \\ 4 & 1 & -1 \\ 4 & 1 & -1 \\ 0 & 3 & 3 \end{bmatrix} = \begin{bmatrix} \frac{1}{2} & -\frac{1}{2} & -\frac{1}{2} & \frac{1}{2} \\ \frac{1}{2} & \frac{1}{2} & \frac{1}{2} & \frac{1}{2} \\ \frac{1}{2} & \frac{1}{2} & -\frac{1}{2} & -\frac{1}{2} \\ \frac{1}{2} & -\frac{1}{2} & \frac{1}{2} & -\frac{1}{2} \end{bmatrix} \begin{bmatrix} 6 & 0 & 0 \\ 0 & 6 & 0 \\ 0 & 0 & 0 \\ 0 & 0 & 0 \end{bmatrix} \begin{bmatrix} \frac{2}{3} & \frac{2}{3} & \frac{1}{3} \\ \frac{2}{3} & -\frac{1}{3} & -\frac{2}{3} \\ \frac{1}{3} & -\frac{2}{3} & \frac{2}{3} \end{bmatrix}$$

a) Give orthonormal bases for $\mathcal{R}(\boldsymbol{A}), \mathcal{R}(\boldsymbol{A}^{\mathrm{T}}), \mathcal{N}(\boldsymbol{A})$ and $\mathcal{N}(\boldsymbol{A}^{\mathrm{T}})$.

Solution. Write $\boldsymbol{U} = [\boldsymbol{u}_1, \boldsymbol{u}_2, \boldsymbol{u}_3, \boldsymbol{u}_4]$ and $\boldsymbol{V} = [\boldsymbol{v}_1, \boldsymbol{v}_2, \boldsymbol{v}_3]$. A direct application of Theorem 7.3 with $r = \operatorname{rank}(\boldsymbol{A}) = 2$ gives

$\{\boldsymbol{u}_1, \boldsymbol{u}_2\}$ is an orthonormal basis for $\operatorname{span}(\boldsymbol{A})$,
$\{\boldsymbol{u}_3, \boldsymbol{u}_4\}$ is an orthonormal basis for $\ker(\boldsymbol{A}^{\mathrm{T}})$,
$\{\boldsymbol{v}_1, \boldsymbol{v}_2\}$ is an orthonormal basis for $\operatorname{span}(\boldsymbol{A}^{\mathrm{T}})$,
$\{\boldsymbol{v}_3\}$ is an orthonormal basis for $\ker(\boldsymbol{A})$.

Since $\boldsymbol{U}$ is orthogonal, $\{\boldsymbol{u}_1, \boldsymbol{u}_2, \boldsymbol{u}_3, \boldsymbol{u}_4\}$ is an orthonormal basis for $\mathbb{R}^4$. In particular $\boldsymbol{u}_3, \boldsymbol{u}_4$ are orthogonal to $\boldsymbol{u}_1, \boldsymbol{u}_2$, so that they span the orthogonal complement $\operatorname{span}(\boldsymbol{A})^\perp$ to $\operatorname{span}(\boldsymbol{A}) = \operatorname{span}\{\boldsymbol{u}_1, \boldsymbol{u}_2\}$.

b) Explain why for all matrices $\boldsymbol{B} \in \mathbb{R}^{4,3}$ of rank one we have $\|\boldsymbol{A} - \boldsymbol{B}\|_F \geq 6$.

Solution. Applying Theorem 7.6 with $r = 1$ yields

$$\|\boldsymbol{A} - \boldsymbol{B}\|_F \geq \sqrt{\sigma_2^2 + \sigma_3^2} = \sqrt{6^2 + 0^2} = 6.$$

c) Give a matrix A_1 of rank one such that $\|A - A_1\|_F = 6$.

Solution. Following Section 7.4.2, with $D' := \operatorname{diag}(\sigma_1, 0, \ldots, 0) \in \mathbb{R}^{n,n}$, take

$$A_1 = A' := U \begin{bmatrix} D' \\ 0 \end{bmatrix} V^{\mathrm{T}} = \begin{bmatrix} 2 & 2 & 1 \\ 2 & 2 & 1 \\ 2 & 2 & 1 \\ 2 & 2 & 1 \end{bmatrix}.$$

Exercise 7.18: Another rank example

Let A be the $n \times n$ matrix that for $n = 4$ takes the form

$$A = \begin{bmatrix} 1 & -1 & -1 & -1 \\ 0 & 1 & -1 & -1 \\ 0 & 0 & 1 & -1 \\ 0 & 0 & 0 & 1 \end{bmatrix}.$$

Thus A is upper triangular with diagonal elements one and all elements above the diagonal equal to -1. Let B be the matrix obtained from A by changing the $(n, 1)$ element from zero to -2^{2-n}.

a) Show that $Bx = 0$, where $x := [2^{n-2}, 2^{n-3}, \ldots, 2^0, 1]^{\mathrm{T}}$. Conclude that B is singular, $\det(A) = 1$, and $\|A - B\|_F = 2^{2-n}$. Thus even if $\det(A)$ is not small the Frobenius norm of $A - B$ is small for large n, and the matrix A is very close to being singular for large n.

Solution. The matrix $B = (b_{ij})_{ij} \in \mathbb{R}^{n,n}$ is defined by

$$b_{ij} = \begin{cases} 1 & \text{if } i = j; \\ -1 & \text{if } i < j; \\ -2^{2-n} & \text{if } (i,j) = (n,1); \\ 0 & \text{otherwise.} \end{cases}$$

while the column vector $x = (x_j)_j \in \mathbb{R}^n$ is given by

$$x_j = \begin{cases} 1 & \text{if } j = n; \\ 2^{n-1-j} & \text{otherwise.} \end{cases}$$

For the final entry in the matrix product Bx one finds that

$$(Bx)_n = \sum_{j=1}^{n} b_{nj} x_j = b_{n1} x_1 + b_{nn} x_n = -2^{2-n} \cdot 2^{n-2} + 1 \cdot 1 = 0.$$

For any of the remaining indices $i \neq n$, the i-th entry of the matrix product Bx can be expressed as

$$
\begin{aligned}
(\boldsymbol{Bx})_i &= \sum_{j=1}^{n} b_{ij}x_j = b_{in} + \sum_{j=1}^{n-1} 2^{n-1-j} b_{ij} \\
&= -1 + 2^{n-1-i} b_{ii} + \sum_{j=i+1}^{n-1} 2^{n-1-j} b_{ij} \\
&= -1 + 2^{n-1-i} - \sum_{j=i+1}^{n-1} 2^{n-1-j} \\
&= -1 + 2^{n-1-i} - 2^{n-2-i} \sum_{j'=0}^{n-2-i} \left(\frac{1}{2}\right)^{j'} \\
&= -1 + 2^{n-1-i} - 2^{n-2-i} \frac{1 - \left(\frac{1}{2}\right)^{n-1-i}}{1 - \frac{1}{2}} \\
&= -1 + 2^{n-1-i} - 2^{n-1-i} \left(1 - 2^{-(n-1-i)}\right) \\
&= 0.
\end{aligned}
$$

As $\boldsymbol{B}$ has a nonzero kernel, it must be singular. The matrix $\boldsymbol{A}$, on the other hand, is nonsingular, as its determinant is $(-1)^n \neq 0$. The matrices $\boldsymbol{A}$ and $\boldsymbol{B}$ differ only in their $(n,1)$-th entry, so one has $\|\boldsymbol{A} - \boldsymbol{B}\|_F = \sqrt{|a_{n1} - b_{n1}|^2} = 2^{2-n}$. In other words, *the tiniest perturbation can make a large matrix with large determinant singular.*

b) Use Theorem 7.6 to show that the smallest singular vale σ_n of $\boldsymbol{A}$ is bounded above by 2^{2-n}.

Solution. Let $\sigma_1 \geq \cdots \geq \sigma_n \geq 0$ be the singular values of $\boldsymbol{A}$. Applying Theorem 7.6 for $r = \operatorname{rank}(\boldsymbol{B}) < n$, we obtain

$$
\sigma_n \leq \sqrt{\sigma_{r+1}^2 + \cdots + \sigma_n^2} = \min_{\substack{\boldsymbol{C} \in \mathbb{R}^{n,n} \\ \operatorname{rank}(\boldsymbol{C}) = r}} \|\boldsymbol{A} - \boldsymbol{C}\|_F \leq \|\boldsymbol{A} - \boldsymbol{B}\|_F = 2^{2-n}.
$$

We conclude that the smallest singular value σ_n can be at most 2^{2-n}.

Exercise 7.19: Norms, Cholesky and SVD (Exam exercise 2016-1)

a) Let $\boldsymbol{A}$ be the matrix

$$
\boldsymbol{A} = \begin{bmatrix} 3 & 1 \\ 2 & 3 \\ -1 & 5 \end{bmatrix}.
$$

Compute $\|\boldsymbol{A}\|_1$, $\|\boldsymbol{A}\|_\infty$ and $\|\boldsymbol{A}\|_F$.

Solution. We have that $\|\boldsymbol{A}\|_F = \sqrt{3^2+1^2+2^2+3^2+(-1)^2+5^2} = \sqrt{49} = 7$. The (absolute) column sums are 6 and 9. $\|\boldsymbol{A}\|_1$ is the biggest of these, which is 9. The (absolute) row sums are 4, 5, and 6. $\|\boldsymbol{A}\|_\infty$ is the biggest of these, which is 6.

b) Let $\boldsymbol{T}$ be the matrix

$$\boldsymbol{T} = \begin{bmatrix} 2 & -1 \\ -1 & 2 \end{bmatrix}.$$

Show that $\boldsymbol{T}$ is symmetric positive definite, and find the Cholesky factorization $\boldsymbol{T} = \boldsymbol{L}\boldsymbol{L}^{\mathrm{T}}$ of $\boldsymbol{T}$.

Solution. The matrix $\boldsymbol{T}$ is clearly symmetric. The characteristic equation is

$$0 = (2-\lambda)^2 - 1 = \lambda^2 - 4\lambda + 3 = (\lambda-1)(\lambda-3),$$

so that the eigenvalues are 1 and 3, implying that $\boldsymbol{T}$ is in fact symmetric positive definite. In Chapter 4 one first finds an $\boldsymbol{L}\boldsymbol{D}\boldsymbol{L}^*$-factorization for the principal submatrices incrementally. Since $\boldsymbol{L}\boldsymbol{D}\boldsymbol{L}^* = \boldsymbol{L}\sqrt{\boldsymbol{D}}\sqrt{\boldsymbol{D}}\boldsymbol{L}^*$ (where the square root means that we take the positive square root of the diagonal entries), a Cholesky factorization is easily found from this. One can also find an $\boldsymbol{L}\boldsymbol{L}^*$ factorization directly from the principal submatrices as follows. For the principal 1×1-submatrix we have $2 = [\sqrt{2}][\sqrt{2}]$. We then have that

$$\begin{bmatrix} 2 & -1 \\ -1 & 2 \end{bmatrix} = \begin{bmatrix} \sqrt{2} & 0 \\ l & d \end{bmatrix} \begin{bmatrix} \sqrt{2} & l \\ 0 & d \end{bmatrix} = \begin{bmatrix} 2 & l\sqrt{2} \\ l\sqrt{2} & l^2+d^2 \end{bmatrix},$$

where l and d are unknowns. We see that $l\sqrt{2} = -1$, so that $l = -\sqrt{2}/2$. Then we obtain $l^2 + d^2 = 1/2 + d^2 = 2$, so that $d = \sqrt{3/2}$, so that

$$\begin{bmatrix} 2 & -1 \\ -1 & 2 \end{bmatrix} = \begin{bmatrix} \sqrt{2} & 0 \\ -\sqrt{2}/2 & \sqrt{3/2} \end{bmatrix} \begin{bmatrix} \sqrt{2} & -\sqrt{2}/2 \\ 0 & \sqrt{3/2} \end{bmatrix}.$$

This gives the Cholesky factorization.

c) Let $\boldsymbol{A} = \boldsymbol{U}\boldsymbol{\Sigma}\boldsymbol{V}^*$ be a singular value decomposition of the $m\times n$-matrix $\boldsymbol{A}$ with $m \geq n$, and let $\boldsymbol{A}' = \sum_{i=1}^r \sigma_i \boldsymbol{u}_i \boldsymbol{v}_i^*$, where $1 \leq r \leq n$, σ_i are the singular values of $\boldsymbol{A}$, and where $\boldsymbol{u}_i, \boldsymbol{v}_i$ are the columns of $\boldsymbol{U}$ and $\boldsymbol{V}$. Prove that

$$\|\boldsymbol{A} - \boldsymbol{A}'\|_F^2 = \sigma_{r+1}^2 + \cdots + \sigma_n^2.$$

Solution. We have that

$$\boldsymbol{A} - \boldsymbol{A}' = \sum_{i=1}^n \sigma_i \boldsymbol{u}_i \boldsymbol{v}_i^* - \sum_{i=1}^r \sigma_i \boldsymbol{u}_i \boldsymbol{v}_i^* = \sum_{i=r+1}^n \sigma_i \boldsymbol{u}_i \boldsymbol{v}_i^* = \boldsymbol{U}\boldsymbol{\Sigma}'\boldsymbol{V}^*,$$

where $\boldsymbol{\Sigma}'$ has $0,\ldots,0,\sigma_{r+1},\ldots,\sigma_n$ on the diagonal. We now obtain

$$\|\boldsymbol{A} - \boldsymbol{A}'\|_F^2 = \|\boldsymbol{U}\boldsymbol{\Sigma}'\boldsymbol{V}^*\|_F^2 = \|\boldsymbol{\Sigma}'\|_F^2 = \sigma_{r+1}^2 + \cdots + \sigma_n^2,$$

where we used the unitary invariance of $\|\cdot\|_F$.

Chapter 8
Matrix Norms and Perturbation Theory for Linear Systems

Exercises section 8.1

Exercise 8.1: An A-norm inequality(Exam exercise 1982-4)
Given a symmetric positive definite matrix $\boldsymbol{A} \in \mathbb{R}^{n\times n}$ with eigenvalues $0 < \lambda_n \le \cdots \le \lambda_1$. Show that

$$\|\boldsymbol{x}\|_{\boldsymbol{A}} \le \|\boldsymbol{y}\|_{\boldsymbol{A}} \implies \|\boldsymbol{x}\|_2 \le \sqrt{\frac{\lambda_1}{\lambda_n}}\|\boldsymbol{y}\|_2,$$

where

$$\|\boldsymbol{x}\|_{\boldsymbol{A}} := \sqrt{\boldsymbol{x}^{\mathrm{T}}\boldsymbol{A}\boldsymbol{x}}, \quad \boldsymbol{x} \in \mathbb{R}^n.$$

Solution. Let

$$\boldsymbol{A}\boldsymbol{u}_j = \lambda_j \boldsymbol{u}_j, \qquad \boldsymbol{u}_i^{\mathrm{T}}\boldsymbol{u}_j = \delta_{i,j}, \qquad i,j = 1,\ldots,n.$$

If $\boldsymbol{x} = \sum_{j=1}^n c_j \boldsymbol{u}_j$ then $\boldsymbol{A}\boldsymbol{x} = \sum_{j=1}^n c_j \lambda_j \boldsymbol{u}_j$. Moreover,

$$\|\boldsymbol{x}\|_{\boldsymbol{A}}^2 = \sum_{j=1}^n c_j^2 \lambda_j, \qquad \|\boldsymbol{x}\|_2^2 = \sum_{j=1}^n c_j^2.$$

We find

$$\|\boldsymbol{x}\|_{\boldsymbol{A}}^2 = \sum_{j=1}^n c_j^2 \lambda_j \le \lambda_1 \sum_{j=1}^n c_j^2 = \lambda_1 \|\boldsymbol{x}\|_2^2,$$

$$\|\boldsymbol{x}\|_{\boldsymbol{A}}^2 \ge \lambda_n \sum_{j=1}^n c_j^2 = \lambda_n \|\boldsymbol{x}\|_2^2,$$

and

T. Lyche et al., *Exercises in Numerical Linear Algebra and Matrix Factorizations*, Texts in Computational Science and Engineering 23, https://doi.org/10.1007/978-3-030-59789-4_8

$$\|\boldsymbol{x}\|_2 \le \sqrt{\frac{1}{\lambda_n}}\|\boldsymbol{x}\|_{\boldsymbol{A}} \le \sqrt{\frac{1}{\lambda_n}}\|\boldsymbol{y}\|_{\boldsymbol{A}} \le \sqrt{\frac{\lambda_1}{\lambda_n}}\|\boldsymbol{y}\|_2.$$

Exercise 8.2: A-orthogonal bases (Exam exercise 1995-4)

Let $\boldsymbol{A} \in \mathbb{R}^{n\times n}$ be a symmetric and positive definite matrix and assume $\boldsymbol{b}_1,\dots,\boldsymbol{b}_n$ is a basis for $\mathbb{R}^n$. We define $\boldsymbol{B}_k := [\boldsymbol{b}_1,\dots,\boldsymbol{b}_k] \in \mathbb{R}^{n\times k}$ for $k = 1,\dots,n$. We consider in this exercise the inner product $\langle\cdot,\cdot\rangle$ defined by $\langle\boldsymbol{x},\boldsymbol{y}\rangle := \boldsymbol{x}^{\mathrm{T}}\boldsymbol{A}\boldsymbol{y}$ for $\boldsymbol{x},\boldsymbol{y} \in \mathbb{R}^n$ and the corresponding norm $\|\boldsymbol{x}\|_{\boldsymbol{A}} := \langle\boldsymbol{x},\boldsymbol{x}\rangle^{1/2}$. We define $\tilde{\boldsymbol{b}}_1 := \boldsymbol{b}_1$ and

$$\tilde{\boldsymbol{b}}_k := \boldsymbol{b}_k - \boldsymbol{B}_{k-1}\big(\boldsymbol{B}_{k-1}^{\mathrm{T}}\boldsymbol{A}\boldsymbol{B}_{k-1}\big)^{-1}\boldsymbol{B}_{k-1}^{\mathrm{T}}\boldsymbol{A}\boldsymbol{b}_k, \quad k = 2,\dots,n.$$

a) Show that $\boldsymbol{B}_k^{\mathrm{T}}\boldsymbol{A}\boldsymbol{B}_k$ is positive definite for $k = 1,\dots,n$.

Solution. Since $\boldsymbol{A}$ is symmetric and $\big(\boldsymbol{B}_k^{\mathrm{T}}\boldsymbol{A}\boldsymbol{B}_k\big)^{\mathrm{T}} = \boldsymbol{B}_k^{\mathrm{T}}\boldsymbol{A}^{\mathrm{T}}\boldsymbol{B}_k = \boldsymbol{B}_k^{\mathrm{T}}\boldsymbol{A}\boldsymbol{B}_k$ we see that $\boldsymbol{B}_k^{\mathrm{T}}\boldsymbol{A}\boldsymbol{B}_k$ is symmetric. For $\boldsymbol{x} \in \mathbb{R}^k$ we have

$$\boldsymbol{x}^{\mathrm{T}}\boldsymbol{B}_k^{\mathrm{T}}\boldsymbol{A}\boldsymbol{B}_k\boldsymbol{x} = (\boldsymbol{B}_k\boldsymbol{x})^{\mathrm{T}}\boldsymbol{A}(\boldsymbol{B}_k\boldsymbol{x}) \ge 0$$

since $\boldsymbol{A}$ is positive definite. If $\boldsymbol{x}^{\mathrm{T}}\boldsymbol{B}_k^{\mathrm{T}}\boldsymbol{A}\boldsymbol{B}_k\boldsymbol{x} = 0$ then $\boldsymbol{B}_k\boldsymbol{x} = 0$ and since $\boldsymbol{B}_k$ has linearly independent columns we conclude that $\boldsymbol{x} = 0$ and $\boldsymbol{B}_k^{\mathrm{T}}\boldsymbol{A}\boldsymbol{B}_k$ is positive definite.

b) Show that for $k = 2,\dots,n$ we have
(i) $\langle\tilde{\boldsymbol{b}}_k,\boldsymbol{b}_j\rangle = 0$ for $j = 1,\dots,k-1$ and
(ii) $\tilde{\boldsymbol{b}}_k - \boldsymbol{b}_k \in \operatorname{span}(\boldsymbol{b}_1,\dots,\boldsymbol{b}_{k-1})$.

Solution. For $1 \le j < k$

$$\begin{aligned}\langle\boldsymbol{b}_j,\tilde{\boldsymbol{b}}_k\rangle &= \boldsymbol{b}_j^{\mathrm{T}}\boldsymbol{A}\tilde{\boldsymbol{b}}_k = \boldsymbol{b}_j^{\mathrm{T}}\boldsymbol{A}\boldsymbol{b}_k - \boldsymbol{e}_j^{\mathrm{T}}(\boldsymbol{B}_{k-1}^{\mathrm{T}}\boldsymbol{A}\boldsymbol{B}_{k-1})\big(\boldsymbol{B}_{k-1}^{\mathrm{T}}\boldsymbol{A}\boldsymbol{B}_{k-1}\big)^{-1}\boldsymbol{B}_{k-1}^{\mathrm{T}}\boldsymbol{A}\boldsymbol{b}_k\\ &= \boldsymbol{b}_j^{\mathrm{T}}\boldsymbol{A}\boldsymbol{b}_k - (\boldsymbol{e}_j^{\mathrm{T}}\boldsymbol{B}_{k-1}^{\mathrm{T}})\boldsymbol{A}\boldsymbol{b}_k = 0.\end{aligned}$$

Since $\boldsymbol{b}_k - \tilde{\boldsymbol{b}}_k = \boldsymbol{B}_{k-1}\boldsymbol{z}$, where $\boldsymbol{z} := \big(\boldsymbol{B}_{k-1}^{\mathrm{T}}\boldsymbol{A}\boldsymbol{B}_{k-1}\big)^{-1}\boldsymbol{B}_{k-1}^{\mathrm{T}}\boldsymbol{A}\boldsymbol{b}_k$, we see that $\boldsymbol{b}_k - \tilde{\boldsymbol{b}}_k$ is a linear combination of the columns of $\boldsymbol{B}_{k-1}$.

c) Explain why $\tilde{\boldsymbol{b}}_1,\dots,\tilde{\boldsymbol{b}}_n$ is a basis for $\mathbb{R}^n$ which in addition is $\boldsymbol{A}$-orthogonal, i.e., $\langle\tilde{\boldsymbol{b}}_i,\tilde{\boldsymbol{b}}_j\rangle = 0$ for all $i,j \le n$, $i \ne j$.

Solution. Since $\tilde{\boldsymbol{b}}_k \in \operatorname{span}(\boldsymbol{b}_1,\dots,\boldsymbol{b}_k)$ we have $\tilde{\boldsymbol{b}}_k = \sum_{j=1}^k c_j\boldsymbol{b}_j$ for some $c_j \in \mathbb{R}$. But then for $i > k$ by **b)**

$$\langle\tilde{\boldsymbol{b}}_i,\tilde{\boldsymbol{b}}_k\rangle = \langle\tilde{\boldsymbol{b}}_i,\sum_{j=1}^k c_j\boldsymbol{b}_j\rangle = \sum_{j=1}^k c_j\langle\tilde{\boldsymbol{b}}_i,\boldsymbol{b}_j\rangle = 0.$$

Since $\tilde{\boldsymbol{b}}_k - \boldsymbol{b}_k \in \operatorname{span}(\boldsymbol{b}_1, \ldots, \boldsymbol{b}_{k-1})$ we see that $\tilde{\boldsymbol{b}}_k \neq \boldsymbol{0}$ and $\tilde{\boldsymbol{b}}_1, \ldots, \tilde{\boldsymbol{b}}_n$ is an $\boldsymbol{A}$-orthogonal basis for $\mathbb{R}^n$.

d) Define $\tilde{\boldsymbol{B}}_n := [\tilde{\boldsymbol{b}}_1, \ldots, \tilde{\boldsymbol{b}}_n]$. Show that there is an upper triangular matrix $\boldsymbol{T} \in \mathbb{R}^{n\times n}$ with ones on the diagonal and satisfies $\boldsymbol{B}_n = \tilde{\boldsymbol{B}}_n \boldsymbol{T}$.

Solution. Since $\tilde{\boldsymbol{b}}_k - \boldsymbol{b}_k \in \operatorname{span}(\boldsymbol{b}_1, \ldots, \boldsymbol{b}_{k-1})$ we have $\tilde{\boldsymbol{b}}_k = \boldsymbol{b}_k + \sum_{j=1}^{k-1} s_{j,k}\boldsymbol{b}_j$ for some $s_{j,k} \in \mathbb{R}$. But this implies that $\tilde{\boldsymbol{B}}_n = \boldsymbol{B}_n\boldsymbol{S}$, where $\boldsymbol{S} \in \mathbb{R}^{n\times n}$ is unit upper triangular. But then $\boldsymbol{B}_n = \tilde{\boldsymbol{B}}_n\boldsymbol{T}$, where $\boldsymbol{T} = \boldsymbol{S}^{-1}$ is also unit upper triangular (cf. Lemma 2.5).

e) Assume that the matrix $\boldsymbol{T}$ in **d)** is such that $|t_{ij}| \leq \frac{1}{2}$ for all $i, j \leq n$, $i \neq j$. Assume also that $\|\tilde{\boldsymbol{b}}_k\|_{\boldsymbol{A}}^2 \leq 2\|\tilde{\boldsymbol{b}}_{k+1}\|_{\boldsymbol{A}}^2$ for $k = 1, \ldots, n-1$ and that $\det(\boldsymbol{B}_n) = 1$. Show that then

$$\|\boldsymbol{b}_1\|_{\boldsymbol{A}}\|\boldsymbol{b}_2\|_{\boldsymbol{A}} \cdots \|\boldsymbol{b}_n\|_{\boldsymbol{A}} \leq 2^{n(n-1)/4}\sqrt{\det(\boldsymbol{A})}.$$

Hint.
Show that $\|\tilde{\boldsymbol{b}}_1\|_{\boldsymbol{A}}^2 \cdots \|\tilde{\boldsymbol{b}}_n\|_{\boldsymbol{A}}^2 = \det(\boldsymbol{A})$

Solution. We first show that $\|\tilde{\boldsymbol{b}}_1\|_{\boldsymbol{A}}^2 \cdots \|\tilde{\boldsymbol{b}}_n\|_{\boldsymbol{A}}^2 = \det(\boldsymbol{A})$. By $\boldsymbol{A}$-orthogonality for all i, j

$$(\tilde{\boldsymbol{B}}_n^{\mathrm{T}}\boldsymbol{A}\tilde{\boldsymbol{B}}_n)(i,j) = \tilde{\boldsymbol{b}}_i^{\mathrm{T}}\boldsymbol{A}\tilde{\boldsymbol{b}}_j = \delta_{i,j}\tilde{\boldsymbol{b}}_i^{\mathrm{T}}\boldsymbol{A}\tilde{\boldsymbol{b}}_i.$$

But then, by construction of $\boldsymbol{T}$ and assumption of $\boldsymbol{B}_n$,

$$\det\left(\boldsymbol{T}^{-1}\right) = \det\left((\boldsymbol{T}^{-1})^{\mathrm{T}}\right) = 1 = \det(\boldsymbol{B}_n) = \det\left((\boldsymbol{B}_n)^{\mathrm{T}}\right)$$

and $\tilde{\boldsymbol{B}}_n = \boldsymbol{B}_n\boldsymbol{T}^{-1}$, it follows that

$$\begin{aligned}\|\tilde{\boldsymbol{b}}_1\|_{\boldsymbol{A}}^2 \cdots \|\tilde{\boldsymbol{b}}_n\|_{\boldsymbol{A}}^2 &= \det\left(\operatorname{diag}(\tilde{\boldsymbol{b}}_1^{\mathrm{T}}\boldsymbol{A}\tilde{\boldsymbol{b}}_1, \ldots, \tilde{\boldsymbol{b}}_n^{\mathrm{T}}\boldsymbol{A}\tilde{\boldsymbol{b}}_n)\right)\\ &= \det\left(\tilde{\boldsymbol{B}}_n^{\mathrm{T}}\boldsymbol{A}\tilde{\boldsymbol{B}}_n\right) = \det\left((\boldsymbol{T}^{-1})^{\mathrm{T}}\boldsymbol{B}_n^{\mathrm{T}}\boldsymbol{A}\boldsymbol{B}_n\boldsymbol{T}^{-1}\right) = \det(\boldsymbol{A}).\end{aligned}$$

Now $\boldsymbol{b}_k = \boldsymbol{B}_n\boldsymbol{e}_k = \tilde{\boldsymbol{B}}_n\boldsymbol{T}\boldsymbol{e}_k = \tilde{\boldsymbol{b}}_k + \sum_{j=1}^{k-1} t_{j,k}\tilde{\boldsymbol{b}}_j$, and by $\boldsymbol{A}$-orthogonality this implies

$$\|\boldsymbol{b}_k\|_{\boldsymbol{A}}^2 = \|\tilde{\boldsymbol{b}}_k\|_{\boldsymbol{A}}^2 + \sum_{j=1}^{k-1}|t_{j,k}|^2\|\tilde{\boldsymbol{b}}_j\|_{\boldsymbol{A}}^2.$$

Since $\|\tilde{\boldsymbol{b}}_k\|_{\boldsymbol{A}}^2 \leq 2\|\tilde{\boldsymbol{b}}_{k+1}\|_{\boldsymbol{A}}^2$ we obtain $\|\tilde{\boldsymbol{b}}_j\|_{\boldsymbol{A}}^2 \leq 2^{k-j}\|\tilde{\boldsymbol{b}}_k\|_{\boldsymbol{A}}^2$ for $j = 1, \ldots, k-1$. Moreover, since $|t_{j,k}| \leq 1/2$ we find

$$\|\boldsymbol{b}_k\|_{\boldsymbol{A}}^2 \leq \|\tilde{\boldsymbol{b}}_k\|_{\boldsymbol{A}}^2 + \frac{1}{4}\sum_{j=1}^{k-1} 2^{k-j}\|\tilde{\boldsymbol{b}}_k\|_{\boldsymbol{A}}^2 \leq \|\tilde{\boldsymbol{b}}_k\|_{\boldsymbol{A}}^2(1+2^{k-2}) \leq \|\tilde{\boldsymbol{b}}_k\|_{\boldsymbol{A}}^2 2^{k-1}.$$

But then

$$\|\boldsymbol{b}_1\|_{\boldsymbol{A}}^2 \cdots \|\boldsymbol{b}_n\|_{\boldsymbol{A}}^2 \leq \prod_{k=1}^{n} \|\tilde{\boldsymbol{b}}_k\|_{\boldsymbol{A}}^2 2^{k-1} = \det(\boldsymbol{A}) \prod_{k=1}^{n} 2^{k-1} = \det(\boldsymbol{A}) 2^{n(n-1)/2}.$$

Taking square roots the result follows.

Exercises section 8.2

Exercise 8.3: Consistency of sum norm?

Show that the sum norm is consistent.

Solution. Observe that the sum norm is a matrix norm. This follows since it is equal to the l_1-norm of the vector $\boldsymbol{v} = \text{vec}(\boldsymbol{A})$ obtained by stacking the columns of a matrix $\boldsymbol{A}$ on top of each other.

Let $\boldsymbol{A} = (a_{ij})_{ij}$ and $\boldsymbol{B} = (b_{ij})_{ij}$ be matrices for which the product $\boldsymbol{AB}$ is defined. Then

$$\begin{aligned} \|\boldsymbol{AB}\|_S &= \sum_{i,j}\left|\sum_k a_{ik}b_{kj}\right| \leq \sum_{i,j,k} |a_{ik}| \cdot |b_{kj}| \\ &\leq \sum_{i,j,k,l} |a_{ik}| \cdot |b_{lj}| = \sum_{i,k} |a_{ik}| \sum_{l,j} |b_{lj}| = \|\boldsymbol{A}\|_S\|\boldsymbol{B}\|_S, \end{aligned}$$

where the first inequality follows from the triangle inequality and multiplicative property of the absolute value $|\cdot|$. Since $\boldsymbol{A}$ and $\boldsymbol{B}$ were arbitrary, this proves that the sum norm is consistent.

Exercise 8.4: Consistency of max norm?

Show that the max norm is not consistent by considering $\left[\begin{smallmatrix}1 & 1\\ 1 & 1\end{smallmatrix}\right]$.

Solution. Observe that the max norm is a matrix norm. This follows since it is equal to the l_∞-norm of the vector $\boldsymbol{v} = \text{vec}(\boldsymbol{A})$ obtained by stacking the columns of a matrix $\boldsymbol{A}$ on top of each other.

To show that the max norm is not consistent we use a counter example. Let $\boldsymbol{A} = \boldsymbol{B} = \begin{bmatrix}1 & 1\\ 1 & 1\end{bmatrix}$. Then

$$\left\|\begin{bmatrix}1 & 1\\ 1 & 1\end{bmatrix}\begin{bmatrix}1 & 1\\ 1 & 1\end{bmatrix}\right\|_M = \left\|\begin{bmatrix}2 & 2\\ 2 & 2\end{bmatrix}\right\|_M = 2 > 1 = \left\|\begin{bmatrix}1 & 1\\ 1 & 1\end{bmatrix}\right\|_M \left\|\begin{bmatrix}1 & 1\\ 1 & 1\end{bmatrix}\right\|_M,$$

contradicting $\|\boldsymbol{AB}\|_M \le \|\boldsymbol{A}\|_M\|\boldsymbol{B}\|_M$.

Exercise 8.5: Consistency of modified max norm

Exercise 8.4 shows that the max norm is not consistent. In this exercise we show that the max norm can be modified so as to define a consistent matrix norm.

a) Show that the norm

$$\|\boldsymbol{A}\| := \sqrt{mn}\|\boldsymbol{A}\|_M, \quad \boldsymbol{A} \in \mathbb{C}^{m\times n}$$

is a consistent matrix norm.

Solution. To show that $\|\cdot\|$ defines a consistent matrix norm, we have to show that it fulfills the three matrix norm properties and that it is submultiplicative. Let $\boldsymbol{A}, \boldsymbol{B} \in \mathbb{C}^{m,n}$ be any matrices and α any scalar.

1. *Positivity*. Clearly $\|\boldsymbol{A}\| := \sqrt{mn}\|\boldsymbol{A}\|_M \ge 0$. Moreover,

$$\|\boldsymbol{A}\| = 0 \iff a_{i,j} = 0\ \forall i,j \iff \boldsymbol{A} = 0.$$

2. *Homogeneity*. $\|\alpha\boldsymbol{A}\| = \sqrt{mn}\|\alpha\boldsymbol{A}\|_M = |\alpha|\sqrt{mn}\|\boldsymbol{A}\|_M = |\alpha|\|\boldsymbol{A}\|$.
3. *Subadditivity*. One has

$$\|\boldsymbol{A}+\boldsymbol{B}\| = \sqrt{nm}\|\boldsymbol{A}+\boldsymbol{B}\|_M \le \sqrt{nm}\Big(\|\boldsymbol{A}\|_M + \|\boldsymbol{B}\|_M\Big) = \|\boldsymbol{A}\| + \|\boldsymbol{B}\|.$$

4. *Submultiplicativity*. One has

$$\begin{aligned}
\|\boldsymbol{AB}\| &= \sqrt{mn}\max_{\substack{1\le i\le m\\ 1\le j\le n}}\left|\sum_{k=1}^{q} a_{i,k}b_{k,j}\right| \\
&\le \sqrt{mn}\max_{\substack{1\le i\le m\\ 1\le j\le n}}\sum_{k=1}^{q}|a_{i,k}||b_{k,j}| \\
&\le \sqrt{mn}\max_{1\le i\le m}\left(\max_{\substack{1\le k\le q\\ 1\le j\le n}}|b_{k,j}|\sum_{k=1}^{q}|a_{i,k}|\right) \\
&\le q\sqrt{mn}\left(\max_{\substack{1\le i\le m\\ 1\le k\le q}}|a_{i,k}|\right)\left(\max_{\substack{1\le k\le q\\ 1\le j\le n}}|b_{k,j}|\right) \\
&= \|\boldsymbol{A}\|\|\boldsymbol{B}\|.
\end{aligned}$$

b) Show that the constant $\sqrt{mn}$ can be replaced by m and by n.

Solution. For any $\boldsymbol{A} \in \mathbb{C}^{m,n}$, let

$$\|\boldsymbol{A}\|^{(1)} := m\|\boldsymbol{A}\|_M \qquad \text{and} \qquad \|\boldsymbol{A}\|^{(2)} := n\|\boldsymbol{A}\|_M.$$

Comparing with the solution of part **a)** we see, that the points of positivity, homogeneity and subadditivity are fulfilled here as well, making $\|\boldsymbol{A}\|^{(1)}$ and $\|\boldsymbol{A}\|^{(2)}$ valid matrix norms. Furthermore, for any $\boldsymbol{A} \in \mathbb{C}^{m,q}, \boldsymbol{B} \in \mathbb{C}^{q,n}$,

$$\begin{aligned}\|\boldsymbol{AB}\|^{(1)} &= m \max_{\substack{1\le i\le m\\1\le j\le n}} \left|\sum_{k=1}^{q} a_{i,k}b_{k,j}\right| \le m\left(\max_{\substack{1\le i\le m\\1\le k\le q}} |a_{i,k}|\right) q \left(\max_{\substack{1\le k\le q\\1\le j\le n}} |b_{k,j}|\right)\\ &= \|\boldsymbol{A}\|^{(1)}\|\boldsymbol{B}\|^{(1)},\\ \|\boldsymbol{AB}\|^{(2)} &= n \max_{\substack{1\le i\le m\\1\le j\le n}} |\sum_{k=1}^{q} a_{i,k}b_{k,j}| \le q\left(\max_{\substack{1\le i\le m\\1\le k\le q}} |a_{i,k}|\right) n \left(\max_{\substack{1\le k\le q\\1\le j\le n}} |b_{k,j}|\right)\\ &= \|\boldsymbol{A}\|^{(2)}\|\boldsymbol{B}\|^{(2)},\end{aligned}$$

which proves the submultiplicativity of both norms.

Exercise 8.6: What is the sum norm subordinate to?

Show that the sum norm is subordinate to the l_1-norm.

Solution. For any matrix $\boldsymbol{A} = [a_{ij}]_{ij} \in \mathbb{C}^{m,n}$ and column vector $\boldsymbol{x} = [x_j]_j \in \mathbb{C}^n$, one has

$$\begin{aligned}\|\boldsymbol{Ax}\|_1 &= \sum_{i=1}^{m}\left|\sum_{j=1}^{n} a_{ij}x_j\right| \le \sum_{i=1}^{m}\sum_{j=1}^{n} |a_{ij}|\cdot|x_j| \le \sum_{i=1}^{m}\sum_{j=1}^{n} |a_{ij}| \sum_{k=1}^{n} |x_k|\\ &= \|\boldsymbol{A}\|_S\|\boldsymbol{x}\|_1,\end{aligned}$$

which shows that the matrix norm $\|\cdot\|_S$ is subordinate to the vector norm $\|\cdot\|_1$.

Exercise 8.7: What is the max norm subordinate to?

Let $\boldsymbol{A} = [a_{ij}]_{ij} \in \mathbb{C}^{m,n}$ be a matrix and $\boldsymbol{x} = [x_j]_j \in \mathbb{C}^n$ a column vector.

a) Show that the max norm is subordinate to the ∞ and 1 norm, i.e., $\|\boldsymbol{Ax}\|_\infty \le \|\boldsymbol{A}\|_M\|\boldsymbol{x}\|_1$ holds for all $\boldsymbol{A} \in \mathbb{C}^{m\times n}$ and all $\boldsymbol{x} \in \mathbb{C}^n$.

Solution. One has

$$\begin{aligned}\|\boldsymbol{Ax}\|_\infty &= \max_{i=1,\ldots,m}\left|\sum_{j=1}^{n} a_{ij}x_j\right| \le \max_{i=1,\ldots,m}\sum_{j=1}^{n} |a_{ij}|\cdot|x_j| \le \max_{\substack{i=1,\ldots,m\\j=1,\ldots,n}} |a_{ij}|\sum_{j=1}^{n}|x_j|\\ &= \|\boldsymbol{A}\|_M\|\boldsymbol{x}\|_1.\end{aligned}$$

b) Show that if $\|\boldsymbol{A}\|_M = |a_{kl}|$, then $\|\boldsymbol{A}\boldsymbol{e}_l\|_\infty = \|\boldsymbol{A}\|_M \|\boldsymbol{e}_l\|_1$.

Solution. Assume that the maximum in the definition of $\|\boldsymbol{A}\|_M$ is attained in column l, implying that $\|\boldsymbol{A}\|_M = |a_{k,l}|$ for some k. Let $\boldsymbol{e}_l$ be the lth standard basis vector. Then $\|\boldsymbol{e}_l\|_1 = 1$ and

$$\|\boldsymbol{A}\boldsymbol{e}_l\|_\infty = \max_{i=1,\dots,m} |a_{i,l}| = |a_{k,l}| = |a_{k,l}| \cdot 1 = \|\boldsymbol{A}\|_M \cdot \|\boldsymbol{e}_l\|_1,$$

which is what needed to be shown.

c) Show that $\|\boldsymbol{A}\|_M = \max_{\boldsymbol{x}\neq\boldsymbol{0}} \frac{\|\boldsymbol{A}\boldsymbol{x}\|_\infty}{\|\boldsymbol{x}\|_1}$.

Solution. By **a)**, $\|\boldsymbol{A}\|_M \geq \|\boldsymbol{A}\boldsymbol{x}\|_\infty / \|\boldsymbol{x}\|_1$ for all nonzero vectors $\boldsymbol{x}$, implying that

$$\|\boldsymbol{A}\|_M \geq \max_{\boldsymbol{x}\neq\boldsymbol{0}} \frac{\|\boldsymbol{A}\boldsymbol{x}\|_\infty}{\|\boldsymbol{x}\|_1}.$$

By **b)**, equality is attained for any standard basis vector $\boldsymbol{e}_l$ for which there exists a k such that $\|\boldsymbol{A}\|_M = |a_{k,l}|$. We conclude that

$$\|\boldsymbol{A}\|_M = \max_{\boldsymbol{x}\neq\boldsymbol{0}} \frac{\|\boldsymbol{A}\boldsymbol{x}\|_\infty}{\|\boldsymbol{x}\|_1},$$

which means that $\|\cdot\|_M$ is the $(\infty, 1)$-operator norm (see Definition 8.6).

Exercise 8.8: Spectral norm

Let $m, n \in \mathbb{N}$ and $\boldsymbol{A} \in \mathbb{C}^{m\times n}$. Show that

$$\|\boldsymbol{A}\|_2 = \max_{\|\boldsymbol{x}\|_2 = \|\boldsymbol{y}\|_2 = 1} |\boldsymbol{y}^*\boldsymbol{A}\boldsymbol{x}|.$$

Solution. Let $\boldsymbol{A} = \boldsymbol{U}\boldsymbol{\Sigma}\boldsymbol{V}^*$ be a singular value decomposition of $\boldsymbol{A}$, and write $\sigma_1 := \|\boldsymbol{A}\|_2$ for the biggest singular value of $\boldsymbol{A}$. Since the orthogonal matrices $\boldsymbol{U}$ and $\boldsymbol{V}$ leave the Euclidean norm invariant,

$$\begin{aligned}
\max_{\|\boldsymbol{x}\|_2=1=\|\boldsymbol{y}\|_2} |\boldsymbol{y}^*\boldsymbol{A}\boldsymbol{x}| &= \max_{\|\boldsymbol{x}\|_2=1=\|\boldsymbol{y}\|_2} |\boldsymbol{y}^*\boldsymbol{U}\boldsymbol{\Sigma}\boldsymbol{V}^*\boldsymbol{x}| = \max_{\|\boldsymbol{x}\|_2=1=\|\boldsymbol{y}\|_2} |\boldsymbol{y}^*\boldsymbol{\Sigma}\boldsymbol{x}| \\
&= \max_{\|\boldsymbol{x}\|_2=1=\|\boldsymbol{y}\|_2} \left| \sum_{i=1}^n \sigma_i x_i \overline{y}_i \right| \leq \max_{\|\boldsymbol{x}\|_2=1=\|\boldsymbol{y}\|_2} \sum_{i=1}^n \sigma_i |x_i \overline{y}_i| \\
&\leq \sigma_1 \max_{\|\boldsymbol{x}\|_2=1=\|\boldsymbol{y}\|_2} \sum_{i=1}^n |x_i \overline{y}_i| \leq \sigma_1 \max_{\|\boldsymbol{x}\|_2=1=\|\boldsymbol{y}\|_2} \||\boldsymbol{y}|\|_2 \||\boldsymbol{x}|\|_2 \\
&= \sigma_1 \max_{\|\boldsymbol{x}\|_2=1=\|\boldsymbol{y}\|_2} \|\boldsymbol{y}\|_2 \|\boldsymbol{x}\|_2 = \sigma_1,
\end{aligned}$$

where $|\boldsymbol{x}| = (|x_1|, \ldots, |x_n|)$. Moreover, this maximum is achieved for $\boldsymbol{x} = \boldsymbol{v}_1$ and $\boldsymbol{y} = \boldsymbol{u}_1$, and we conclude

$$\|\boldsymbol{A}\|_2 = \sigma_1 = \max_{\|\boldsymbol{x}\|_2 = 1 = \|\boldsymbol{y}\|_2} |\boldsymbol{y}^* \boldsymbol{A}\boldsymbol{x}|.$$

Exercise 8.9: Spectral norm of the inverse

Suppose $\boldsymbol{A} \in \mathbb{C}^{n\times n}$ is nonsingular. Show that $\|\boldsymbol{A}\boldsymbol{x}\|_2 \geq \sigma_n$ for all $\boldsymbol{x} \in \mathbb{C}^n$ with $\|\boldsymbol{x}\|_2 = 1$. Show also that

$$\|\boldsymbol{A}^{-1}\|_2 = \max_{\boldsymbol{x}\neq \boldsymbol{0}} \frac{\|\boldsymbol{x}\|_2}{\|\boldsymbol{A}\boldsymbol{x}\|_2}.$$

Solution. Since $\boldsymbol{A}$ is nonsingular we find

$$\|\boldsymbol{A}^{-1}\|_2 = \max_{\boldsymbol{x}\neq \boldsymbol{0}} \frac{\|\boldsymbol{A}^{-1}\boldsymbol{x}\|_2}{\|\boldsymbol{x}\|_2} = \max_{\boldsymbol{x}\neq \boldsymbol{0}} \frac{\|\boldsymbol{x}\|_2}{\|\boldsymbol{A}\boldsymbol{x}\|_2}.$$

This proves the second claim. For the first claim, let $\sigma_1 \geq \cdots \geq \sigma_n$ be the singular values of $\boldsymbol{A}$. Again since $\boldsymbol{A}$ is nonsingular, σ_n must be nonzero, and Equation (8.19) states that $\frac{1}{\sigma_n} = \|\boldsymbol{A}^{-1}\|_2$. From this and what we just proved we have that $\frac{1}{\sigma_n} \geq \frac{1}{\|\boldsymbol{A}\boldsymbol{x}\|_2}$ for any $\boldsymbol{x}$ so that $\|\boldsymbol{x}\|_2 = 1$, so that also $\|\boldsymbol{A}\boldsymbol{x}\|_2 \geq \sigma_n$ for such $\boldsymbol{x}$.

Exercise 8.10: p-norm example

Let

$$\boldsymbol{A} = \begin{bmatrix} 2 & -1 \\ -1 & 2 \end{bmatrix}.$$

Compute $\|\boldsymbol{A}\|_p$ and $\|\boldsymbol{A}^{-1}\|_p$ for $p = 1, 2, \infty$.

Solution. We have

$$\boldsymbol{A} = \begin{bmatrix} 2 & -1 \\ -1 & 2 \end{bmatrix}, \qquad \boldsymbol{A}^{-1} = \frac{1}{3}\begin{bmatrix} 2 & 1 \\ 1 & 2 \end{bmatrix}.$$

Using Theorem 8.3, one finds $\|\boldsymbol{A}\|_1 = \|\boldsymbol{A}\|_\infty = 3$ and $\|\boldsymbol{A}^{-1}\|_1 = \|\boldsymbol{A}^{-1}\|_\infty = 1$. The singular values $\sigma_1 \geq \sigma_2$ of $\boldsymbol{A}$ are the square roots of the zeros of

$$0 = \det(\boldsymbol{A}^{\mathrm{T}}\boldsymbol{A} - \lambda \boldsymbol{I}) = (5-\lambda)^2 - 16 = \lambda^2 - 10\lambda + 9 = (\lambda - 9)(\lambda - 1).$$

Using Theorem 8.4, we find $\|\boldsymbol{A}\|_2 = \sigma_1 = 3$ and $\|\boldsymbol{A}^{-1}\|_2 = \sigma_2^{-1} = 1$. Alternatively, since $\boldsymbol{A}$ is symmetric positive definite, we know from (8.20) that $\|\boldsymbol{A}\|_2 = \lambda_1$ and $\|\boldsymbol{A}^{-1}\|_2 = 1/\lambda_2$, where $\lambda_1 = 3$ is the biggest eigenvalue of $\boldsymbol{A}$ and $\lambda_2 = 1$ is the smallest.

Exercise 8.11: Unitary invariance of the spectral norm

Show that $\|VA\|_2 = \|A\|_2$ holds even for a rectangular V as long as $V^*V = I$.

Solution. Suppose V is a rectangular matrix satisfying $V^*V = I$. Then

$$\begin{aligned}\|VA\|_2^2 &= \max_{\|x\|_2=1} \|VAx\|_2^2 = \max_{\|x\|_2=1} x^*A^*V^*VAx \\ &= \max_{\|x\|_2=1} x^*A^*Ax = \max_{\|x\|_2=1} \|Ax\|_2^2 = \|A\|_2^2.\end{aligned}$$

The result follows by taking square roots.

Exercise 8.12: $\|AU\|_2$ rectangular A

Find $A \in \mathbb{R}^{2\times 2}$ and $U \in \mathbb{R}^{2\times 1}$ with $U^{\mathrm{T}}U = 1$ such that $\|AU\|_2 < \|A\|_2$. Thus, in general, $\|AU\|_2 = \|A\|_2$ does not hold for a rectangular U even if $U^*U = I$.

Solution. Let $U = [u_1, u_2]^{\mathrm{T}}$ be any 2×1 matrix satisfying $1 = U^{\mathrm{T}}U$. Then AU is a 2×1-matrix, and clearly the operator 2-norm of a 2×1-matrix equals its euclidean norm (when viewed as a vector):

$$\left\| \begin{bmatrix} a_1 \\ a_2 \end{bmatrix} \begin{bmatrix} x \end{bmatrix} \right\|_2 = \left\| \begin{bmatrix} a_1 x \\ a_2 x \end{bmatrix} \right\|_2 = |x| \left\| \begin{bmatrix} a_1 \\ a_2 \end{bmatrix} \right\|_2 .$$

In order for $\|AU\|_2 < \|A\|_2$ to hold, we need to find a vector v with $\|v\|_2 = 1$ so that $\|AU\|_2 < \|Av\|_2$. In other words, we need to pick a matrix A that scales more in the direction v than in the direction U. For instance, if

$$A = \begin{bmatrix} 2 & 0 \\ 0 & 1 \end{bmatrix}, \qquad U = \begin{bmatrix} 0 \\ 1 \end{bmatrix}, \qquad v = \begin{bmatrix} 1 \\ 0 \end{bmatrix},$$

then

$$\|A\|_2 = \max_{\|x\|_2=1} \|Ax\|_2 \geq \|Av\|_2 = 2 > 1 = \|AU\|_2.$$

Exercise 8.13: p-norm of diagonal matrix

Show that $\|A\|_p = \rho(A) := \max |\lambda_i|$ (the largest eigenvalue of A), $1 \leq p \leq \infty$, when A is a diagonal matrix.

Solution. $A = \mathrm{diag}(\lambda_1, \ldots, \lambda_n)$ has eigenpairs $(\lambda_1, e_1), \ldots, (\lambda_n, e_n)$. For $\rho(A) = \max\{|\lambda_1|, \ldots, |\lambda_n|\}$, one has

$$\|A\|_p = \max_{(x_1,\ldots,x_n)\neq 0} \frac{(|\lambda_1 x_1|^p + \cdots + |\lambda_n x_n|^p)^{1/p}}{(|x_1|^p + \cdots + |x_n|^p)^{1/p}}$$

$$\leq \max_{(x_1,\dots,x_n)\neq \mathbf{0}} \frac{\left(\rho(\boldsymbol{A})^p|x_1|^p + \cdots + \rho(\boldsymbol{A})^p|x_n|^p\right)^{1/p}}{(|x_1|^p + \cdots + |x_n|^p)^{1/p}} = \rho(\boldsymbol{A}).$$

On the other hand, for j such that $\rho(\boldsymbol{A}) = |\lambda_j|$, one finds

$$\|\boldsymbol{A}\|_p = \max_{\boldsymbol{x}\neq 0} \frac{\|\boldsymbol{A}\boldsymbol{x}\|_p}{\|\boldsymbol{x}\|_p} \geq \frac{\|\boldsymbol{A}\boldsymbol{e}_j\|_p}{\|\boldsymbol{e}_j\|_p} = \rho(\boldsymbol{A}).$$

Together, the above two statements imply that $\|\boldsymbol{A}\|_p = \rho(\boldsymbol{A})$ for any diagonal matrix $\boldsymbol{A}$ and any p satisfying $1 \leq p \leq \infty$.

Exercise 8.14: Spectral norm of a column vector

A vector $\boldsymbol{a} \in \mathbb{C}^m$ can also be considered as a matrix $\boldsymbol{A} \in \mathbb{C}^{m,1}$.

a) Show that the spectral matrix norm (2-norm) of $\boldsymbol{A}$ equals the Euclidean vector norm of $\boldsymbol{a}$.

b) Show that $\|\boldsymbol{A}\|_p = \|\boldsymbol{a}\|_p$ for $1 \leq p \leq \infty$.

Solution. We write $\boldsymbol{A} \in \mathbb{C}^{m,1}$ for the matrix corresponding to the column vector $\boldsymbol{a} \in \mathbb{C}^m$. Write $\|\boldsymbol{A}\|_p$ for the operator p-norm of $\boldsymbol{A}$ and $\|\boldsymbol{a}\|_p$ for the vector p-norm of $\boldsymbol{a}$. In particular $\|\boldsymbol{A}\|_2$ is the spectral norm of $\boldsymbol{A}$ and $\|\boldsymbol{a}\|_2$ is the Euclidean norm of $\boldsymbol{a}$. Then

$$\|\boldsymbol{A}\|_p = \max_{x\neq 0} \frac{\|\boldsymbol{A}x\|_p}{|x|} = \max_{x\neq 0} \frac{|x|\|\boldsymbol{a}\|_p}{|x|} = \|\boldsymbol{a}\|_p,$$

proving **b)**. Note that **a)** follows as the special case $p = 2$.

Exercise 8.15: Norm of absolute value matrix

If $\boldsymbol{A} \in \mathbb{C}^{m\times n}$ has elements a_{ij}, let $|\boldsymbol{A}| \in \mathbb{R}^{m\times n}$ be the matrix with elements $|a_{ij}|$.

a) Compute $|\boldsymbol{A}|$ if $\boldsymbol{A} = \begin{bmatrix} 1{+}i & -2 \\ 1 & 1{-}i \end{bmatrix}$, $i = \sqrt{-1}$.

Solution. One finds

$$|\boldsymbol{A}| = \begin{bmatrix} |1+i| & |-2| \\ |1| & |1-i| \end{bmatrix} = \begin{bmatrix} \sqrt{2} & 2 \\ 1 & \sqrt{2} \end{bmatrix}.$$

b) Show that for any $\boldsymbol{A} \in \mathbb{C}^{m\times n}$,

$$\|\boldsymbol{A}\|_F = \|\,|\boldsymbol{A}|\,\|_F, \qquad \|\boldsymbol{A}\|_p = \|\,|\boldsymbol{A}|\,\|_p, \qquad \text{for } p = 1, \infty.$$

Solution. Let $b_{i,j}$ denote the entries of $|\boldsymbol{A}|$. Observe that $b_{i,j} = |a_{i,j}| = |b_{i,j}|$. Together with Theorem 8.3, these relations yield

$$\|A\|_F = \left(\sum_{i=1}^{m}\sum_{j=1}^{n}|a_{i,j}|^2\right)^{\frac{1}{2}} = \left(\sum_{i=1}^{m}\sum_{j=1}^{n}|b_{i,j}|^2\right)^{\frac{1}{2}} = \|\,|A|\,\|_F,$$

$$\|A\|_1 = \max_{1\le j\le n}\left(\sum_{i=1}^{m}|a_{i,j}|\right) = \max_{1\le j\le n}\left(\sum_{i=1}^{m}|b_{i,j}|\right) = \|\,|A|\,\|_1,$$

$$\|A\|_\infty = \max_{1\le i\le m}\left(\sum_{j=1}^{n}|a_{i,j}|\right) = \max_{1\le i\le m}\left(\sum_{j=1}^{n}|b_{i,j}|\right) = \|\,|A|\,\|_\infty,$$

which is what needed to be shown.

c) Show that for any $A \in \mathbb{C}^{m\times n}$ $\|A\|_2 \le \|\,|A|\,\|_2$.

Solution. To show this relation between the operator 2-norms of A and $|A|$, we first examine the connection between the l_2-norms of Ax and $|A||x|$, where $x = (x_1,\ldots,x_n)$ and $|x| = (|x_1|,\ldots,|x_n|)$. We find

$$\|Ax\|_2 = \left(\sum_{i=1}^{m}\left|\sum_{j=1}^{n}a_{i,j}x_j\right|^2\right)^{\frac{1}{2}} \le \left(\sum_{i=1}^{m}\left(\sum_{j=1}^{n}|a_{i,j}||x_j|\right)^2\right)^{\frac{1}{2}} = \|\,|A||x|\,\|_2.$$

Now let x^* with $\|x^*\|_2 = 1$ be a vector for which $\|A\|_2 = \|Ax^*\|_2$. That is, let x^* be a unit vector for which the maximum in the definition of the 2-norm is attained. Observe that $|x^*|$ is then a unit vector as well, $\|\,|x^*|\,\|_2 = 1$. Then, by the above estimate of l_2-norms and definition of the 2-norm,

$$\|A\|_2 = \|Ax^*\|_2 \le \|\,|A||x^*|\,\|_2 \le \|\,|A|\,\|_2.$$

d) Find a real symmetric 2×2 matrix A such that $\|A\|_2 < \|\,|A|\,\|_2$.

Solution. By Theorem 8.3, we can solve this exercise by finding a matrix A for which A and $|A|$ have different largest singular values. As A is real and symmetric, there exist $a, b, c \in \mathbb{R}$ such that

$$A = \begin{bmatrix} a & b \\ b & c \end{bmatrix}, \qquad |A| = \begin{bmatrix} |a| & |b| \\ |b| & |c| \end{bmatrix},$$

$$A^{\mathrm{T}}A = \begin{bmatrix} a^2+b^2 & ab+bc \\ ab+bc & b^2+c^2 \end{bmatrix}, \qquad |A|^{\mathrm{T}}|A| = \begin{bmatrix} a^2+b^2 & |ab|+|bc| \\ |ab|+|bc| & b^2+c^2 \end{bmatrix}.$$

To simplify these equations we first try the case $a + c = 0$. Eliminating c we get

$$A^{\mathrm{T}}A = \begin{bmatrix} a^2+b^2 & 0 \\ 0 & a^2+b^2 \end{bmatrix}, \qquad |A|^{\mathrm{T}}|A| = \begin{bmatrix} a^2+b^2 & 2|ab| \\ 2|ab| & a^2+b^2 \end{bmatrix}.$$

To get different norms we have to choose a, b in such a way that the maximal eigenvalues of $\boldsymbol{A}^{\mathrm{T}}\boldsymbol{A}$ and $|\boldsymbol{A}|^{\mathrm{T}}|\boldsymbol{A}|$ are different. Clearly $\boldsymbol{A}^{\mathrm{T}}\boldsymbol{A}$ has a unique eigenvalue $\lambda := a^2+b^2$ and putting the characteristic polynomial $\pi(\mu) = (a^2+b^2-\mu)^2-4|ab|^2$ of $|\boldsymbol{A}|^{\mathrm{T}}|\boldsymbol{A}|$ to zero yields eigenvalues $\mu_\pm := a^2+b^2 \pm 2|ab|$. Hence $|\boldsymbol{A}|^{\mathrm{T}}|\boldsymbol{A}|$ has maximal eigenvalue $\mu_+ = a^2+b^2+2|ab| = \lambda + 2|ab|$. The spectral norms of $\boldsymbol{A}$ and $|\boldsymbol{A}|$ therefore differ whenever both a and b are nonzero. For example, when $a = b = -c = 1$ we find

$$\boldsymbol{A} = \begin{bmatrix} 1 & 1 \\ 1 & -1 \end{bmatrix}, \qquad \|\boldsymbol{A}\|_2 = \sqrt{2}, \qquad \|\,|\boldsymbol{A}|\,\|_2 = 2.$$

Exercise 8.16: An iterative method (Exam exercise 2017-3)

Assume that $\boldsymbol{A} \in \mathbb{C}^{n\times n}$ is non-singular and nondefective (the eigenvectors of $\boldsymbol{A}$ form a basis for $\mathbb{C}^n$). We wish to solve $\boldsymbol{A}\boldsymbol{x} = \boldsymbol{b}$. Assume that we have a list of the eigenvalues $\{\lambda_1, \lambda_2, \ldots, \lambda_m\}$, in no particular order. We have that $m \le n$, since some of the eigenvalues may have multiplicity larger than one. Given $\boldsymbol{x}_0 \in \mathbb{C}^n$, and $k \ge 0$, we define the sequence $\{\boldsymbol{x}_k\}_{k=0}^{m-1}$ by

$$\boldsymbol{x}_{k+1} = \boldsymbol{x}_k + \frac{1}{\lambda_{k+1}}\boldsymbol{r}_k, \text{ where } \boldsymbol{r}_k = \boldsymbol{b} - \boldsymbol{A}\boldsymbol{x}_k.$$

a) Let the coefficients c_{ik} be defined by

$$\boldsymbol{r}_k = \sum_{i=1}^{n} c_{ik}\boldsymbol{u}_i,$$

where $\{(\sigma_i, \boldsymbol{u}_i)\}_{i=1}^{n}$ are the eigenpairs of $\boldsymbol{A}$. Show that

$$c_{i,k+1} = \begin{cases} 0 & \text{if } \sigma_i = \lambda_{k+1}, \\ c_{i,k}\left(1 - \frac{\sigma_i}{\lambda_{k+1}}\right) & \text{otherwise.} \end{cases}$$

Solution. Observe that $\boldsymbol{A}\boldsymbol{u}_j = \sigma_j\boldsymbol{u}_j$, where $\sigma_j \in \{\lambda_1, \ldots, \lambda_m\}$. We have that

$$\begin{aligned} \boldsymbol{r}_{k+1} &= \boldsymbol{b} - \boldsymbol{A}\boldsymbol{x}_{k+1} \\ &= \boldsymbol{b} - \boldsymbol{A}\left(\boldsymbol{x}_k + \frac{1}{\lambda_{k+1}}\boldsymbol{r}_k\right) \\ &= \boldsymbol{r}_k - \frac{1}{\lambda_{k+1}}\boldsymbol{A}\boldsymbol{r}_k \\ &= \sum_i c_{ik}\left(1 - \frac{\sigma_i}{\lambda_{k+1}}\right)\boldsymbol{u}_i. \end{aligned}$$

Hence

$$c_{i,k+1} = c_{i,k}\left(1 - \frac{\sigma_i}{\lambda_{k+1}}\right),$$

which is zero when $\sigma_i = \lambda_{k+1}$.

b) Show that for some $l \le m$, we have that $\boldsymbol{x}_l = \boldsymbol{x}_{l+1} = \cdots = \boldsymbol{x}_m = \boldsymbol{x}$, where $\boldsymbol{A}\boldsymbol{x} = \boldsymbol{b}$.

Solution. After at most m iterations we will have $c_{i,k} = 0$ for all i, and thus $\boldsymbol{x}_k = \boldsymbol{x}$.

c) Consider this iteration for the $n \times n$ matrix $\boldsymbol{T} = \text{tridiag}(c, d, c)$, where d and c are positive real numbers and $d > 2c$. The eigenvalues of $\boldsymbol{T}$ are

$$\lambda_j = d + 2c\cos\left(\frac{j\pi}{n+1}\right), \qquad j = 1, \dots, n.$$

What is the operation count for solving $\boldsymbol{T}\boldsymbol{x} = \boldsymbol{b}$ using the iterative algorithm above?

Solution. Each iterative step consists in finding $\boldsymbol{x} + (\boldsymbol{b} - \boldsymbol{T}\boldsymbol{x})/\lambda$. The matrix multiplication $\boldsymbol{T}\boldsymbol{x}$ requires $\mathcal{O}(5n)$ arithmetic operations, we have to add $\boldsymbol{b}$ (n operations), divide by λ (n operations), and add $\boldsymbol{x}$ (n operations). Altogether $\mathcal{O}(5n)+n+n+n = \mathcal{O}(8n)$ operations. The iteration reaches a fixed point in at most n steps, so the total count is $\mathcal{O}(8n^2)$.

d) Let now $\boldsymbol{B}$ be a symmetric $n \times n$ matrix which is zero on the "tridiagonal", i.e., $b_{ij} = 0$ if $|i - j| \le 1$. Set $\boldsymbol{A} = \boldsymbol{T} + \boldsymbol{B}$, where $\boldsymbol{T}$ is the tridiagonal matrix above. We wish to solve $\boldsymbol{A}\boldsymbol{x} = \boldsymbol{b}$ by the iterative scheme

$$\boldsymbol{T}\boldsymbol{x}_{k+1} = \boldsymbol{b} - \boldsymbol{B}\boldsymbol{x}_k. \tag{8.41}$$

Recall that if $\boldsymbol{E} \in \mathbb{R}^{n\times n}$ has eigenvalues $\lambda_1, \dots, \lambda_n$ then $\rho(\boldsymbol{E}) := \max_i|\lambda_i|$ is the spectral radius of $\boldsymbol{E}$. Show that $\rho(\boldsymbol{T}^{-1}\boldsymbol{B}) \le \rho(\boldsymbol{T}^{-1})\rho(\boldsymbol{B})$.

Solution. Choose $\boldsymbol{x} \in \mathbb{R}^n$ so that

$$\boldsymbol{T}^{-1}\boldsymbol{B}\boldsymbol{x} = \lambda\boldsymbol{x}, \qquad |\lambda| = \rho(\boldsymbol{T}^{-1}\boldsymbol{B}), \qquad \boldsymbol{x}^*\boldsymbol{x} = 1.$$

Multiplying $\boldsymbol{T}^{-1}\boldsymbol{B}\boldsymbol{x} = \lambda\boldsymbol{x}$ by $\boldsymbol{x}^*\boldsymbol{T}$ and taking absolute values we find

$$\rho(\boldsymbol{T}^{-1}\boldsymbol{B}) = \frac{|\boldsymbol{x}^*\boldsymbol{B}\boldsymbol{x}|}{|\boldsymbol{x}^*\boldsymbol{T}\boldsymbol{x}|}. \tag{8.i}$$

Since $\boldsymbol{T}$ is symmetric positive definite it has positive eigenvalues and orthonormal eigenvectors

$$\boldsymbol{T}\boldsymbol{u}_j = \lambda_j\boldsymbol{u}_j, \qquad j = 1, \dots, n, \qquad \lambda_1 \ge \cdots \ge \lambda_n > 0, \qquad \boldsymbol{u}_j^*\boldsymbol{u}_k = \delta_{jk},$$

Since $\boldsymbol{B}$ is symmetric it has real eigenvalues and orthonormal eigenvectors

$$\boldsymbol{B}\boldsymbol{v}_j = \mu_j \boldsymbol{v}_j, \qquad j = 1, \ldots, n, \qquad 0 < |\mu_1| \le \cdots \le |\mu_n|, \qquad \boldsymbol{v}_j^* \boldsymbol{v}_k = \delta_{jk}.$$

Moreover,

$$\rho(\boldsymbol{T}^{-1}) = \frac{1}{\lambda_n}, \qquad \rho(\boldsymbol{B}) = |\mu_n|. \tag{8.ii}$$

For some $c_j, d_j \in \mathbb{R}$

$$\boldsymbol{x} = \sum_{j=1}^{n} c_j \boldsymbol{u}_j = \sum_{j=1}^{n} d_j \boldsymbol{v}_j, \qquad \sum_{j=1}^{n} c_j^2 = \sum_{j=1}^{n} d_j^2 = 1.$$

We find by (8.ii)

$$\boldsymbol{x}^* \boldsymbol{T} \boldsymbol{x} = \sum_{j=1}^{n} c_j^2 \lambda_j \ge \lambda_n \sum_{j=1}^{n} c_j^2 = \lambda_n = 1/\rho(\boldsymbol{T}^{-1}),$$

$$|\boldsymbol{x}^* \boldsymbol{B} \boldsymbol{x}| = \left| \sum_{j=1}^{n} d_j^2 \mu_j \right| \le |\mu_n| \sum_{j=1}^{n} d_j^2 = \rho(\boldsymbol{B}).$$

But then $\rho(\boldsymbol{T}^{-1}\boldsymbol{B}) \le \rho(\boldsymbol{T}^{-1})\rho(\boldsymbol{B})$ follows from (8.i).

e) Show that the iteration (8.41) will converge if

$$\min\left\{ \max_i \sum_{j=1}^{n} |b_{ij}|, \max_j \sum_{i=1}^{n} |b_{ij}| \right\} < d - 2c.$$

Hint.
Use Gershgorin's theorem.

Solution. The iterative scheme can be written

$$\boldsymbol{x}_{k+1} = -\boldsymbol{T}^{-1}\boldsymbol{B}\boldsymbol{x}_k + \boldsymbol{T}^{-1}\boldsymbol{b} =: \boldsymbol{G}\boldsymbol{x}_k + \mathbf{c}.$$

The iteration will converge if $\rho(\boldsymbol{G}) < 1$ (see Chapter 12). By **d)** we have that $\rho(\boldsymbol{G}) \le \rho(\boldsymbol{T}^{-1})\rho(\boldsymbol{B})$. The eigenvalues of $\boldsymbol{T}$ satisfy

$$\lambda_j > d - 2c, \qquad \frac{1}{\lambda_j} < \frac{1}{d - 2c}.$$

Hence $\rho(\boldsymbol{T}^{-1}) < 1/(d - 2c) < \infty$. Regarding the eigenvalues μ_j of $\boldsymbol{B}$, by Gershgorin's circle theorem, since $b_{ii} = 0$

$$|\mu_j| \le \min\left\{\max_i \sum_{\substack{j\\|i-j|>1}} |b_{ij}|, \max_j \sum_{\substack{i\\|i-j|>1}} |b_{ij}|\right\} =: C_{\boldsymbol{B}}.$$

Hence the algorithm will converge if

$$C_{\boldsymbol{B}} \le d - 2c.$$

Exercises section 8.3

Exercise 8.17: Perturbed linear equation (Exam exercise 1981-2)
Given the systems $\boldsymbol{A}\boldsymbol{x} = \boldsymbol{b}$, $\boldsymbol{A}\boldsymbol{y} = \boldsymbol{b} + \boldsymbol{e}$, where

$$\boldsymbol{A} := \begin{bmatrix} 1.1 & 1 \\ 1 & 1 \end{bmatrix}, \quad \boldsymbol{b} := \begin{bmatrix} b_1 \\ b_2 \end{bmatrix} = \begin{bmatrix} 2.1 \\ 2.0 \end{bmatrix}, \quad \boldsymbol{e} := \begin{bmatrix} e_1 \\ e_2 \end{bmatrix}, \quad \|\boldsymbol{e}\|_2 = 0.1.$$

We define $\delta = \|\boldsymbol{x} - \boldsymbol{y}\|_2 / \|\boldsymbol{x}\|_2$.

a) Determine $K_2(\boldsymbol{A}) = \|\boldsymbol{A}\|_2 \|\boldsymbol{A}^{-1}\|_2$. Give an upper bound and a positive lower bound for δ without computing $\boldsymbol{x}$ and $\boldsymbol{y}$.

Solution. Since $\boldsymbol{A}^{\mathrm{T}} = \boldsymbol{A}$ we have $K_2(\boldsymbol{A}) = |\lambda_{\max}|/|\lambda_{\min}|$. We find $\det(\boldsymbol{A} - \lambda\boldsymbol{I}) = \lambda^2 - 2.1\lambda + 0.1$ giving $\lambda = \frac{2.1\pm\sqrt{4.01}}{2} = \frac{21\pm\sqrt{401}}{20}$. We then get

$$\frac{\lambda_{\max}}{\lambda_{\min}} = \frac{21+\sqrt{401}}{21-\sqrt{401}} = \frac{(21+\sqrt{401})^2}{40} \approx 42.08.$$

Thus $K_2(\boldsymbol{A}) \approx 42.08$ and by Theorem 8.7

$$\delta \le K_2(\boldsymbol{A}) \frac{\|\boldsymbol{e}\|_2}{\|\boldsymbol{b}\|_2} \le 42.08 \frac{0.1}{2.9} < 1.46$$

$$\delta \ge \frac{1}{K_2(\boldsymbol{A})} \frac{\|\boldsymbol{e}\|_2}{\|\boldsymbol{b}\|_2} \ge \frac{1}{42.08} \frac{0.1}{2.9} > 0.0008.$$

b) Suppose as before that $b_2 = 2.0$ and $\|\boldsymbol{e}\|_2 = 0.1$. Determine b_1 and $\boldsymbol{e}$ which maximize δ.

Solution. We have

$$\delta_{\max} = \frac{\max\limits_{\|\boldsymbol{e}\|_2 = 0.1} \|\boldsymbol{A}^{-1}\boldsymbol{e}\|_2}{\min\limits_{b_1 \in \mathbb{R}} \|\boldsymbol{A}^{-1}\boldsymbol{b}\|_2}.$$

Now $\boldsymbol{A}^{-1} = \begin{bmatrix} 10 & -10 \\ -10 & 11 \end{bmatrix}$ so that

$$\|\boldsymbol{A}^{-1}\boldsymbol{b}\|_2^2 = (10b_1 - 10b_2)^2 + (10b_1 - 11b_2)^2 = (10b_1 - 20)^2 + (10b_1 - 22)^2,$$

which achieves its minimum $\|\boldsymbol{A}^{-1}\boldsymbol{b}\|_2^2 = 2$ at $b_1 = 2.1$.

Since $\boldsymbol{A}^{-1}$ is symmetric, it has an orthonormal basis of eigenvectors $\{\boldsymbol{v}_1, \boldsymbol{v}_2\}$. If $\boldsymbol{e} = e_1\boldsymbol{v}_1 + e_2\boldsymbol{v}_2$ and corresponding eigenvalues λ_1, λ_2 of $\boldsymbol{A}$ are ordered such that $|\lambda_1| \geq |\lambda_2|$, we have that

$$\|\boldsymbol{A}^{-1}\boldsymbol{e}\|_2^2 = \left\|\frac{e_1}{\lambda_1}\boldsymbol{v}_1 + \frac{e_2}{\lambda_2}\boldsymbol{v}_2\right\|^2 = \frac{e_1^2}{\lambda_1^2} + \frac{e_2^2}{\lambda_2^2} \leq \frac{0.01}{\lambda_2^2},$$

with equality if $e_1 = 0$, $e_2 = 0.1$. The maximum $\|\boldsymbol{A}^{-1}\boldsymbol{e}\|_2$ is therefore obtained for $\boldsymbol{e} = 0.1\boldsymbol{v}_2$, and is $\frac{0.1}{\lambda_2} = \frac{2}{21-\sqrt{401}}$. The maximum for δ is thus $\frac{2}{21-\sqrt{401}}\frac{1}{\sqrt{2}} \approx 1.45$. To find $\boldsymbol{e}$ we need to find an eigenvector for λ_2:

$$\begin{aligned}
&\begin{bmatrix} 1.1 & 1 \\ 1 & 1 \end{bmatrix} - \frac{21 - \sqrt{401}}{20}\boldsymbol{I} \\
&= \begin{bmatrix} 0.05 + \sqrt{401}/20 & 1 \\ 1 & -0.05 + \sqrt{401}/20 \end{bmatrix} \\
&\sim \begin{bmatrix} 20 & \sqrt{401} - 1 \\ 0 & 0 \end{bmatrix}.
\end{aligned}$$

Hence $\begin{bmatrix} 1 - \sqrt{401} \\ 20 \end{bmatrix}$ is an eigenvector for λ_2. Normalizing this to a vector with length 0.1 we get $\boldsymbol{e} \approx \begin{bmatrix} -0.069 \\ 0.072 \end{bmatrix}$.

Exercise 8.18: Sharpness of perturbation bounds

The upper and lower bounds for $\|\boldsymbol{y} - \boldsymbol{x}\|/\|\boldsymbol{x}\|$ given by Equation (8.22), i.e.,

$$\frac{1}{K(\boldsymbol{A})}\frac{\|\boldsymbol{e}\|}{\|\boldsymbol{b}\|} \leq \frac{\|\boldsymbol{y} - \boldsymbol{x}\|}{\|\boldsymbol{x}\|} \leq K(\boldsymbol{A})\frac{\|\boldsymbol{e}\|}{\|\boldsymbol{b}\|},$$

can be attained for any matrix $\boldsymbol{A}$, but only for special choices of $\boldsymbol{b}$. Suppose $\boldsymbol{y}_{\boldsymbol{A}}$ and $\boldsymbol{y}_{\boldsymbol{A}^{-1}}$ are vectors with $\|\boldsymbol{y}_{\boldsymbol{A}}\| = \|\boldsymbol{y}_{\boldsymbol{A}^{-1}}\| = 1$ and $\|\boldsymbol{A}\| = \|\boldsymbol{A}\boldsymbol{y}_{\boldsymbol{A}}\|$ and $\|\boldsymbol{A}^{-1}\| = \|\boldsymbol{A}^{-1}\boldsymbol{y}_{\boldsymbol{A}^{-1}}\|$.

a) Show that the upper bound in Equation (8.22) is attained if $\boldsymbol{b} = \boldsymbol{A}\boldsymbol{y}_{\boldsymbol{A}}$ and $\boldsymbol{e} = \boldsymbol{y}_{\boldsymbol{A}^{-1}}$.

b) Show that the lower bound is attained if $\boldsymbol{b} = \boldsymbol{y}_{\boldsymbol{A}^{-1}}$ and $\boldsymbol{e} = \boldsymbol{A}\boldsymbol{y}_{\boldsymbol{A}}$.

Solution. Suppose $\boldsymbol{A}\boldsymbol{x} = \boldsymbol{b}$ and $\boldsymbol{A}\boldsymbol{y} = \boldsymbol{b} + \boldsymbol{e}$. Let $K = K(\boldsymbol{A}) = \|\boldsymbol{A}\|\|\boldsymbol{A}^{-1}\|$ be the condition number of $\boldsymbol{A}$. Let $\boldsymbol{y}_{\boldsymbol{A}}$ and $\boldsymbol{y}_{\boldsymbol{A}^{-1}}$ be unit vectors for which the maxima in the definition of the operator norms of $\boldsymbol{A}$ and $\boldsymbol{A}^{-1}$ are attained. That is, $\|\boldsymbol{y_A}\| = 1 = \|\boldsymbol{y}_{\boldsymbol{A}^{-1}}\|$, $\|\boldsymbol{A}\| = \|\boldsymbol{A}\boldsymbol{y_A}\|$, and $\|\boldsymbol{A}^{-1}\| = \|\boldsymbol{A}^{-1}\boldsymbol{y}_{\boldsymbol{A}^{-1}}\|$. If $\boldsymbol{b} = \boldsymbol{A}\boldsymbol{y_A}$ and $\boldsymbol{e} = \boldsymbol{y}_{\boldsymbol{A}^{-1}}$, then

$$\frac{\|\boldsymbol{y}-\boldsymbol{x}\|}{\|\boldsymbol{x}\|} = \frac{\|\boldsymbol{A}^{-1}\boldsymbol{e}\|}{\|\boldsymbol{A}^{-1}\boldsymbol{b}\|} = \frac{\|\boldsymbol{A}^{-1}\boldsymbol{y}_{\boldsymbol{A}^{-1}}\|}{\|\boldsymbol{y_A}\|}$$
$$= \|\boldsymbol{A}^{-1}\| = \|\boldsymbol{A}\|\|\boldsymbol{A}^{-1}\|\frac{\|\boldsymbol{y}_{\boldsymbol{A}^{-1}}\|}{\|\boldsymbol{A}\boldsymbol{y_A}\|} = K\frac{\|\boldsymbol{e}\|}{\|\boldsymbol{b}\|},$$

showing that the upper bound is sharp. If $\boldsymbol{b} = \boldsymbol{y}_{\boldsymbol{A}^{-1}}$ and $\boldsymbol{e} = \boldsymbol{A}\boldsymbol{y_A}$, then

$$\frac{\|\boldsymbol{y}-\boldsymbol{x}\|}{\|\boldsymbol{x}\|} = \frac{\|\boldsymbol{A}^{-1}\boldsymbol{e}\|}{\|\boldsymbol{A}^{-1}\boldsymbol{b}\|} = \frac{\|\boldsymbol{y_A}\|}{\|\boldsymbol{A}^{-1}\boldsymbol{y}_{\boldsymbol{A}^{-1}}\|}$$
$$= \frac{1}{\|\boldsymbol{A}^{-1}\|} = \frac{1}{\|\boldsymbol{A}\|\|\boldsymbol{A}^{-1}\|}\frac{\|\boldsymbol{A}\boldsymbol{y_A}|}{\|\boldsymbol{y}_{\boldsymbol{A}^{-1}}\|} = \frac{1}{K}\frac{\|\boldsymbol{e}\|}{\|\boldsymbol{b}\|},$$

showing that the lower bound is sharp.

Exercise 8.19: Condition number of 2. derivative matrix

In this exercise we will show that for $m \geq 1$

$$\frac{4}{\pi^2}(m+1)^2 - 2/3 < \mathrm{cond}_p(\boldsymbol{T}) \leq \frac{1}{2}(m+1)^2, \qquad p = 1, 2, \infty, \quad (8.42)$$

where $\boldsymbol{T} := \mathrm{tridiag}(-1, 2, -1) \in \mathbb{R}^{m\times m}$ and $\mathrm{cond}_p(\boldsymbol{T}) := \|\boldsymbol{T}\|_p\|\boldsymbol{T}^{-1}\|_p$ is the p-norm condition number of $\boldsymbol{T}$. The p matrix norm is given by (8.17). You will need the explicit inverse of $\boldsymbol{T}$ given by Equation (2.39) and the eigenvalues given in Lemma 2.2. As usual we define $h := 1/(m+1)$.

a) Show that for $m \geq 3$

$$\mathrm{cond}_1(\boldsymbol{T}) = \mathrm{cond}_\infty(\boldsymbol{T}) = \frac{1}{2} \times \begin{cases} h^{-2}, & m \text{ odd}, \\ h^{-2} - 1, & m \text{ even}. \end{cases}$$

and that $\mathrm{cond}_1(\boldsymbol{T}) = \mathrm{cond}_\infty(\boldsymbol{T}) = 3$ for $m = 2$.

Solution. Equation (2.39) said that $\boldsymbol{T}^{-1}$ has components

$$\left(\boldsymbol{T}^{-1}\right)_{ij} = \left(\boldsymbol{T}^{-1}\right)_{ji} = (1 - ih)j > 0, \quad 1 \leq j \leq i \leq m, \quad h = \frac{1}{m+1}.$$

From Theorems 8.3 and 8.4, we have the following explicit expressions for the 1-, 2- and ∞-norms

$$\|\boldsymbol{A}\|_1 = \max_{1\leq j\leq n}\sum_{i=1}^m |a_{i,j}|,\ \|\boldsymbol{A}\|_2 = \sigma_1,\ \|\boldsymbol{A}^{-1}\|_2 = \frac{1}{\sigma_m},\ \|\boldsymbol{A}\|_\infty = \max_{1\leq i\leq m}\sum_{j=1}^n |a_{i,j}|$$

for any matrix $\boldsymbol{A} \in \mathbb{C}^{m,n}$, where σ_1 is the largest singular value of $\boldsymbol{A}$, σ_m the smallest singular value of $\boldsymbol{A}$, and we assumed $\boldsymbol{A}$ to be nonsingular in the third equation.

For the matrix $\boldsymbol{T}$ one obtains $\|\boldsymbol{T}\|_1 = \|\boldsymbol{T}\|_\infty = m+1$ for $m = 1, 2$ and $\|\boldsymbol{T}\|_1 = \|\boldsymbol{T}\|_\infty = 4$ for $m \geq 3$. For the inverse we get $\|\boldsymbol{T}^{-1}\|_1 = \|\boldsymbol{T}^{-1}\|_\infty = \frac{1}{2} = \frac{1}{8}h^{-2}$ for $m = 1$ and

$$\|\boldsymbol{T}^{-1}\|_1 = \left\| \frac{1}{3} \begin{bmatrix} 2 & 1 \\ 1 & 2 \end{bmatrix} \right\|_1 = 1 = \left\| \frac{1}{3} \begin{bmatrix} 2 & 1 \\ 1 & 2 \end{bmatrix} \right\|_\infty = \|\boldsymbol{T}^{-1}\|_\infty$$

for $m = 2$. For $m > 2$, one obtains

$$\sum_{i=1}^{m} \left| \left(\boldsymbol{T}^{-1}\right)_{ij} \right| = \sum_{i=1}^{j-1} (1 - jh)i + \sum_{i=j}^{m} (1 - ih)j$$

It is easy to commit an error when simplifying this sum. It can be computed symbolically using Symbolic Math Toolbox using the following code.

code/symbolic_example.m

```
syms i j m
simplify(symsum((1-j/(m+1))*i,i,1,j-1) + symsum( (1-i/(m+1))*
    j,i,j,m))
```

Listing 8.1: Simplify a symbolic sum using the Symbolic Math Toolbox in MATLAB.

Here we have substituted $h = 1/(m+1)$. This produces the result $\frac{j}{2}(m+1-j)$. To arrive at this ourselves, we can first rewrite the expression as

$$\sum_{i=1}^{j-1} (1 - jh)i + \sum_{i=1}^{m} (1 - ih)j - \sum_{i=1}^{j-1} (1 - ih)j.$$

The first sum here equals $(1 - jh)\frac{(j-1)j}{2}$. The second sum equals

$$\sum_{i=1}^{m} j - jh \sum_{i=1}^{m} i = mj - jh\frac{m(m+1)}{2} = mj - mj/2 = mj/2.$$

The third sum equals

$$\sum_{i=1}^{j-1} j - jh \sum_{i=1}^{j-1} i = j(j-1) - hj^2\frac{j-1}{2} = j(j-1)(2 - hj)/2.$$

Combining the three sums we get

$$\frac{j}{2}\left((j-1)(1-jh) + m - (j-1)(2-hj)\right) = \frac{j}{2}(m+1-j),$$

which we also arrived at above. This can also be written as $\frac{1}{2h}j - \frac{1}{2}j^2$, which is a quadratic function in j that attains its maximum at $j = \frac{1}{2h} = \frac{m+1}{2}$. For odd $m > 1$, this function takes its maximum at integral j, yielding $\|\boldsymbol{T}^{-1}\|_1 = \frac{1}{8}h^{-2}$. For even $m > 2$, on the other hand, the maximum over all integral j is attained at $j = \frac{m}{2} = \frac{1-h}{2h}$ or $j = \frac{m+2}{2} = \frac{1+h}{2h}$, which both give $\|\boldsymbol{T}^{-1}\|_1 = \frac{1}{8}(h^{-2} - 1)$.

Since $\boldsymbol{T}^{-1}$ is symmetric, the row sums equal the column sums, so that $\|\boldsymbol{T}^{-1}\|_\infty = \|\boldsymbol{T}^{-1}\|_1$. We conclude that the 1- and ∞-condition numbers of $\boldsymbol{T}$ are

$$\operatorname{cond}_1(\boldsymbol{T}) = \operatorname{cond}_\infty(\boldsymbol{T}) = \frac{1}{2}\begin{cases} 2, & m = 1, \\ 6, & m = 2, \\ h^{-2}, & m \text{ odd}, \ m > 1, \\ h^{-2} - 1, & m \text{ even}, \ m > 2. \end{cases}$$

b) Show that for $p = 2$ and $m \geq 1$ we have

$$\operatorname{cond}_2(\boldsymbol{T}) = \cot^2\left(\frac{\pi h}{2}\right) = 1/\tan^2\left(\frac{\pi h}{2}\right).$$

Solution. Since the matrix $\boldsymbol{T}$ is symmetric, $\boldsymbol{T}^{\mathrm{T}}\boldsymbol{T} = \boldsymbol{T}^2$ and the eigenvalues of $\boldsymbol{T}^{\mathrm{T}}\boldsymbol{T}$ are the squares of the eigenvalues $\lambda_1, \ldots, \lambda_n$ of $\boldsymbol{T}$. As all eigenvalues of $\boldsymbol{T}$ are positive, each singular value of $\boldsymbol{T}$ is equal to an eigenvalue. Using that $\lambda_i = 2 - 2\cos(i\pi h)$, we find

$$\sigma_1 = |\lambda_m| = 2 - 2\cos(m\pi h) = 2 + 2\cos(\pi h),$$

$$\sigma_m = |\lambda_1| = 2 - 2\cos(\pi h).$$

It follows that

$$\operatorname{cond}_2(\boldsymbol{T}) = \frac{\sigma_1}{\sigma_m} = \frac{1 + \cos(\pi h)}{1 - \cos(\pi h)} = \cot^2\left(\frac{\pi h}{2}\right).$$

c) Show the bounds

$$\frac{4}{\pi^2}h^{-2} - \frac{2}{3} < \operatorname{cond}_2(\boldsymbol{T}) < \frac{4}{\pi^2}h^{-2}.$$

Hint.
For the upper bound use the inequality $\tan x > x$, which is valid for $0 < x < \pi/2$. For the lower bound we use (without proof) the inequality $\cot^2 x > \frac{1}{x^2} - \frac{2}{3}$ for $x > 0$.

Solution. From $\tan x > x$ we obtain $\cot^2 x = \frac{1}{\tan^2 x} < \frac{1}{x^2}$. Using this and $\cot^2 x > x^{-2} - \frac{2}{3}$ we find

$$\frac{4}{\pi^2 h^2} - \frac{2}{3} < \text{cond}_2(\boldsymbol{T}) < \frac{4}{\pi^2 h^2}.$$

d) Prove Equation (8.42).

Solution. For $p = 2$, substitute $h = 1/(m+1)$ in **c)** and use that $4/\pi^2 < 1/2$. For $p = 1, \infty$ we need to show due to **a)** that

$$\frac{4}{\pi^2} h^{-2} - 2/3 < \frac{1}{2} h^{-2} \leq \frac{1}{2} h^{-2}.$$

when m is odd, and that

$$\frac{4}{\pi^2} h^{-2} - 2/3 < \frac{1}{2}(h^{-2} - 1) \leq \frac{1}{2} h^{-2}.$$

when m is even. The right hand sides in these equations are obvious. The left equation for m odd is also obvious since $4/\pi^2 < 1/2$. The left equation for m even is also obvious since $-2/3 < -1/2$.

Exercise 8.20: Perturbation of the Identity matrix

Let $\boldsymbol{E}$ be a square matrix.

a) Show that if $\boldsymbol{I} - \boldsymbol{E}$ is nonsingular then

$$\frac{\|(\boldsymbol{I} - \boldsymbol{E})^{-1} - \boldsymbol{I}\|}{\|(\boldsymbol{I} - \boldsymbol{E})^{-1}\|} \leq \|\boldsymbol{E}\|.$$

Solution. Since $\boldsymbol{I} - \boldsymbol{E}$ is nonsingular, we can write $\boldsymbol{E} = (\boldsymbol{I} - \boldsymbol{E})\big((\boldsymbol{I} - \boldsymbol{E})^{-1} - \boldsymbol{I}\big)$. Multiplying this equation from the left by $(\boldsymbol{I} - \boldsymbol{E})^{-1}$ gives $(\boldsymbol{I} - \boldsymbol{E})^{-1}\boldsymbol{E} = \big((\boldsymbol{I} - \boldsymbol{E})^{-1} - \boldsymbol{I}\big)$. Consistency now gives $\|((\boldsymbol{I} - \boldsymbol{E})^{-1} - \boldsymbol{I})\| \leq \|(\boldsymbol{I} - \boldsymbol{E})^{-1}\| \cdot \|\boldsymbol{E}\|$, from which the result follows.

b) If $\|\boldsymbol{E}\| < 1$ then $(\boldsymbol{I} - \boldsymbol{E})^{-1}$ exists and

$$\frac{1}{1 + \|\boldsymbol{E}\|} \leq \|(\boldsymbol{I} - \boldsymbol{E})^{-1}\| \leq \frac{1}{1 - \|\boldsymbol{E}\|}$$

Show the lower bound. Show the upper bound if $\|\boldsymbol{I}\| = 1$. In general for a consistent matrix norm (such as the Frobenius norm) the upper bound follows from Theorem 12.14 using Neumann series.

Solution. From **a)** it follows that $\frac{\|(\boldsymbol{I}-\boldsymbol{E})^{-1} - \boldsymbol{I}\|}{\|\boldsymbol{E}\|} \leq \|(\boldsymbol{I}-\boldsymbol{E})^{-1}\|$. On the left side now multiply with $\|\boldsymbol{I} - \boldsymbol{E}\|$ in the numerator and the denominator. For the numerator we have that

$$\|\boldsymbol{E}\| = \|(\boldsymbol{I}-\boldsymbol{E})\big((\boldsymbol{I}-\boldsymbol{E})^{-1}-\boldsymbol{I}\big)\| \le \|\boldsymbol{I}-\boldsymbol{E}\|\|(\boldsymbol{I}-\boldsymbol{E})^{-1}-\boldsymbol{I}\|,$$

and it follows from the triangle inequality that

$$\frac{1}{1+\|\boldsymbol{E}\|} \le \frac{1}{\|\boldsymbol{I}-\boldsymbol{E}\|} = \frac{\|\boldsymbol{E}\|}{\|\boldsymbol{I}-\boldsymbol{E}\|\|\boldsymbol{E}\|} \le \frac{\|(\boldsymbol{I}-\boldsymbol{E})^{-1}-\boldsymbol{I}\|}{\|\boldsymbol{E}\|} \le \|(\boldsymbol{I}-\boldsymbol{E})^{-1}\|,$$

where the final inequality follows from **a)**.

For the upper bound, recall the inverse triangle inequality, $\big|\|\boldsymbol{a}\|-\|\boldsymbol{b}\|\big| \le \|\boldsymbol{a}-\boldsymbol{b}\|$. With $\boldsymbol{a} = (\boldsymbol{I}-\boldsymbol{E})^{-1}$ and $\boldsymbol{b} = \boldsymbol{I}$, and applied to **a)**, this yields

$$1 - \frac{\|\boldsymbol{I}\|}{\|(\boldsymbol{I}-\boldsymbol{E})^{-1}\|} \le \|\boldsymbol{E}\|.$$

Reorganizing this we obtain $1 - \|\boldsymbol{E}\| \le \frac{\|\boldsymbol{I}\|}{\|(\boldsymbol{I}-\boldsymbol{E})^{-1}\|}$. Setting $\|\boldsymbol{I}\| = 1$ and moving things around again gives the upper bound.

c) Show that if $\|\boldsymbol{E}\| < 1$ then

$$\|(\boldsymbol{I}-\boldsymbol{E})^{-1}-\boldsymbol{I}\| \le \frac{\|\boldsymbol{E}\|}{1-\|\boldsymbol{E}\|}.$$

Solution. From **a)** and **b)** it follows that

$$\|(\boldsymbol{I}-\boldsymbol{E})^{-1}-\boldsymbol{I}\| \le \|(\boldsymbol{I}-\boldsymbol{E})^{-1}\|\|\boldsymbol{E}\| \le \frac{\|\boldsymbol{E}\|}{1-\|\boldsymbol{E}\|}.$$

Exercise 8.21: Lower bounds in Equations (8.27), (8.29)

a) Solve for $\boldsymbol{E}$ in

$$\boldsymbol{B}^{-1}-\boldsymbol{A}^{-1} = -\boldsymbol{A}^{-1}\boldsymbol{E}\boldsymbol{B}^{-1} = -\boldsymbol{B}^{-1}\boldsymbol{E}\boldsymbol{A}^{-1}, \tag{8.30}$$

and show that

$$K(\boldsymbol{B})^{-1}\frac{\|\boldsymbol{E}\|}{\|\boldsymbol{A}\|} \le \frac{\|\boldsymbol{B}^{-1}-\boldsymbol{A}^{-1}\|}{\|\boldsymbol{B}^{-1}\|}.$$

Solution. The statement is the same as $\|\boldsymbol{E}\| \le \|\boldsymbol{B}^{-1}-\boldsymbol{A}^{-1}\|\|\boldsymbol{A}\|\|\boldsymbol{B}\|$. Solving for $\boldsymbol{E}$ in Equation (8.30) gives $\boldsymbol{E} = -\boldsymbol{B}(\boldsymbol{B}^{-1}-\boldsymbol{A}^{-1})\boldsymbol{A}$, so the result follows by applying norms on both sides, and using consistency on the right hand side.

b) Equation (8.28) said that $\frac{1}{1+r} \le \frac{\|\boldsymbol{B}^{-1}\|}{\|\boldsymbol{A}^{-1}\|} \le \frac{1}{1-r}$. Use this and **a)** to show that

$$\frac{K(\boldsymbol{B})^{-1}}{1+r}\frac{\|\boldsymbol{E}\|}{\|\boldsymbol{A}\|} \le \frac{\|\boldsymbol{B}^{-1}-\boldsymbol{A}^{-1}\|}{\|\boldsymbol{A}^{-1}\|}.$$

Solution. Dividing with $1+r$ on both sides in the expression obtained in **a)**, and using the left inequality in Equation (8.28),we obtain

$$\begin{aligned}\frac{K(\boldsymbol{B})^{-1}}{1+r}\frac{\|\boldsymbol{E}\|}{\|\boldsymbol{A}\|} &\leq \frac{1}{1+r}\frac{\|\boldsymbol{B}^{-1}-\boldsymbol{A}^{-1}\|}{\|\boldsymbol{B}^{-1}\|}\\ &\leq \frac{\|\boldsymbol{B}^{-1}\|}{\|\boldsymbol{A}^{-1}\|}\frac{\|\boldsymbol{B}^{-1}-\boldsymbol{A}^{-1}\|}{\|\boldsymbol{B}^{-1}\|}\\ &= \frac{\|\boldsymbol{B}^{-1}-\boldsymbol{A}^{-1}\|}{\|\boldsymbol{A}^{-1}\|},\end{aligned}$$

which proves the claim.

Exercise 8.22: Periodic spline interpolation (Exam exercise 1993-2)

Let the components of $\boldsymbol{x} = [x_0, \ldots, x_n]^{\mathrm{T}} \in \mathbb{R}^{n+1}$ define a partition of the interval $[a, b]$,

$$a = x_0 < x_1 < \cdots < x_n = b,$$

and given a dataset $\boldsymbol{y} := [y_0, \ldots, y_n]^{\mathrm{T}} \in \mathbb{R}^{n+1}$, where we assume $y_0 = y_n$. The periodic cubic spline interpolation problem is defined by finding a cubic spline function g satisfying the conditions

$$\begin{aligned} g(x_i) &= y_i, \qquad i = 0, 1, \ldots, n, \\ g'(a) &= g'(b), \quad g''(a) = g''(b). \end{aligned}$$

(Recall that g is a cubic polynomial on each interval (x_{i-1}, x_i), for $i = 1, \ldots, n$ with smoothness $C^2[a, b]$.)

We define $s_i := g'(x_i)$, $i = 0, \ldots, n$. It can be shown that the vector $\boldsymbol{s} := [s_1, \ldots, s_n]^{\mathrm{T}}$ is determined from a linear system

$$\boldsymbol{A}\boldsymbol{s} = \boldsymbol{b}, \tag{8.45}$$

where $\boldsymbol{b} \in \mathbb{R}^n$ is a given vector determined by $\boldsymbol{x}$ and $\boldsymbol{y}$. The matrix $\boldsymbol{A} \in \mathbb{R}^{n \times n}$ is given by

$$\boldsymbol{A} := \begin{bmatrix} 2 & \mu_1 & 0 & \cdots & 0 & \lambda_1 \\ \lambda_2 & 2 & \mu_2 & \ddots & & 0 \\ 0 & \ddots & \ddots & \ddots & \ddots & \vdots \\ \vdots & \ddots & \ddots & \ddots & \ddots & 0 \\ 0 & & \ddots & \lambda_{n-1} & 2 & \mu_{n-1} \\ \mu_n & 0 & \cdots & 0 & \lambda_n & 2 \end{bmatrix},$$

where

$$\lambda_i := \frac{h_i}{h_{i-1} + h_i}, \quad \mu_i := \frac{h_{i-1}}{h_{i-1} + h_i}, \quad , i = 1, \ldots, n,$$

and

$$h_i = x_{i+1} - x_i, \quad i = 0, \ldots, n-1, \text{ and } h_n = h_0.$$

You shall not argue or prove the system (8.45). Throughout this exercise we assume that

$$\frac{1}{2} \le \frac{h_i}{h_{i-1}} \le 2, \quad i = 1, \ldots, n.$$

a) Show that

$$\|\boldsymbol{A}\|_\infty = 3 \quad \text{and that} \quad \|\boldsymbol{A}\|_1 \le \frac{10}{3}.$$

Solution. We have

$$\|\boldsymbol{A}\|_\infty = \max_i \sum_j |a_{ij}| = \max_i (2 + \mu_i + \lambda_i) = 2 + 1 = 3.$$

We have

$$\mu_i = \frac{1}{1 + h_i/h_{i-1}} \le \frac{1}{1 + 1/2} = 2/3, \qquad i = 1, \dots, n,$$

and since also $1/2 \le h_{i-1}/h_i \le 2$ we find

$$\lambda_i = \frac{1}{1 + h_{i-1}/h_i} \le \frac{1}{1 + 1/2} = 2/3, \qquad i = 1, \dots, n.$$

But then with $\mu_0 := \mu_n$ and $\lambda_{n+1} := \lambda_1$

$$\|\boldsymbol{A}\|_1 = \max_j \sum_i |a_{ij}| = \max_j (2 + \mu_{j-1} + \lambda_{j+1}) \le 2 + \frac{2}{3} + \frac{2}{3} = \frac{10}{3}.$$

b) Show that $\|\boldsymbol{A}^{-1}\|_\infty \le 1$.

Solution. We have

$$\sigma_i := |a_{ii}| - \sum_{j \ne i} |a_{ij}| = 2 - \lambda_i - \mu_i = 1, \qquad i = 1, \dots, n.$$

Hence $\boldsymbol{A}$ is strictly diagonally dominant, so that $\|\boldsymbol{A}^{-1}\boldsymbol{b}\|_\infty \le \|\boldsymbol{b}\|_\infty$ for all $\boldsymbol{b} \in \mathbb{R}^n$ by Theorem 2.2. But then

$$\|\boldsymbol{A}^{-1}\|_\infty = \max_{\boldsymbol{b} \ne 0} \frac{\|\boldsymbol{A}^{-1}\boldsymbol{b}\|_\infty}{\|\boldsymbol{b}\|_\infty} \le \max_{\boldsymbol{b} \ne 0} \frac{\|\boldsymbol{b}\|_\infty}{\|\boldsymbol{b}\|_\infty} = 1.$$

c) Show that $\|\boldsymbol{A}^{-1}\|_1 \le \frac{3}{2}$.

Solution. Since

$$|a_{jj}| - \sum_{i \ne j} |a_{ij}| = 2 - \lambda_{i+1} - \mu_{i-1} \ge 2 - \frac{4}{3} = \frac{2}{3}, \qquad i = 1, \dots, n,$$

it follows that $\boldsymbol{A}^{\mathrm{T}}$ also is strictly diagonally dominant, so that $\|\boldsymbol{A}^{-\mathrm{T}}\boldsymbol{b}\|_\infty \le \frac{3}{2}\|\boldsymbol{b}\|_\infty$ for all $\boldsymbol{b} \in \mathbb{R}^n$ by Theorem 2.2. But then

$$\|\boldsymbol{A}^{-1}\|_1 = \|\boldsymbol{A}^{-\mathrm{T}}\|_\infty = \max_{\boldsymbol{b} \ne 0} \frac{\|\boldsymbol{A}^{-\mathrm{T}}\boldsymbol{b}\|_\infty}{\|\boldsymbol{b}\|_\infty} \le \frac{3}{2} \max_{\boldsymbol{b} \ne 0} \frac{\|\boldsymbol{b}\|_\infty}{\|\boldsymbol{b}\|_\infty} = \frac{3}{2}.$$

d) Let $\boldsymbol{s}$ and $\boldsymbol{b}$ be as in (8.45), where we assume $\boldsymbol{b} \neq \boldsymbol{0}$. Let $\boldsymbol{e} \in \mathbb{R}^n$ be such that $\|\boldsymbol{e}\|_p / \|\boldsymbol{b}\|_p \leq 0.01$, $p = 1, \infty$. Suppose $\hat{\boldsymbol{s}}$ satisfies

$$\boldsymbol{A}\hat{\boldsymbol{s}} = \boldsymbol{b} + \boldsymbol{e}.$$

Give estimates for

$$\frac{\|\hat{\boldsymbol{s}} - \boldsymbol{s}\|_\infty}{\|\boldsymbol{s}\|_\infty} \quad \text{and} \quad \frac{\|\hat{\boldsymbol{s}} - \boldsymbol{s}\|_1}{\|\boldsymbol{s}\|_1}.$$

Solution. By Theorem 8.7 for $p = 1$ and $p = \infty$,

$$\frac{\|\hat{\boldsymbol{s}} - \boldsymbol{s}\|_p}{\|\boldsymbol{s}\|_p} \leq K_p(\boldsymbol{A}) \frac{\|\boldsymbol{e}\|_p}{\|\boldsymbol{b}\|_p} \leq 0.01 K_p(\boldsymbol{A}),$$

where

$$K_p(\boldsymbol{A}) := \|\boldsymbol{A}\|_p \|\boldsymbol{A}^{-1}\|_p \leq \begin{cases} \frac{10}{3} \cdot \frac{3}{2} = 5, & \text{if } p = 1, \\ 3 \cdot 1 = 3, & \text{if } p = \infty. \end{cases}$$

We obtain

$$\frac{\|\hat{\boldsymbol{s}} - \boldsymbol{s}\|_1}{\|\boldsymbol{s}\|_1} \leq 0.05 \qquad \text{and} \qquad \frac{\|\hat{\boldsymbol{s}} - \boldsymbol{s}\|_\infty}{\|\boldsymbol{s}\|_\infty} \leq 0.03.$$

Exercise 8.23: LSQ MATLAB program (Exam exercise 2013-4)

Suppose $\boldsymbol{A} \in \mathbb{R}^{m \times n}, \boldsymbol{b} \in \mathbb{R}^m$, where $\boldsymbol{A}$ has rank n and let $\boldsymbol{A} = \boldsymbol{U}\boldsymbol{\Sigma}\boldsymbol{V}^{\mathrm{T}}$ be a singular value factorization of $\boldsymbol{A}$. Thus $\boldsymbol{U} \in \mathbb{R}^{m \times n}$ and $\boldsymbol{\Sigma}, \boldsymbol{V} \in \mathbb{R}^{n \times n}$. Write a MATLAB `function [x,K]=lsq(A,b)` that uses the singular value factorization of $\boldsymbol{A}$ to calculate a least squares solution $\boldsymbol{x} = \boldsymbol{V}\boldsymbol{\Sigma}^{-1}\boldsymbol{U}^{\mathrm{T}}\boldsymbol{b}$ to the system $\boldsymbol{A}\boldsymbol{x} = \boldsymbol{b}$ and the spectral (2-norm) condition number of $\boldsymbol{A}$. The MATLAB command `[U,Sigma,V]=svd(A,0)` computes the singular value factorization of $\boldsymbol{A}$.

Solution. The matrix $\boldsymbol{\Sigma}$ is a diagonal matrix with the singular values on the diagonal ordered so that $\sigma_1 \geq \cdots \geq \sigma_n$. Moreover, $\sigma_n > 0$ since $\boldsymbol{A}$ has rank n. The spectral condition number is $K = \sigma_1/\sigma_n$. We also use the MATLAB function `diag(Sigma)` that extracts the diagonal of $\boldsymbol{\Sigma}$. This leads to the following program

code/lsq.m

```
function [x,K]=lsq(A,b)
[U,Sigma,V]=svd(A,0);
s=diag(Sigma);
x=V*((U'*b)./s);
```

```
K=s(1)/s(length(s));
```

Listing 8.2: For a full rank matrix $\boldsymbol{A}$, use its singular value factorization to compute the least square solution to $\boldsymbol{A}\boldsymbol{x} = \boldsymbol{b}$ and its spectral condition number K.

Exercises section 8.4

Exercise 8.24: When is a complex norm an inner product norm?

Given a vector norm in a complex vector space $\mathcal{V}$, and suppose (8.35) holds for all $\boldsymbol{x}, \boldsymbol{y} \in \mathcal{V}$. Show that

$$\langle \boldsymbol{x}, \boldsymbol{y} \rangle := \frac{1}{4}\left(\|\boldsymbol{x}+\boldsymbol{y}\|^2 - \|\boldsymbol{x}-\boldsymbol{y}\|^2 + i\|\boldsymbol{x}+i\boldsymbol{y}\|^2 - i\|\boldsymbol{x}-i\boldsymbol{y}\|^2\right),$$

defines an inner product on $\mathcal{V}$, where $i = \sqrt{-1}$. This identity is called the *polarization identity*.

Hint.
We have $\langle \boldsymbol{x}, \boldsymbol{y} \rangle = s(\boldsymbol{x}, \boldsymbol{y}) + is(\boldsymbol{x}, i\boldsymbol{y})$, where $s(\boldsymbol{x}, \boldsymbol{y}) := \frac{1}{4}\left(\|\boldsymbol{x}+\boldsymbol{y}\|^2 - \|\boldsymbol{x}-\boldsymbol{y}\|^2\right)$.

Solution. As in the exercise, we let

$$\langle \boldsymbol{x}, \boldsymbol{y} \rangle = s(\boldsymbol{x}, \boldsymbol{y}) + is(\boldsymbol{x}, i\boldsymbol{y}), \qquad s(\boldsymbol{x}, \boldsymbol{y}) = \frac{\|\boldsymbol{x}+\boldsymbol{y}\|^2 - \|\boldsymbol{x}-\boldsymbol{y}\|^2}{4}.$$

We need to verify the three properties that define an inner product. Let $\boldsymbol{x}, \boldsymbol{y}, \boldsymbol{z}$ be arbitrary vectors in $\mathbb{C}^m$ and $a \in \mathbb{C}$ be an arbitrary scalar.

1. *Positive-definiteness*. One has $s(\boldsymbol{x}, \boldsymbol{x}) = \|\boldsymbol{x}\|^2 \geq 0$ and

$$\begin{aligned} s(\boldsymbol{x}, i\boldsymbol{x}) &= \frac{\|\boldsymbol{x}+i\boldsymbol{x}\|^2 - \|\boldsymbol{x}-i\boldsymbol{x}\|^2}{4} = \frac{\|(1+i)\boldsymbol{x}\|^2 - \|(1-i)\boldsymbol{x}\|^2}{4} \\ &= \frac{(|1+i| - |1-i|)\|\boldsymbol{x}\|^2}{4} = 0, \end{aligned}$$

   so that $\langle \boldsymbol{x}, \boldsymbol{x} \rangle = \|\boldsymbol{x}\|^2 \geq 0$, with equality holding precisely when $\boldsymbol{x} = \boldsymbol{0}$.
2. *Conjugate symmetry*. Since $s(\boldsymbol{x}, \boldsymbol{y})$ is real, $s(\boldsymbol{x}, \boldsymbol{y}) = s(\boldsymbol{y}, \boldsymbol{x})$, $s(a\boldsymbol{x}, a\boldsymbol{y}) = |a|^2 s(\boldsymbol{x}, \boldsymbol{y})$, and $s(\boldsymbol{x}, -\boldsymbol{y}) = -s(\boldsymbol{x}, \boldsymbol{y})$,

$$\begin{aligned} \overline{\langle \boldsymbol{y}, \boldsymbol{x} \rangle} &= s(\boldsymbol{y}, \boldsymbol{x}) - is(\boldsymbol{y}, i\boldsymbol{x}) = s(\boldsymbol{x}, \boldsymbol{y}) - is(i\boldsymbol{x}, \boldsymbol{y}) \\ &= s(\boldsymbol{x}, \boldsymbol{y}) - is(\boldsymbol{x}, -i\boldsymbol{y}) = \langle \boldsymbol{x}, \boldsymbol{y} \rangle. \end{aligned}$$

3. *Linearity in the first argument*. Assuming the parallelogram identity,

$$
\begin{aligned}
&2s(\boldsymbol{x},\boldsymbol{z}) + 2s(\boldsymbol{y},\boldsymbol{z})\\
&= \frac{1}{2}\|\boldsymbol{x}+\boldsymbol{z}\|^2 - \frac{1}{2}\|\boldsymbol{z}-\boldsymbol{x}\|^2 + \frac{1}{2}\|\boldsymbol{y}+\boldsymbol{z}\|^2 - \frac{1}{2}\|\boldsymbol{z}-\boldsymbol{y}\|^2\\
&= \frac{1}{2}\left\|\boldsymbol{z}+\frac{\boldsymbol{x}+\boldsymbol{y}}{2}+\frac{\boldsymbol{x}-\boldsymbol{y}}{2}\right\|^2 - \frac{1}{2}\left\|\boldsymbol{z}-\frac{\boldsymbol{x}+\boldsymbol{y}}{2}-\frac{\boldsymbol{x}-\boldsymbol{y}}{2}\right\|^2 +\\
&\frac{1}{2}\left\|\boldsymbol{z}+\frac{\boldsymbol{x}+\boldsymbol{y}}{2}-\frac{\boldsymbol{x}-\boldsymbol{y}}{2}\right\|^2 - \frac{1}{2}\left\|\boldsymbol{z}-\frac{\boldsymbol{x}+\boldsymbol{y}}{2}+\frac{\boldsymbol{x}-\boldsymbol{y}}{2}\right\|^2\\
&= \left\|\boldsymbol{z}+\frac{\boldsymbol{x}+\boldsymbol{y}}{2}\right\|^2 + \left\|\frac{\boldsymbol{x}-\boldsymbol{y}}{2}\right\|^2 - \left\|\boldsymbol{z}-\frac{\boldsymbol{x}+\boldsymbol{y}}{2}\right\|^2 - \left\|\frac{\boldsymbol{x}-\boldsymbol{y}}{2}\right\|^2\\
&= \left\|\boldsymbol{z}+\frac{\boldsymbol{x}+\boldsymbol{y}}{2}\right\|^2 - \left\|\boldsymbol{z}-\frac{\boldsymbol{x}+\boldsymbol{y}}{2}\right\|^2\\
&= 4s\left(\frac{\boldsymbol{x}+\boldsymbol{y}}{2},\boldsymbol{z}\right),
\end{aligned}
$$

so that $s(\boldsymbol{x},\boldsymbol{z}) + s(\boldsymbol{y},\boldsymbol{z}) = 2s\left(\frac{\boldsymbol{x}+\boldsymbol{y}}{2},\boldsymbol{z}\right)$. For $\boldsymbol{y} = \boldsymbol{0}$ we have clearly that $s(\boldsymbol{y},\boldsymbol{z}) = 0$, and it follows that $s(\boldsymbol{x},\boldsymbol{z}) = 2s\left(\frac{\boldsymbol{x}}{2},\boldsymbol{z}\right)$. It follows that $s(\boldsymbol{x}+\boldsymbol{y},\boldsymbol{z}) = s(\boldsymbol{x},\boldsymbol{z}) + s(\boldsymbol{y},\boldsymbol{z})$. Finally

$$
\begin{aligned}
\langle \boldsymbol{x}+\boldsymbol{y},\boldsymbol{z}\rangle &= s(\boldsymbol{x}+\boldsymbol{y},\boldsymbol{z}) + is(\boldsymbol{x}+\boldsymbol{y},i\boldsymbol{z})\\
&= s(\boldsymbol{x},\boldsymbol{z}) + s(\boldsymbol{y},\boldsymbol{z}) + is(\boldsymbol{x},i\boldsymbol{z}) + is(\boldsymbol{y},i\boldsymbol{z})\\
&= s(\boldsymbol{x},\boldsymbol{z}) + is(\boldsymbol{x},i\boldsymbol{z}) + s(\boldsymbol{y},\boldsymbol{z}) + is(\boldsymbol{y},i\boldsymbol{z})\\
&= \langle \boldsymbol{x},\boldsymbol{z}\rangle + \langle \boldsymbol{y},\boldsymbol{z}\rangle.
\end{aligned}
$$

That $\langle a\boldsymbol{x},\boldsymbol{y}\rangle = a\langle \boldsymbol{x},\boldsymbol{y}\rangle$ follows, *mutatis mutandis*, from the proof of Theorem 8.15.

Exercise 8.25: p norm for $p = 1$ and $p = \infty$

Show that $\|\cdot\|_p$ is a vector norm in $\mathbb{R}^n$ for $p = 1$, $p = \infty$.

Solution. We need to verify the three properties that define a norm. Consider arbitrary vectors $\boldsymbol{x} = [x_1,\ldots,x_n]^{\mathrm{T}}$ and $\boldsymbol{y} = [y_1,\ldots,y_n]^{\mathrm{T}}$ in $\mathbb{R}^n$ and a scalar $a \in \mathbb{R}$. First we verify that $\|\cdot\|_1$ is a norm.

1. *Positivity*. Clearly $\|\boldsymbol{x}\|_1 = |x_1| + \cdots + |x_n| \geq 0$, with equality holding precisely when $|x_1| = \cdots = |x_n| = 0$, which happens if and only if $\boldsymbol{x}$ is the zero vector.
2. *Homogeneity*. One has

$$\|a\boldsymbol{x}\|_1 = |ax_1| + \cdots + |ax_n| = |a|(|x_1| + \cdots + |x_n|) = |a|\|\boldsymbol{x}\|_1.$$

3. *Subadditivity*. Using the triangle inequality for the absolute value,

$$\|\boldsymbol{x}+\boldsymbol{y}\|_1 = |x_1+y_1| + \cdots + |x_n+y_n|$$

$$\leq |x_1| + |y_1| + \cdots + |x_n| + |y_n| = \|\boldsymbol{x}\|_1 + \|\boldsymbol{y}\|_1.$$

Next we verify that $\|\cdot\|_\infty$ is a norm.

1. *Positivity*. Clearly $\|\boldsymbol{x}\|_\infty := \max\{|x_1|, \ldots, |x_n|\} \geq 0$, with equality holding precisely when $|x_1| = \cdots = |x_n| = 0$, which happens if and only if $\boldsymbol{x}$ is the zero vector.
2. *Homogeneity*. One has

$$\|a\boldsymbol{x}\|_\infty = \max\{|a||x_1|, \ldots, |a||x_n|\} = |a| \max\{|x_1|, \ldots, |x_n|\} = |a|\|\boldsymbol{x}\|_\infty.$$

3. *Subadditivity*. Using the triangle inequality for the absolute value,

$$\begin{aligned}
\|\boldsymbol{x} + \boldsymbol{y}\|_\infty &= \max\{|x_1 + y_1|, \ldots, |x_n + y_n|\} \\
&\leq \max\{|x_1| + |y_1|, \ldots, |x_n| + |y_n|\} \\
&\leq \max\{|x_1|, \ldots, |x_n|\} + \max\{|y_1|, \ldots, |y_n|\} = \|\boldsymbol{x}\|_\infty + \|\boldsymbol{y}\|_\infty.
\end{aligned}$$

Exercise 8.26: The p-norm unit sphere

The set

$$S_p = \{\boldsymbol{x} \in \mathbb{R}^n : \|\boldsymbol{x}\|_p = 1\}$$

is called the unit sphere in $\mathbb{R}^n$ with respect to p. Draw S_p for $p = 1, 2, \infty$ for $n = 2$.

Solution. In the plane, unit spheres for the 1-norm, 2-norm, and ∞-norm are

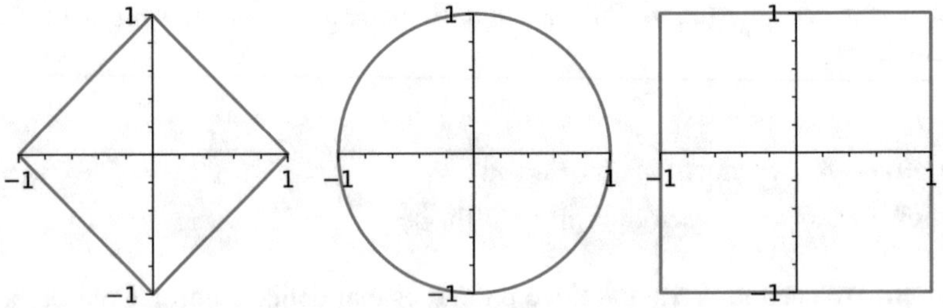

Exercise 8.27: Sharpness of p-norm inequality

For $p \geq 1$, and any $\boldsymbol{x} \in \mathbb{C}^n$ we have $\|\boldsymbol{x}\|_\infty \leq \|\boldsymbol{x}\|_p \leq n^{1/p}\|\boldsymbol{x}\|_\infty$ (cf. (8.5)).

Produce a vector $\boldsymbol{x}_l$ such that $\|\boldsymbol{x}_l\|_\infty = \|\boldsymbol{x}_l\|_p$ and another vector $\boldsymbol{x}_u$ such that $\|\boldsymbol{x}_u\|_p = n^{1/p}\|\boldsymbol{x}_u\|_\infty$. Thus, these inequalities are sharp.

Solution. Let $1 \leq p \leq \infty$. The vector $\boldsymbol{x}_l = [1, 0, \ldots, 0]^{\mathrm{T}} \in \mathbb{R}^n$ satisfies

$$\|\boldsymbol{x}_l\|_p = (|1|^p + |0|^p + \cdots + |0|^p)^{1/p} = 1 = \max\{|1|, |0|, \ldots, |0|\} = \|\boldsymbol{x}_l\|_\infty,$$

and the vector $\boldsymbol{x}_u = [1, 1, \ldots, 1]^{\mathrm{T}} \in \mathbb{R}^n$ satisfies

$$\|\boldsymbol{x}_u\|_p = (|1|^p + \cdots + |1|^p)^{1/p} = n^{1/p} = n^{1/p} \max\{|1|, \ldots, |1|\} = n^{1/p}\|\boldsymbol{x}_u\|_\infty.$$

Exercise 8.28: ***p*-norm inequalities for arbitrary *p***

If $1 \le q \le p \le \infty$ then

$$\|\boldsymbol{x}\|_p \le \|\boldsymbol{x}\|_q \le n^{1/q-1/p}\|\boldsymbol{x}\|_p, \quad \boldsymbol{x} \in \mathbb{C}^n.$$

Hint.

For the rightmost inequality use Jensen's inequality. Cf. Theorem 8.13 with $f(z) = z^{p/q}$ and $z_i = |x_i|^q$. For the left inequality consider first $y_i = x_i/\|\boldsymbol{x}\|_\infty$, $i = 1, 2, \ldots, n$.

Solution. Let p and q be integers satisfying $1 \le q \le p$, and let $\boldsymbol{x} = [x_1, \ldots, x_n]^{\mathrm{T}} \in \mathbb{C}^n$. Since $p/q \ge 1$, the function $f(z) = z^{p/q}$ is convex on $[0, \infty)$. For any $z_1, \ldots, z_n \in [0, \infty)$ and $\lambda_1, \ldots, \lambda_n \ge 0$ satisfying $\lambda_1 + \cdots + \lambda_n = 1$, Jensen's inequality gives

$$\left(\sum_{i=1}^n \lambda_i z_i\right)^{p/q} = f\left(\sum_{i=1}^n \lambda_i z_i\right) \le \sum_{i=1}^n \lambda_i f(z_i) = \sum_{i=1}^n \lambda_i z_i^{p/q}.$$

In particular for $z_i = |x_i|^q$ and $\lambda_1 = \cdots = \lambda_n = 1/n$,

$$n^{-p/q}\left(\sum_{i=1}^n |x_i|^q\right)^{p/q} = \left(\sum_{i=1}^n \frac{1}{n}|x_i|^q\right)^{p/q} \le \sum_{i=1}^n \frac{1}{n}\left(|x_i|^q\right)^{p/q} = n^{-1}\sum_{i=1}^n |x_i|^p.$$

Since the function $x \longmapsto x^{1/p}$ is monotone, we obtain

$$n^{-1/q}\|\boldsymbol{x}\|_q = n^{-1/q}\left(\sum_{i=1}^n |x_i|^q\right)^{1/q} \le n^{-1/p}\left(\sum_{i=1}^n |x_i|^p\right)^{1/p} = n^{-1/p}\|\boldsymbol{x}\|_p,$$

from which the right inequality in the exercise follows.

The left inequality clearly holds for $\boldsymbol{x} = \boldsymbol{0}$, so assume $\boldsymbol{x} \ne \boldsymbol{0}$. Without loss of generality we can then assume $\|\boldsymbol{x}\|_\infty = 1$, since $\|a\boldsymbol{x}\|_p \le \|a\boldsymbol{x}\|_q$ if and only if $\|\boldsymbol{x}\|_p \le \|\boldsymbol{x}\|_q$ for any nonzero scalar a. Then, for any $i = 1, \ldots, n$, one has $|x_i| \le 1$, implying that $|x_i|^p \le |x_i|^q$. Moreover, since $|x_i| = 1$ for some i, one has $|x_1|^q + \cdots + |x_n|^q \ge 1$, so that

$$\|\boldsymbol{x}\|_p = \left(\sum_{i=1}^n |x_i|^p\right)^{1/p} \le \left(\sum_{i=1}^n |x_i|^q\right)^{1/p} \le \left(\sum_{i=1}^n |x_i|^q\right)^{1/q} = \|\boldsymbol{x}\|_q.$$

Finally we consider the case $p = \infty$. The statement is obvious for $q = p$, so assume that q is an integer. Then

$$\|\boldsymbol{x}\|_q = \left(\sum_{i=1}^{n} |x_i|^q\right)^{1/q} \leq \left(\sum_{i=1}^{n} \|\boldsymbol{x}\|_\infty^q\right)^{1/q} = n^{1/q}\|\boldsymbol{x}\|_\infty,$$

proving the right inequality. Using that the map $x \longmapsto x^{1/q}$ is monotone, the left inequality follows from

$$\|\boldsymbol{x}\|_\infty^q = (\max_i |x_i|)^q \leq \sum_{i=1}^{n} |x_i|^q = \|\boldsymbol{x}\|_q^q.$$

Chapter 9
Least Squares

Exercises section 9.1

Exercise 9.1: Fitting a circle to points

In this exercise we derive an algorithm to fit a circle $(t-c_1)^2+(y-c_2)^2=r^2$ to $m \geq 3$ given points $(t_i, y_i)_{i=1}^m$ in the (t, y)-plane. We obtain the overdetermined system

$$(t_i - c_1)^2 + (y_i - c_2)^2 = r^2, \qquad i = 1, \ldots, m,$$

(System (9.22)) of m equations in the three unknowns c_1, c_2 and r. This system is nonlinear, but it can be solved from the linear system

$$t_i x_1 + y_i x_2 + x_3 = t_i^2 + y_i^2, \qquad i = 1, \ldots, m,$$

(System (9.23)), and then setting $c_1 = x_1/2$, $c_2 = x_2/2$ and $r^2 = c_1^2+c_2^2+x_3$.

a) Derive (9.23) from (9.22). Explain how we can find c_1, c_2, r once $[x_1, x_2, x_3]$ is determined.

Solution. Let $c_1 = x_1/2$, $c_2 = x_2/2$, and $r^2 = x_3 + c_1^2 + c_2^2$ as in the exercise. Then, for $i = 1, \ldots, m$,

$$\begin{aligned} 0 &= (t_i - c_1)^2 + (y_i - c_2)^2 - r^2 \\ &= \left(t_i - \frac{x_1}{2}\right)^2 + \left(y_i - \frac{x_2}{2}\right)^2 - x_3 - \left(\frac{x_1}{2}\right)^2 - \left(\frac{x_2}{2}\right)^2 \\ &= t_i^2 + y_i^2 - t_i x_1 - y_i x_2 - x_3, \end{aligned}$$

from which Equation (9.23) follows immediately. Once x_1, x_2, and x_3 are determined, we can compute

T. Lyche et al., *Exercises in Numerical Linear Algebra and Matrix Factorizations*, Texts in Computational Science and Engineering 23, https://doi.org/10.1007/978-3-030-59789-4_9

$$c_1 = \frac{x_1}{2}, \qquad c_2 = \frac{x_2}{2}, \qquad r = \sqrt{\frac{1}{4}x_1^2 + \frac{1}{4}x_2^2 + x_3}.$$

b) Formulate (9.23) as a linear least squares problem for suitable $\boldsymbol{A}$ and $\boldsymbol{b}$.

Solution. The linear least square problem is to minimize $\|\boldsymbol{Ax} - \boldsymbol{b}\|_2^2$, with

$$\boldsymbol{A} = \begin{bmatrix} t_1 & y_1 & 1 \\ \vdots & \vdots & \vdots \\ t_m & y_m & 1 \end{bmatrix}, \qquad \boldsymbol{b} = \begin{bmatrix} t_1^2 + y_1^2 \\ \vdots \\ t_m^2 + y_m^2 \end{bmatrix}, \qquad \boldsymbol{x} = \begin{bmatrix} x_1 \\ x_2 \\ x_3 \end{bmatrix}.$$

c) Does the matrix $\boldsymbol{A}$ in **b)** have linearly independent columns?

Solution. Whether or not $\boldsymbol{A}$ has independent columns depends on the data t_i, y_i. For instance, if $t_i = y_i = 1$ for all i, then the columns of $\boldsymbol{A}$ are clearly dependent. In general, $\boldsymbol{A}$ has independent columns whenever we can find three points (t_i, y_i) not on a straight line.

d) Use (9.23) to find the circle passing through the three points $(1, 4)$, $(3, 2)$, and $(1, 0)$.

Solution. For these points the matrix $\boldsymbol{A}$ becomes

$$\boldsymbol{A} = \begin{bmatrix} 1 & 4 & 1 \\ 3 & 2 & 1 \\ 1 & 0 & 1 \end{bmatrix},$$

which clearly is invertible. We find

$$\boldsymbol{x} = \begin{bmatrix} x_1 \\ x_2 \\ x_3 \end{bmatrix} = \begin{bmatrix} 1 & 4 & 1 \\ 3 & 2 & 1 \\ 1 & 0 & 1 \end{bmatrix}^{-1} \begin{bmatrix} 17 \\ 13 \\ 1 \end{bmatrix} = \begin{bmatrix} 2 \\ 4 \\ -1 \end{bmatrix}.$$

It follows that $c_1 = 1$, $c_2 = 2$, and $r = 2$. The points $(t, y) = (1, 4), (3, 2), (1, 0)$ therefore all lie on the circle

$$(t-1)^2 + (y-2)^2 = 4,$$

as shown in Figure 9.1.

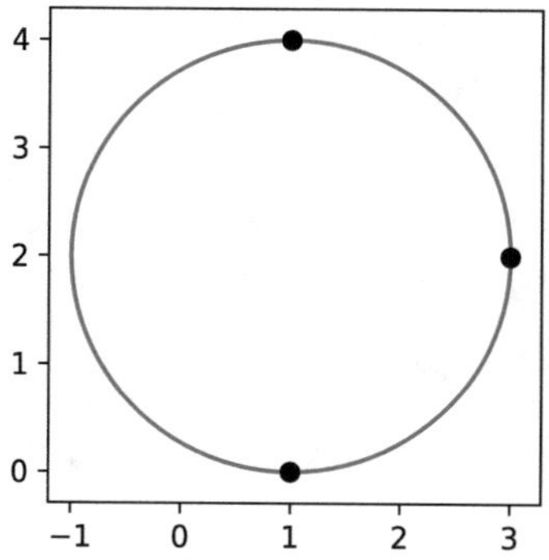

Figure 9.1: The circle fit to the points in Exercise 9.1.

Exercise 9.2: Least square fit (Exam exercise 2018-1)

a) Let $\boldsymbol{A}$ be the matrix $\begin{bmatrix} \sqrt{2} & \sqrt{2} \\ 0 & \sqrt{3} \end{bmatrix}$. Find the singular values of $\boldsymbol{A}$, and compute $\|\boldsymbol{A}\|_2$.

Solution. We first compute $\boldsymbol{A}^*\boldsymbol{A} = \begin{bmatrix} 2 & 2 \\ 2 & 5 \end{bmatrix}$, which has eigenvalues 6 and 1. The corresponding singular values are thus $\sigma_1 = \sqrt{6}$ and $\sigma_2 = 1$. In particular $\|\boldsymbol{A}\|_2 = \sigma_1 = \sqrt{6}$.

b) Consider the matrix $\boldsymbol{A} = \begin{bmatrix} 3 & \alpha \\ \alpha & 1 \end{bmatrix}$, where α is a real number. For which values of α is $\boldsymbol{A}$ positive definite?

Solution. Since $\boldsymbol{A}$ is symmetric and $a_{11} = 3 > 0$ the matrix is positive definite if and only if $\det(\boldsymbol{A}) = 3 - \alpha^2 > 0$ or $\alpha \in (-\sqrt{3}, \sqrt{3})$.

Alternatively, the matrix is positive definite if and only if the eigenvalues are positive. The eigenvalues of $\boldsymbol{A}$ are solutions to

$$0 = \det(\boldsymbol{A} - \lambda \boldsymbol{I}) = (3 - \lambda)(1 - \lambda) - \alpha^2 = \lambda^2 - 4\lambda + 3 - \alpha^2.$$

We find the eigenvalues

$$\lambda_{\pm} = \frac{4 \pm \sqrt{16 - 4(3 - \alpha^2)}}{2} = 2 \pm \sqrt{1 + \alpha^2},$$

and, since α is real, these values are both bigger than zero if and only if $1 + \alpha^2 < 4$, which happens precisely when $\alpha \in (-\sqrt{3}, \sqrt{3})$.

c) We would like to fit the points $\boldsymbol{p}_1 = (0, 1)$, $\boldsymbol{p}_2 = (1, 0)$, $\boldsymbol{p}_3 = (2, 1)$ to a straight line in the plane. Find a line $p(x) = mx + b$ which minimizes

$$\sum_{i=1}^{3} \|p(x_i) - y_i\|^2,$$

where $\boldsymbol{p}_i = (x_i, y_i)$. Is this solution unique?

Solution. We would like to find a least squares solution to

$$\begin{bmatrix} 1 & 0 \\ 1 & 1 \\ 1 & 2 \end{bmatrix} \begin{bmatrix} b \\ m \end{bmatrix} = \begin{bmatrix} 1 \\ 0 \\ 1 \end{bmatrix}.$$

Denote the coefficient matrix by $\boldsymbol{A}$, and the right hand side by $\boldsymbol{b}$. We have that

$$\boldsymbol{A}^{\mathrm{T}}\boldsymbol{A} = \begin{bmatrix} 3 & 3 \\ 3 & 5 \end{bmatrix}, \qquad \boldsymbol{A}^{\mathrm{T}}\boldsymbol{b} = \begin{bmatrix} 2 \\ 2 \end{bmatrix}.$$

Applying row reduction to the augmented matrix yields

$$[\boldsymbol{A}^{\mathrm{T}}\boldsymbol{A}|\boldsymbol{A}^{\mathrm{T}}\boldsymbol{b}] = \left[\begin{array}{cc|c} 3 & 3 & 2 \\ 3 & 5 & 2 \end{array}\right] \sim \left[\begin{array}{cc|c} 1 & 0 & 2/3 \\ 0 & 1 & 0 \end{array}\right],$$

so that $b = 2/3$, $m = 0$. It follows that the horizontal line $y = p(x) = 2/3$ is the least squares fit. The least squares solution is unique since $\boldsymbol{A}$ has linearly independent columns.

Exercises section 9.2

Exercise 9.3: A least squares problem (Exam exercise 1983-2)

Suppose $\boldsymbol{A} \in \mathbb{R}^{m \times n}$ and let $\boldsymbol{I} \in \mathbb{R}^{n \times n}$ be the identity matrix. We define $F : \mathbb{R}^n \to \mathbb{R}$ by

$$F(\boldsymbol{x}) := \|\boldsymbol{A}\boldsymbol{x} - \boldsymbol{b}\|_2^2 + \|\boldsymbol{x}\|_2^2.$$

a) Show that the matrix $\boldsymbol{B} := \boldsymbol{I} + \boldsymbol{A}^{\mathrm{T}}\boldsymbol{A}$ is symmetric and positive definite.

Solution. For any $\boldsymbol{x} \in \mathbb{R}^n$ we have

$$\boldsymbol{x}^{\mathrm{T}}(\boldsymbol{I} + \boldsymbol{A}^{\mathrm{T}}\boldsymbol{A})\boldsymbol{x} = \boldsymbol{x}^{\mathrm{T}}\boldsymbol{x} + \boldsymbol{x}^{\mathrm{T}}\boldsymbol{A}^{\mathrm{T}}\boldsymbol{A}\boldsymbol{x} = \|\boldsymbol{x}\|_2^2 + \|\boldsymbol{A}\boldsymbol{x}\|_2^2 \geq 0.$$

Moreover, if equality holds then $\|\boldsymbol{x}\|_2 = 0$ so that $\boldsymbol{x} = 0$. Finally,

$$\boldsymbol{B}^{\mathrm{T}} = (\boldsymbol{I} + \boldsymbol{A}^{\mathrm{T}}\boldsymbol{A})^{\mathrm{T}} = \boldsymbol{I}^{\mathrm{T}} + (\boldsymbol{A}^{\mathrm{T}}\boldsymbol{A})^{\mathrm{T}} = \boldsymbol{B}$$

so that $\boldsymbol{B}$ is symmetric.

b) Show that

$$F(\boldsymbol{x}) = \boldsymbol{x}^{\mathrm{T}}\boldsymbol{B}\boldsymbol{x} - 2\boldsymbol{c}^{\mathrm{T}}\boldsymbol{x} + \boldsymbol{b}^{\mathrm{T}}\boldsymbol{b}, \quad \text{where} \quad \boldsymbol{c} = \boldsymbol{A}^{\mathrm{T}}\boldsymbol{b}.$$

Solution. We find

$$\begin{aligned}
F(\boldsymbol{x}) &= (\boldsymbol{A}\boldsymbol{x} - \boldsymbol{b})^{\mathrm{T}}(\boldsymbol{A}\boldsymbol{x} - \boldsymbol{b}) + \boldsymbol{x}^{\mathrm{T}}\boldsymbol{x} \\
&= \boldsymbol{x}^{\mathrm{T}}\boldsymbol{A}^{\mathrm{T}}\boldsymbol{A}\boldsymbol{x} - (\boldsymbol{A}\boldsymbol{x})^{\mathrm{T}}\boldsymbol{b} - \boldsymbol{b}^{\mathrm{T}}\boldsymbol{A}\boldsymbol{x} + \boldsymbol{b}^{\mathrm{T}}\boldsymbol{b} + \boldsymbol{x}^{\mathrm{T}}\boldsymbol{I}\boldsymbol{x} \\
&= \boldsymbol{x}^{\mathrm{T}}(\boldsymbol{I} + \boldsymbol{A}^{\mathrm{T}}\boldsymbol{A})\boldsymbol{x} - \boldsymbol{x}^{\mathrm{T}}\boldsymbol{A}^{\mathrm{T}}\boldsymbol{b} - (\boldsymbol{A}^{\mathrm{T}}\boldsymbol{b})^{\mathrm{T}}\boldsymbol{x} + \boldsymbol{b}^{\mathrm{T}}\boldsymbol{b} \\
&= \boldsymbol{x}^{\mathrm{T}}\boldsymbol{B}\boldsymbol{x} - 2\boldsymbol{c}^{\mathrm{T}}\boldsymbol{x} + \boldsymbol{b}^{\mathrm{T}}\boldsymbol{b}.
\end{aligned}$$

c) Show that to every $\boldsymbol{b} \in \mathbb{R}^m$ there is a unique $\boldsymbol{x}$ which minimizes F. Moreover, $\boldsymbol{x}$ is the unique solution of the linear system $(\boldsymbol{I} + \boldsymbol{A}^{\mathrm{T}}\boldsymbol{A})\boldsymbol{x} = \boldsymbol{A}^{\mathrm{T}}\boldsymbol{b}$.

Solution. Since $\boldsymbol{B}$ is positive definite it follows from Lemma 13.1 that there is a unique $\boldsymbol{x} \in \mathbb{R}^n$ which minimizes F. Moreover, by the same lemma $\boldsymbol{x}$ is the unique solution of $\boldsymbol{B}\boldsymbol{x} = \boldsymbol{c}$ or $(\boldsymbol{I} + \boldsymbol{A}^{\mathrm{T}}\boldsymbol{A})\boldsymbol{x} = \boldsymbol{A}^{\mathrm{T}}\boldsymbol{b}$.

Exercise 9.4: Weighted least squares (Exam exercise 1977-2) ?

For $m \geq n$ we are given $\boldsymbol{A} \in \mathbb{R}^{m\times n}$ with linearly independent columns, $\boldsymbol{b} \in \mathbb{R}^m$, and $\boldsymbol{D} := \mathrm{diag}(d_1, d_2, \ldots, d_m) \in \mathbb{R}^{m\times m}$, where $d_i > 0$, $i = 1, 2, \ldots, m$. We want to minimize

$$\|\boldsymbol{r}(\boldsymbol{x})\|_D^2 := \sum_{i=1}^{m} r_i(x)^2 d_i, \qquad \boldsymbol{x} \in \mathbb{R}^n, \tag{9.24}$$

where $r_i = r_i(\boldsymbol{x})$, $i = 1, 2, \ldots, m$ are the components of the vector

$$\boldsymbol{r} = \boldsymbol{r}(\boldsymbol{x}) = \boldsymbol{b} - \boldsymbol{A}\boldsymbol{x}.$$

a) Show that $\|\boldsymbol{r}(\boldsymbol{x})\|_D^2$ in (9.24) obtains a unique minimum when $\boldsymbol{x} = \boldsymbol{x}_{\min}$ is the solution of the system

$$\boldsymbol{A}^{\mathrm{T}}\boldsymbol{D}\boldsymbol{A}\boldsymbol{x} = \boldsymbol{A}^{\mathrm{T}}\boldsymbol{D}\boldsymbol{b}.$$

Solution. With $\boldsymbol{D}^{1/2} := \mathrm{diag}\left(\sqrt{d_1}, \sqrt{d_2}, \ldots, \sqrt{d_m}\right)$ we find

$$\|\boldsymbol{r}(\boldsymbol{x})\|_D^2 = \|\boldsymbol{D}^{1/2}\boldsymbol{r}(\boldsymbol{x})\|_2^2 = \|\boldsymbol{D}^{1/2}\boldsymbol{A}\boldsymbol{x} - \boldsymbol{D}^{1/2}\boldsymbol{b}\|_2^2.$$

We obtain an ordinary least squares problem

$$\min\|\hat{\boldsymbol{A}}\boldsymbol{x} - \hat{\boldsymbol{b}}\|_2, \qquad \hat{\boldsymbol{A}} := \boldsymbol{D}^{1/2}\boldsymbol{A}, \qquad \hat{\boldsymbol{b}} := \boldsymbol{D}^{1/2}\boldsymbol{b}.$$

Since $\boldsymbol{D}$ is nonsingular $\hat{\boldsymbol{A}}$ has linearly independent columns, and by Theorems 9.1, 9.2, 9.3, $\boldsymbol{x} = \boldsymbol{x}_{\min}$ is unique and the solution of $\hat{\boldsymbol{A}}^{\mathrm{T}}\hat{\boldsymbol{A}}\boldsymbol{x} = \hat{\boldsymbol{A}}^{\mathrm{T}}\hat{\boldsymbol{b}}$, which is the same as $\boldsymbol{A}^{\mathrm{T}}\boldsymbol{D}\boldsymbol{A}\boldsymbol{x} = \boldsymbol{A}^{\mathrm{T}}\boldsymbol{D}\boldsymbol{b}$.

b) Show that

$$K_2(\boldsymbol{A}^{\mathrm{T}}\boldsymbol{D}\boldsymbol{A}) \le K_2(\boldsymbol{A}^{\mathrm{T}}\boldsymbol{A})K_2(\boldsymbol{D}),$$

where for any nonsingular matrix $K_2(\boldsymbol{B}) := \|\boldsymbol{B}\|_2\|\boldsymbol{B}^{-1}\|_2$.

Solution. Since $\boldsymbol{A}^{\mathrm{T}}\boldsymbol{D}\boldsymbol{A} = \hat{\boldsymbol{A}}^{\mathrm{T}}\hat{\boldsymbol{A}}$ and $\hat{\boldsymbol{A}}$ has linearly independent columns it follows from Lemma 4.2 that both $\boldsymbol{A}^{\mathrm{T}}\boldsymbol{A}$ and $\boldsymbol{A}^{\mathrm{T}}\boldsymbol{D}\boldsymbol{A}$ are positive definite. Let μ_1, μ_n and λ_1, λ_n be the largest and smallest eigenvalues of $\boldsymbol{A}^{\mathrm{T}}\boldsymbol{A}$ and $\boldsymbol{A}^{\mathrm{T}}\boldsymbol{D}\boldsymbol{A}$, respectively. Suppose $(\lambda, \boldsymbol{v})$ is an eigenpair of $\boldsymbol{A}^{\mathrm{T}}\boldsymbol{D}\boldsymbol{A}$,

$$\boldsymbol{A}^{\mathrm{T}}\boldsymbol{D}\boldsymbol{A}\boldsymbol{v} = \lambda\boldsymbol{v}, \quad \|\boldsymbol{v}\|_2 = 1.$$

Then with $\boldsymbol{w} := \boldsymbol{A}\boldsymbol{v}$

$$\lambda = \lambda\boldsymbol{v}^{\mathrm{T}}\boldsymbol{v} = \boldsymbol{v}^{\mathrm{T}}\boldsymbol{A}^{\mathrm{T}}\boldsymbol{D}\boldsymbol{A}\boldsymbol{v} = \boldsymbol{w}^{\mathrm{T}}\boldsymbol{D}\boldsymbol{w} = \sum_{i=1}^{n} d_i w_i^2.$$

With $d_{\max}$ and $d_{\min}$ the largest and smallest diagonal element in $\boldsymbol{D}$, it follows that

$$d_{\min}\|\boldsymbol{A}\boldsymbol{v}\|^2 \le \lambda \le d_{\max}\|\boldsymbol{A}\boldsymbol{v}\|^2.$$

Since also $\mu_n \le \langle \boldsymbol{A}^T\boldsymbol{A}\boldsymbol{v}, \boldsymbol{v}\rangle = \|\boldsymbol{A}\boldsymbol{v}\|_2^2$ and $\|\boldsymbol{A}\boldsymbol{v}\|_2^2 = \langle \boldsymbol{A}^T\boldsymbol{A}\boldsymbol{v}, \boldsymbol{v}\rangle \le \mu_1$, it follows that

$$d_{\min}\mu_n \le d_{\min}\|\boldsymbol{A}\boldsymbol{v}\|_2^2 \le \lambda \le d_{\max}\|\boldsymbol{A}\boldsymbol{v}\|_2^2 \le d_{\max}\mu_1.$$

By Exercise 8.13 we have $\|\boldsymbol{D}\|_2 = d_{\max}$ and $\|\boldsymbol{D}^{-1}\|_2 = 1/d_{\min}$. But then

$$K_2(\boldsymbol{A}^{\mathrm{T}}\boldsymbol{D}\boldsymbol{A}) = \frac{\lambda_1}{\lambda_n} \le \frac{d_{\max}\mu_1}{d_{\min}\mu_n} = K_2(\boldsymbol{D})K_2(\boldsymbol{A}^{\mathrm{T}}\boldsymbol{A}).$$

Exercises section 9.3

Exercise 9.5: Uniqueness of generalized inverse

Given $\boldsymbol{A} \in \mathbb{C}^{m\times n}$, and suppose $\boldsymbol{B}, \boldsymbol{C} \in \mathbb{C}^{n\times m}$ satisfy

$$\begin{array}{rcl} \boldsymbol{ABA} = \boldsymbol{A} & (1) & \boldsymbol{ACA} = \boldsymbol{A}, \\ \boldsymbol{BAB} = \boldsymbol{B} & (2) & \boldsymbol{CAC} = \boldsymbol{C}, \\ (\boldsymbol{AB})^* = \boldsymbol{AB} & (3) & (\boldsymbol{AC})^* = \boldsymbol{AC}, \\ (\boldsymbol{BA})^* = \boldsymbol{BA} & (4) & (\boldsymbol{CA})^* = \boldsymbol{CA}. \end{array}$$

Verify the following proof that $\boldsymbol{B} = \boldsymbol{C}$.

$$\begin{aligned} \boldsymbol{B} &= (\boldsymbol{BA})\boldsymbol{B} = (\boldsymbol{A}^*)\boldsymbol{B}^*\boldsymbol{B} = (\boldsymbol{A}^*\boldsymbol{C}^*)\boldsymbol{A}^*\boldsymbol{B}^*\boldsymbol{B} = \boldsymbol{CA}(\boldsymbol{A}^*\boldsymbol{B}^*)\boldsymbol{B} \\ &= \boldsymbol{CA}(\boldsymbol{BAB}) = (\boldsymbol{C})\boldsymbol{AB} = \boldsymbol{C}(\boldsymbol{AC})\boldsymbol{AB} = \boldsymbol{CC}^*\boldsymbol{A}^*(\boldsymbol{AB}) \\ &= \boldsymbol{CC}^*(\boldsymbol{A}^*\boldsymbol{B}^*\boldsymbol{A}^*) = \boldsymbol{C}(\boldsymbol{C}^*\boldsymbol{A}^*) = \boldsymbol{CAC} = \boldsymbol{C}. \end{aligned}$$

Solution. Denote the properties to the left by $(1_B), (2_B), (3_B), (4_B)$ and the properties to the right by $(1_C), (2_C), (3_C), (4_C)$. Then one uses, in order, (2_B), (4_B), (1_C), (4_C), (4_B), (2_B), (2_C), (3_C), (3_B), (1_B), (3_C), and (2_C).

Exercise 9.6: Verify that a matrix is a generalized inverse

Show that the generalized inverse of $\boldsymbol{A} = \begin{bmatrix} 1 & 1 \\ 1 & 1 \\ 0 & 0 \end{bmatrix}$ is $\boldsymbol{A}^\dagger = \frac{1}{4}\begin{bmatrix} 1 & 1 & 0 \\ 1 & 1 & 0 \end{bmatrix}$ without using the singular value decomposition of $\boldsymbol{A}$.

Solution. Let

$$\boldsymbol{A} = \begin{bmatrix} 1 & 1 \\ 1 & 1 \\ 0 & 0 \end{bmatrix}, \qquad \boldsymbol{B} = \frac{1}{4}\begin{bmatrix} 1 & 1 & 0 \\ 1 & 1 & 0 \end{bmatrix}$$

be as in the exercise. One finds

$$\boldsymbol{AB} = \begin{bmatrix} 1 & 1 \\ 1 & 1 \\ 0 & 0 \end{bmatrix} \frac{1}{4}\begin{bmatrix} 1 & 1 & 0 \\ 1 & 1 & 0 \end{bmatrix} = \frac{1}{2}\begin{bmatrix} 1 & 1 & 0 \\ 1 & 1 & 0 \\ 0 & 0 & 0 \end{bmatrix},$$

$$\boldsymbol{BA} = \frac{1}{4}\begin{bmatrix} 1 & 1 & 0 \\ 1 & 1 & 0 \end{bmatrix}\begin{bmatrix} 1 & 1 \\ 1 & 1 \\ 0 & 0 \end{bmatrix} = \frac{1}{2}\begin{bmatrix} 1 & 1 \\ 1 & 1 \end{bmatrix},$$

so that $(\boldsymbol{AB})^* = \boldsymbol{AB}$ and $(\boldsymbol{BA})^* = \boldsymbol{BA}$. Moreover,

$$\boldsymbol{ABA} = \boldsymbol{A}(\boldsymbol{BA}) = \begin{bmatrix} 1 & 1 \\ 1 & 1 \\ 0 & 0 \end{bmatrix} \frac{1}{2}\begin{bmatrix} 1 & 1 \\ 1 & 1 \end{bmatrix} = \begin{bmatrix} 1 & 1 \\ 1 & 1 \\ 0 & 0 \end{bmatrix} = \boldsymbol{A},$$

$$\boldsymbol{BAB} = (\boldsymbol{BA})\boldsymbol{B} = \frac{1}{2}\begin{bmatrix}1 & 1\\ 1 & 1\end{bmatrix}\frac{1}{4}\begin{bmatrix}1 & 1 & 0\\ 1 & 1 & 0\end{bmatrix} = \frac{1}{4}\begin{bmatrix}1 & 1 & 0\\ 1 & 1 & 0\end{bmatrix} = \boldsymbol{B}.$$

By Theorem 9.5 we conclude that $\boldsymbol{B}$ must be the pseudoinverse of $\boldsymbol{A}$.

Exercise 9.7: Linearly independent columns and generalized inverse

Suppose $\boldsymbol{A} \in \mathbb{C}^{m\times n}$ has linearly independent columns. Show that $\boldsymbol{A}^*\boldsymbol{A}$ is nonsingular and $\boldsymbol{A}^\dagger = (\boldsymbol{A}^*\boldsymbol{A})^{-1}\boldsymbol{A}^*$. If $\boldsymbol{A}$ has linearly independent rows, then show that $\boldsymbol{A}\boldsymbol{A}^*$ is nonsingular and $\boldsymbol{A}^\dagger = \boldsymbol{A}^*(\boldsymbol{A}\boldsymbol{A}^*)^{-1}$.

Solution. If $\boldsymbol{A} \in \mathbb{C}^{m,n}$ has independent columns then both $\boldsymbol{A}$ and $\boldsymbol{A}^*$ have rank $n \le m$. Then, by Exercise 7.11, $\boldsymbol{A}^*\boldsymbol{A}$ must have rank n as well. Since $\boldsymbol{A}^*\boldsymbol{A}$ is an $n\times n$-matrix of maximal rank, it is nonsingular and we can define $\boldsymbol{B} := (\boldsymbol{A}^*\boldsymbol{A})^{-1}\boldsymbol{A}^*$. Let $\boldsymbol{A} = \boldsymbol{U}_1\boldsymbol{\Sigma}_1\boldsymbol{V}_1^*$ be a singular value factorization of $\boldsymbol{A}$. Note that $\boldsymbol{V}_1$ is $n \times n$, so that it is unitary. We have that

$$\begin{aligned}(\boldsymbol{A}^*\boldsymbol{A})^{-1}\boldsymbol{A}^* &= ((\boldsymbol{U}_1\boldsymbol{\Sigma}_1\boldsymbol{V}_1^*)^*\boldsymbol{U}_1\boldsymbol{\Sigma}_1\boldsymbol{V}_1^*)^{-1}(\boldsymbol{U}_1\boldsymbol{\Sigma}_1\boldsymbol{V}_1^*)^*\\ &= (\boldsymbol{V}_1\boldsymbol{\Sigma}_1\boldsymbol{U}_1^*\boldsymbol{U}_1\boldsymbol{\Sigma}_1\boldsymbol{V}_1^*)^{-1}(\boldsymbol{U}_1\boldsymbol{\Sigma}_1\boldsymbol{V}_1^*)^*\\ &= \left(\boldsymbol{V}_1\boldsymbol{\Sigma}_1^2\boldsymbol{V}_1^*\right)^{-1}\boldsymbol{V}_1\boldsymbol{\Sigma}_1\boldsymbol{U}_1^* = \boldsymbol{V}_1\boldsymbol{\Sigma}_1^{-2}\boldsymbol{V}_1^*\boldsymbol{V}_1\boldsymbol{\Sigma}_1\boldsymbol{U}_1^*\\ &= \boldsymbol{V}_1\boldsymbol{\Sigma}_1^{-1}\boldsymbol{U}_1^* = \boldsymbol{A}^\dagger,\end{aligned}$$

where we used that $\left(\boldsymbol{V}_1\boldsymbol{\Sigma}_1^2\boldsymbol{V}_1^*\right)^{-1} = \boldsymbol{V}_1\boldsymbol{\Sigma}_1^{-2}\boldsymbol{V}_1^*$. $\boldsymbol{V}_1\boldsymbol{V}_1^*$ is $m \times m$ ($\boldsymbol{V}_1$ has full rank).

Alternatively, with $\boldsymbol{B} := (\boldsymbol{A}^*\boldsymbol{A})^{-1}\boldsymbol{A}^*$, we can verify that $\boldsymbol{B}$ satisfies the four axioms of Exercise 9.5.

1. $\boldsymbol{ABA} = \boldsymbol{A}(\boldsymbol{A}^*\boldsymbol{A})^{-1}\boldsymbol{A}^*\boldsymbol{A} = \boldsymbol{A}$
2. $\boldsymbol{BAB} = (\boldsymbol{A}^*\boldsymbol{A})^{-1}\boldsymbol{A}^*\boldsymbol{A}(\boldsymbol{A}^*\boldsymbol{A})^{-1}\boldsymbol{A}^* = (\boldsymbol{A}^*\boldsymbol{A})^{-1}\boldsymbol{A}^* = \boldsymbol{B}$
3. $(\boldsymbol{BA})^* = \left((\boldsymbol{A}^*\boldsymbol{A})^{-1}\boldsymbol{A}^*\boldsymbol{A}\right)^* = \boldsymbol{I}_n^* = \boldsymbol{I}_n = (\boldsymbol{A}^*\boldsymbol{A})^{-1}\boldsymbol{A}^*\boldsymbol{A} = \boldsymbol{BA}$
4. $(\boldsymbol{AB})^* = \left(\boldsymbol{A}(\boldsymbol{A}^*\boldsymbol{A})^{-1}\boldsymbol{A}^*\right)^* = \boldsymbol{A}\left((\boldsymbol{A}^*\boldsymbol{A})^{-1}\right)^*\boldsymbol{A}^*$
   $= \boldsymbol{A}(\boldsymbol{A}^*\boldsymbol{A})^{-1}\boldsymbol{A}^* = \boldsymbol{AB}$

It follows that $\boldsymbol{B} = \boldsymbol{A}^\dagger$. The second claim follows similarly.

Alternatively, one can use the fact that the unique solution of the least squares problem is $\boldsymbol{A}^\dagger\boldsymbol{b}$ and compare this with the solution of the normal equation.

Exercise 9.8: More orthogonal projections

Given $m, n \in \mathbb{N}$, $\boldsymbol{A} \in \mathbb{C}^{m\times n}$ of rank r, and let $\mathcal{S}$ be one of the subspaces $\mathcal{R}(\boldsymbol{A}^*), \mathcal{N}(\boldsymbol{A})$. Show that the orthogonal projection of $\boldsymbol{v} \in \mathbb{C}^n$ into $\mathcal{S}$ can be written as a matrix $\boldsymbol{P}_{\mathcal{S}}$ times the vector $\boldsymbol{v}$ in the form $\boldsymbol{P}_{\mathcal{S}}\boldsymbol{v}$, where

$$\begin{aligned} \boldsymbol{P}_{\mathcal{R}(\boldsymbol{A}^*)} &= \boldsymbol{A}^\dagger\boldsymbol{A} = \boldsymbol{V}_1\boldsymbol{V}_1^* = \sum_{j=1}^{r} \boldsymbol{v}_j\boldsymbol{v}_j^* \in \mathbb{C}^{n\times n}, \\ \boldsymbol{P}_{\mathcal{N}(\boldsymbol{A})} &= \boldsymbol{I} - \boldsymbol{A}^\dagger\boldsymbol{A} = \boldsymbol{V}_2\boldsymbol{V}_2^* = \sum_{j=r+1}^{n} \boldsymbol{v}_j\boldsymbol{v}_j^* \in \mathbb{C}^{n\times n}, \end{aligned} \tag{9.28}$$

where $\boldsymbol{A}^\dagger$ is the generalized inverse of $\boldsymbol{A}$ and $\boldsymbol{A} = \boldsymbol{U\Sigma V}^* \in \mathbb{C}^{m\times n}$, is a singular value decomposition of $\boldsymbol{A}$ (cf. (9.7)). Thus (9.12) and (9.28) give the orthogonal projections into the 4 fundamental subspaces.

Hint.

By Theorem 7.3 we have the orthogonal sum $\mathbb{C}^n = \mathcal{R}(\boldsymbol{A}^*)\overset{\perp}{\oplus}\mathcal{N}(\boldsymbol{A})$.

Solution. We know that $\{\boldsymbol{v}_j\}_{j=1}^r$ is an orthonormal basis for $\mathcal{R}(\boldsymbol{A}^*)$, and that $\{\boldsymbol{v}_j\}_{j=r+1}^n$ is an orthonormal basis for $\boldsymbol{P}_{\mathcal{N}(\boldsymbol{A})}$. From this the right hand side expressions $\boldsymbol{V}_1\boldsymbol{V}_1^* = \sum_{j=1}^r \boldsymbol{v}_j\boldsymbol{v}_j^* \in \mathbb{C}^{n\times n}$ and $\boldsymbol{V}_2\boldsymbol{V}_2^* = \sum_{j=r+1}^n \boldsymbol{v}_j\boldsymbol{v}_j^* \in \mathbb{C}^{n\times n}$ follow immediately. We also have that

$$\boldsymbol{A}^\dagger\boldsymbol{A} = \boldsymbol{V}_1\boldsymbol{\Sigma}^{-1}\boldsymbol{U}_1^*\boldsymbol{U}_1\boldsymbol{\Sigma}\boldsymbol{V}_1^* = \boldsymbol{V}_1\boldsymbol{V}_1^*.$$

Also

$$\boldsymbol{I} - \boldsymbol{A}^\dagger\boldsymbol{A} = \boldsymbol{I} - \boldsymbol{P}_{\mathcal{R}(\boldsymbol{A}^*)} = \boldsymbol{P}_{\mathcal{N}(\boldsymbol{A})},$$

and the result follows.

Exercise 9.9: The generalized inverse of a vector

Show that $\boldsymbol{u}^\dagger = (\boldsymbol{u}^*\boldsymbol{u})^{-1}\boldsymbol{u}^*$ if $\boldsymbol{u} \in \mathbb{C}^{n,1}$ is nonzero.

Solution. This is a special case of Exercise 9.7. In particular, if $\boldsymbol{u}$ is a nonzero vector, then $\boldsymbol{u}^*\boldsymbol{u} = \langle \boldsymbol{u}, \boldsymbol{u}\rangle = \|\boldsymbol{u}\|^2$ is a nonzero number and $(\boldsymbol{u}^*\boldsymbol{u})^{-1}\boldsymbol{u}^*$ is defined. One can again check the axioms of Exercise 9.5 to show that this vector must be the pseudoinverse of $\boldsymbol{u}^*$.

Exercise 9.10: The generalized inverse of an outer product

If $\boldsymbol{A} = \boldsymbol{uv}^*$ where $\boldsymbol{u} \in \mathbb{C}^m$, $\boldsymbol{v} \in \mathbb{C}^n$ are nonzero, show that

$$\boldsymbol{A}^\dagger = \frac{1}{\alpha}\boldsymbol{A}^*, \qquad \alpha = \|\boldsymbol{u}\|_2^2\|\boldsymbol{v}\|_2^2.$$

Solution. Let $\boldsymbol{A} = \boldsymbol{u}\boldsymbol{v}^*$ be as in the exercise. Since $\boldsymbol{u}$ and $\boldsymbol{v}$ are nonzero,

$$\boldsymbol{A} = \boldsymbol{U}_1\boldsymbol{\Sigma}_1\boldsymbol{V}_1^* = \frac{\boldsymbol{u}}{\|\boldsymbol{u}\|_2}\left[\|\boldsymbol{u}\|_2\|\boldsymbol{v}\|_2\right]\frac{\boldsymbol{v}^*}{\|\boldsymbol{v}\|_2}$$

is a singular value factorization of $\boldsymbol{A}$. But then

$$\boldsymbol{A}^\dagger = \boldsymbol{V}_1\boldsymbol{\Sigma}_1^{-1}\boldsymbol{U}_1^* = \frac{\boldsymbol{v}}{\|\boldsymbol{v}\|_2}\left[\frac{1}{\|\boldsymbol{u}\|_2\|\boldsymbol{v}\|_2}\right]\frac{\boldsymbol{u}^*}{\|\boldsymbol{u}\|_2} = \frac{1}{\|\boldsymbol{u}\|_2^2\|\boldsymbol{v}\|_2^2}\boldsymbol{v}\boldsymbol{u}^* = \frac{\boldsymbol{A}^*}{\alpha}.$$

Exercise 9.11: The generalized inverse of a diagonal matrix

Show that $\mathrm{diag}(\lambda_1, \ldots, \lambda_n)^\dagger = \mathrm{diag}(\lambda_1^\dagger, \ldots, \lambda_n^\dagger)$ where

$$\lambda_i^\dagger = \begin{cases} 1/\lambda_i, & \lambda_i \neq 0, \\ 0, & \lambda_i = 0. \end{cases}$$

Solution. Let $\boldsymbol{A} := \mathrm{diag}(\lambda_1, \ldots, \lambda_n)$ and $\boldsymbol{B} := \mathrm{diag}(\lambda_1^\dagger, \ldots, \lambda_n^\dagger)$ as in the exercise. Note that, by definition, $\lambda_j^\dagger$ indeed represents the pseudoinverse of the number λ_j for any j. It therefore satisfies the axioms of Exercise 9.5, which we shall use below. We now verify the axioms for $\boldsymbol{B}$ to show that $\boldsymbol{B}$ must be the pseudoinverse of $\boldsymbol{A}$.

1. $\boldsymbol{ABA} = \mathrm{diag}(\lambda_1\lambda_1^\dagger\lambda_1, \ldots, \lambda_n\lambda_n^\dagger\lambda_n) = \mathrm{diag}(\lambda_1, \ldots, \lambda_n) = \boldsymbol{A}$;
2. $\boldsymbol{BAB} = \mathrm{diag}(\lambda_1^\dagger\lambda_1\lambda_1^\dagger, \ldots, \lambda_n^\dagger\lambda_n\lambda_n^\dagger) = \mathrm{diag}(\lambda_1^\dagger, \ldots, \lambda_n^\dagger) = \boldsymbol{B}$;
3. $(\boldsymbol{BA})^* = (\mathrm{diag}(\lambda_1^\dagger\lambda_1, \ldots, \lambda_n^\dagger\lambda_n))^* = \mathrm{diag}(\lambda_1^\dagger\lambda_1, \ldots, \lambda_n^\dagger\lambda_n) = \boldsymbol{BA}$;
4. $(\boldsymbol{AB})^* = (\mathrm{diag}(\lambda_1\lambda_1^\dagger, \ldots, \lambda_n\lambda_n^\dagger))^* = \mathrm{diag}(\lambda_1\lambda_1^\dagger, \ldots, \lambda_n\lambda_n^\dagger) = \boldsymbol{AB}$.

This proves that $\boldsymbol{B}$ is the pseudoinverse of $\boldsymbol{A}$.

Exercise 9.12: Properties of the generalized inverse

Suppose $\boldsymbol{A} \in \mathbb{C}^{m\times n}$. Show that

a) $(\boldsymbol{A}^*)^\dagger = (\boldsymbol{A}^\dagger)^*$.

Solution. Let $\boldsymbol{A} = \boldsymbol{U\Sigma V}^*$ be a singular value decomposition of $\boldsymbol{A}$ and $\boldsymbol{A} = \boldsymbol{U}_1\boldsymbol{\Sigma}_1\boldsymbol{V}_1^*$ the corresponding singular value factorization. By definition of the pseudoinverse, $\boldsymbol{A}^\dagger := \boldsymbol{V}_1\boldsymbol{\Sigma}_1^{-1}\boldsymbol{U}_1^*$.

One has $(\boldsymbol{A}^\dagger)^* = (\boldsymbol{V}_1\boldsymbol{\Sigma}_1^{-1}\boldsymbol{U}_1^*)^* = \boldsymbol{U}_1\boldsymbol{\Sigma}_1^{-*}\boldsymbol{V}_1^*$. On the other hand, the matrix $\boldsymbol{A}^*$ has singular value factorization $\boldsymbol{A}^* = \boldsymbol{V}_1\boldsymbol{\Sigma}_1^*\boldsymbol{U}_1^*$, so that its pseudoinverse is $(\boldsymbol{A}^*)^\dagger := \boldsymbol{U}_1\boldsymbol{\Sigma}_1^{-*}\boldsymbol{V}_1^*$ as well. We conclude that $(\boldsymbol{A}^\dagger)^* = (\boldsymbol{A}^*)^\dagger$.

b) $(\boldsymbol{A}^\dagger)^\dagger = \boldsymbol{A}$.

Solution. Since $\boldsymbol{A}^\dagger := \boldsymbol{V}_1\boldsymbol{\Sigma}_1^{-1}\boldsymbol{U}_1^*$ is a singular value factorization, it has pseudoinverse $(\boldsymbol{A}^\dagger)^\dagger = (\boldsymbol{U}_1^*)^*(\boldsymbol{\Sigma}_1^{-1})^{-1}\boldsymbol{V}_1^* = \boldsymbol{U}_1\boldsymbol{\Sigma}_1\boldsymbol{V}_1^* = \boldsymbol{A}$.

c) $(\alpha\boldsymbol{A})^\dagger = \frac{1}{\alpha}\boldsymbol{A}^\dagger, \quad \alpha \neq 0.$

Solution. Since the matrix $\alpha\boldsymbol{A}$ has singular value factorization $\boldsymbol{U}_1(\alpha\boldsymbol{\Sigma}_1)\boldsymbol{V}_1^*$, it has pseudoinverse

$$(\alpha\boldsymbol{A})^\dagger = \boldsymbol{V}_1(\alpha\boldsymbol{\Sigma}_1)^{-1}\boldsymbol{U}_1^* = \alpha^{-1}\boldsymbol{V}_1\boldsymbol{\Sigma}_1^{-1}\boldsymbol{U}_1^* = \alpha^{-1}\boldsymbol{A}^\dagger.$$

Exercise 9.13: The generalized inverse of a product

Suppose $k, m, n \in \mathbb{N}$, $\boldsymbol{A} \in \mathbb{C}^{m\times n}$, $\boldsymbol{B} \in \mathbb{C}^{n\times k}$. Suppose $\boldsymbol{A}$ has linearly independent columns and $\boldsymbol{B}$ has linearly independent rows.

a) Show that $(\boldsymbol{A}\boldsymbol{B})^\dagger = \boldsymbol{B}^\dagger\boldsymbol{A}^\dagger$.

Hint.

Let $\boldsymbol{E} = \boldsymbol{A}\boldsymbol{B}$, $\boldsymbol{F} = \boldsymbol{B}^\dagger\boldsymbol{A}^\dagger$. Show by using $\boldsymbol{A}^\dagger\boldsymbol{A} = \boldsymbol{B}\boldsymbol{B}^\dagger = \boldsymbol{I}$ that $\boldsymbol{F}$ is the generalized inverse of $\boldsymbol{E}$.

Solution. Let $\boldsymbol{A}$ and $\boldsymbol{B}$ have singular value factorizations $\boldsymbol{A} = \boldsymbol{U}_1\boldsymbol{\Sigma}_1\boldsymbol{V}_1^*$ and $\boldsymbol{B} = \boldsymbol{U}_2\boldsymbol{\Sigma}_2\boldsymbol{V}_2^*$. Since $\boldsymbol{A}$ has full column rank and $\boldsymbol{B}$ has full row rank, $\boldsymbol{V}_1$ and $\boldsymbol{U}_2$ are unitary. We have that $\boldsymbol{A}\boldsymbol{B} = \boldsymbol{U}_1\boldsymbol{\Sigma}_1\boldsymbol{V}_1^*\boldsymbol{U}_2\boldsymbol{\Sigma}_2\boldsymbol{V}_2^*$. Now, let $\boldsymbol{U}_3\boldsymbol{\Sigma}_3\boldsymbol{V}_3^*$ be a singular value factorization of $\boldsymbol{\Sigma}_1\boldsymbol{V}_1^*\boldsymbol{U}_2\boldsymbol{\Sigma}_2$. This matrix is nonsingular, $\boldsymbol{U}_3$ and $\boldsymbol{V}_3$ are unitary, and inversion gives that $\boldsymbol{V}_3\boldsymbol{\Sigma}_3^{-1}\boldsymbol{U}_3^* = \boldsymbol{\Sigma}_2^{-1}\boldsymbol{U}_2^*\boldsymbol{V}_1\boldsymbol{\Sigma}_1^{-1}$. We then have that $\boldsymbol{A}\boldsymbol{B} = \boldsymbol{U}_1\boldsymbol{U}_3\boldsymbol{\Sigma}_3\boldsymbol{V}_3^*\boldsymbol{V}_2^* = \boldsymbol{U}_1\boldsymbol{U}_3\boldsymbol{\Sigma}_3(\boldsymbol{V}_2\boldsymbol{V}_3)^*$ is a singular value factorization of $\boldsymbol{A}\boldsymbol{B}$, so that

$$(\boldsymbol{A}\boldsymbol{B})^\dagger = \boldsymbol{V}_2\boldsymbol{V}_3\boldsymbol{\Sigma}_3^{-1}\boldsymbol{U}_3^*\boldsymbol{U}_1^* = \boldsymbol{V}_2\boldsymbol{\Sigma}_2^{-1}\boldsymbol{U}_2^*\boldsymbol{V}_1\boldsymbol{\Sigma}_1^{-1}\boldsymbol{U}_1^* = \boldsymbol{B}^\dagger\boldsymbol{A}^\dagger.$$

Alternatively we can verify the properties from Exercise 9.5. First, when $\boldsymbol{A}$ has linearly independent columns, $\boldsymbol{B}$ has linearly independent rows, it follows immediately from Exercise 9.7 that $\boldsymbol{A}^\dagger\boldsymbol{A} = \boldsymbol{B}\boldsymbol{B}^\dagger = \boldsymbol{I}$. We know also from Exercise 9.5 that $(\boldsymbol{A}\boldsymbol{A}^\dagger)^* = \boldsymbol{A}\boldsymbol{A}^\dagger$ and $(\boldsymbol{B}^\dagger\boldsymbol{B})^* = \boldsymbol{B}^\dagger\boldsymbol{B}$. We now let $\boldsymbol{E} := \boldsymbol{A}\boldsymbol{B}$ and $\boldsymbol{F} := \boldsymbol{B}^\dagger\boldsymbol{A}^\dagger$. Hence we want to show that $\boldsymbol{E}^\dagger = \boldsymbol{F}$, i.e., that $\boldsymbol{F}$ satisfies the four properties

$$\begin{aligned}
\boldsymbol{E}\boldsymbol{F}\boldsymbol{E} &= \boldsymbol{A}\boldsymbol{B}\boldsymbol{B}^\dagger\boldsymbol{A}^\dagger\boldsymbol{A}\boldsymbol{B} = \boldsymbol{A}\boldsymbol{B} = \boldsymbol{E},\\
\boldsymbol{F}\boldsymbol{E}\boldsymbol{F} &= \boldsymbol{B}^\dagger\boldsymbol{A}^\dagger\boldsymbol{A}\boldsymbol{B}\boldsymbol{B}^\dagger\boldsymbol{A}^\dagger = \boldsymbol{B}^\dagger\boldsymbol{A}^\dagger = \boldsymbol{F},\\
(\boldsymbol{F}\boldsymbol{E})^* &= (\boldsymbol{B}^\dagger\boldsymbol{A}^\dagger\boldsymbol{A}\boldsymbol{B})^* = (\boldsymbol{B}^\dagger\boldsymbol{B})^* = \boldsymbol{B}^\dagger\boldsymbol{B} = \boldsymbol{B}^\dagger\boldsymbol{A}^\dagger\boldsymbol{A}\boldsymbol{B} = \boldsymbol{F}\boldsymbol{E},\\
(\boldsymbol{E}\boldsymbol{F})^* &= (\boldsymbol{A}\boldsymbol{B}\boldsymbol{B}^\dagger\boldsymbol{A}^\dagger)^* = (\boldsymbol{A}\boldsymbol{A}^\dagger)^* = \boldsymbol{A}\boldsymbol{A}^\dagger = \boldsymbol{A}\boldsymbol{B}\boldsymbol{B}^\dagger\boldsymbol{A}^\dagger = \boldsymbol{E}\boldsymbol{F}.
\end{aligned}$$

b) Find $A \in \mathbb{R}^{1,2}$, $B \in \mathbb{R}^{2,1}$ such that $(AB)^\dagger \neq B^\dagger A^\dagger$.

Solution. Let $A = u^*$ and $B = v$, where u and v are column vectors. From exercises 9.9 and 9.10 we have that $A^\dagger = u/\|u\|_2^2$, and $B^\dagger = v^*/\|v\|_2^2$. We have that

$$(AB)^\dagger = (u^*v)^\dagger = \frac{1}{u^*v}, \qquad B^\dagger A^\dagger = \frac{v^*u}{\|v\|_2^2\|u\|_2^2}.$$

If these are to be equal we must have that $(u^*v)^2 = \|v\|_2^2\|u\|_2^2$. We must thus have equality in the Cauchy-Schwarz inequality, and this can happen only if u and v are parallel. It is thus enough to find u and v which are not parallel, in order to produce a counterexample.

Exercise 9.14: The generalized inverse of the conjugate transpose

Show that $A^* = A^\dagger$ if and only if all singular values of A are either zero or one.

Solution. Let A have singular value factorization $A = U_1\Sigma_1V_1^*$, so that $A^* = V_1\Sigma_1^*U_1^*$ and $A^\dagger = V_1\Sigma_1^{-1}U_1^*$. Then $A^* = A^\dagger$ if and only if $\Sigma_1^* = \Sigma_1^{-1}$, which, since the singular values are nonnegative, happens precisely when all nonzero singular values of A are one.

Exercise 9.15: Linearly independent columns

Show that if A has rank n then $A(A^*A)^{-1}A^*b$ is the projection of b into $\mathcal{R}(A)$ (cf. Exercise 9.8).

Solution. By Exercise 9.7, if $A \in \mathbb{C}^{m\times n}$ has rank n, then $A^\dagger = (A^*A)^{-1}A^*$. Then $A(A^*A)^{-1}A^*b = AA^\dagger b$, which is the orthogonal projection of b into $\operatorname{span}(A)$ by Theorem 9.6.

Exercise 9.16: Analysis of the general linear system

Consider the linear system $Ax = b$ where $A \in \mathbb{C}^{n\times n}$ has rank $r > 0$ and $b \in \mathbb{C}^n$. Let

$$U^*AV = \begin{bmatrix} \Sigma_1 & 0 \\ 0 & 0 \end{bmatrix}$$

represent a singular value decomposition of A.

Solution. In this exercise, we can write

$$\Sigma = \begin{bmatrix} \Sigma_1 & 0 \\ 0 & 0 \end{bmatrix}, \qquad \Sigma_1 = \operatorname{diag}(\sigma_1, \ldots, \sigma_r), \qquad \sigma_1 \geq \cdots \geq \sigma_r > 0.$$

a) Let $\boldsymbol{c} = [c_1, \ldots, c_n]^* = \boldsymbol{U}^*\boldsymbol{b}$ and $\boldsymbol{y} = [y_1, \ldots, y_n]^* = \boldsymbol{V}^*\boldsymbol{x}$. Show that $\boldsymbol{Ax} = \boldsymbol{b}$ if and only if

$$\begin{bmatrix} \boldsymbol{\Sigma}_1 & \boldsymbol{0} \\ \boldsymbol{0} & \boldsymbol{0} \end{bmatrix} \boldsymbol{y} = \boldsymbol{c}.$$

Solution. As $\boldsymbol{U}$ is unitary, we have $\boldsymbol{U}^*\boldsymbol{U} = \boldsymbol{I}$. We find the following sequence of equivalences.

$$\boldsymbol{Ax} = \boldsymbol{b} \iff \boldsymbol{U\Sigma V}^*\boldsymbol{x} = \boldsymbol{b} \iff \boldsymbol{U}^*\boldsymbol{U\Sigma}(\boldsymbol{V}^*\boldsymbol{x}) = \boldsymbol{U}^*\boldsymbol{b} \iff \boldsymbol{\Sigma y} = \boldsymbol{c},$$

which is what needed to be shown.

b) Show that $\boldsymbol{Ax} = \boldsymbol{b}$ has a solution $\boldsymbol{x}$ if and only if $c_{r+1} = \cdots = c_n = 0$.

Solution. By **a)**, the linear system $\boldsymbol{Ax} = \boldsymbol{b}$ has a solution if and only if the system

$$\begin{bmatrix} \boldsymbol{\Sigma}_1 & \boldsymbol{0} \\ \boldsymbol{0} & \boldsymbol{0} \end{bmatrix} \boldsymbol{y} = \begin{bmatrix} \sigma_1 y_1 \\ \vdots \\ \sigma_r y_r \\ 0 \\ \vdots \\ 0 \end{bmatrix} = \begin{bmatrix} c_1 \\ \vdots \\ c_r \\ c_{r+1} \\ \vdots \\ c_n \end{bmatrix} = \boldsymbol{c}$$

has a solution $\boldsymbol{y}$. Since $\sigma_1, \ldots, \sigma_r \neq 0$, this system has a solution if and only if $c_{r+1} = \cdots = c_n = 0$. We conclude that $\boldsymbol{Ax} = \boldsymbol{b}$ has a solution if and only if $c_{r+1} = \cdots = c_n = 0$.

c) Deduce that a linear system $\boldsymbol{Ax} = \boldsymbol{b}$ has either no solution, one solution or infinitely many solutions.

Solution. By **a)**, the linear system $\boldsymbol{Ax} = \boldsymbol{b}$ has a solution if and only if the system $\boldsymbol{\Sigma y} = \boldsymbol{c}$ has a solution. Hence we have the following three cases.

$r = n$
: Here $y_i = c_i/\sigma_i$ for $i = 1, \ldots, n$ provides the only solution to the system $\boldsymbol{\Sigma y} = \boldsymbol{b}$, and therefore $\boldsymbol{x} = \boldsymbol{Vy}$ is the only solution to $\boldsymbol{Ax} = \boldsymbol{b}$. It follows that the system has exactly one solution.

$r < n$, $c_i = 0$ for $i = r+1, \ldots, n$
: Here each solution $\boldsymbol{y}$ must satisfy $y_i = c_i/\sigma_i$ for $i = 1, \ldots, r$. The remaining $y_{r+1}, \ldots, y_n$, however, can be chosen arbitrarily. Hence we have infinitely many solutions to $\boldsymbol{\Sigma y} = \boldsymbol{b}$ as well as to $\boldsymbol{Ax} = \boldsymbol{b}$.

$r < n$, $c_i \neq 0$ for some i with $r + 1 \leq i \leq n$
: In this case it is impossible to find a $\boldsymbol{y}$ that satisfies $\boldsymbol{\Sigma y} = \boldsymbol{b}$, and therefore the system $\boldsymbol{Ax} = \boldsymbol{b}$ has no solution at all.

Exercise 9.17: Fredholm's alternative

For any $\boldsymbol{A} \in \mathbb{C}^{m\times n}$, $\boldsymbol{b} \in \mathbb{C}^n$ show that one and only one of the following systems has a solution

$$(1) \quad \boldsymbol{A}\boldsymbol{x} = \boldsymbol{b}, \qquad\qquad (2) \quad \boldsymbol{A}^*\boldsymbol{y} = \boldsymbol{0},\ \boldsymbol{y}^*\boldsymbol{b} \neq 0.$$

In other words either $\boldsymbol{b} \in \mathcal{R}(\boldsymbol{A})$, or we can find $\boldsymbol{y} \in \mathcal{N}(\boldsymbol{A}^*)$ such that $\boldsymbol{y}^*\boldsymbol{b} \neq 0$. This is called *Fredholm's alternative*.

Solution. Suppose that the system $\boldsymbol{A}\boldsymbol{x} = \boldsymbol{b}$ has a solution, i.e., $\boldsymbol{b} \in \operatorname{span}(\boldsymbol{A})$. Suppose in addition that $\boldsymbol{A}^*\boldsymbol{y} = \boldsymbol{0}$ has a solution, i.e., $\boldsymbol{y} \in \ker(\boldsymbol{A}^*)$. Since $\operatorname{span}(\boldsymbol{A})^\perp = \ker(\boldsymbol{A}^*)$, one has $\langle \boldsymbol{y}, \boldsymbol{b}\rangle = \boldsymbol{y}^*\boldsymbol{b} = 0$. Thus if the system $\boldsymbol{A}\boldsymbol{x} = \boldsymbol{b}$ has a solution, then we can not find a solution to $\boldsymbol{A}^*\boldsymbol{y} = \boldsymbol{0}$, $\boldsymbol{y}^*\boldsymbol{b} \neq 0$. Conversely if $\boldsymbol{y} \in \ker(\boldsymbol{A}^*)$ and $\boldsymbol{y}^*\boldsymbol{b} \neq 0$, then $\boldsymbol{b} \notin \ker(\boldsymbol{A}^*)^\perp = \operatorname{span}(\boldsymbol{A})$, implying that the system $\boldsymbol{A}\boldsymbol{x} = \boldsymbol{b}$ does not have a solution.

Exercise 9.18: SVD (Exam exercise 2017-2)

Let $\boldsymbol{A} \in \mathbb{C}^{m\times n}$, with $m \geq n$, be a matrix on the form

$$\boldsymbol{A} = \begin{bmatrix} \boldsymbol{B} \\ \boldsymbol{C} \end{bmatrix}$$

where $\boldsymbol{B}$ is a non-singular $n \times n$ matrix and $\boldsymbol{C}$ is in $\mathbb{C}^{(m-n)\times n}$. Let $\boldsymbol{A}^\dagger$ denote the pseudoinverse of $\boldsymbol{A}$. Show that $\|\boldsymbol{A}^\dagger\|_2 \leq \|\boldsymbol{B}^{-1}\|_2$.

Solution. Since $\boldsymbol{B}$ is nonsingular the matrix $\boldsymbol{A}$ has rank n and singular values $\sigma_1 \geq \cdots \geq \sigma_n > 0$. The SVD of $\boldsymbol{A}$ can be written

$$\boldsymbol{A} = \boldsymbol{U} \begin{bmatrix} \boldsymbol{\Sigma}_1 \\ \boldsymbol{0} \end{bmatrix} \boldsymbol{V}^*,$$

where $\boldsymbol{U} \in \mathbb{C}^{m\times m}$, $\boldsymbol{V} \in \mathbb{C}^{n\times n}$ are unitary and $\boldsymbol{\Sigma}_1 := \operatorname{diag}(\sigma_1, \ldots, \sigma_n)$. Moreover, by the Courant-Fischer theorem for singular values (cf. (9.17))

$$\sigma_n = \min_{\substack{\boldsymbol{x}\in\mathbb{C}^n \\ \|\boldsymbol{x}\|_2=1}} \|\boldsymbol{A}\boldsymbol{x}\|_2.$$

But then, using also (8.19)

$$\boldsymbol{A}^\dagger = \boldsymbol{V}[\boldsymbol{\Sigma}_1^{-1}, \boldsymbol{0}]\boldsymbol{U}^*, \qquad \|\boldsymbol{B}^{-1}\|_2 = \frac{1}{\sigma_{\boldsymbol{B}}}, \tag{9.i}$$

where $\sigma_{\boldsymbol{B}}$ is the smallest singular value of $\boldsymbol{B}$. Now, for any $\boldsymbol{x} \in \mathbb{C}^n$

$$\|\boldsymbol{A}\boldsymbol{x}\|_2^2 = \boldsymbol{x}^*\boldsymbol{A}^*\boldsymbol{A}\boldsymbol{x} = \boldsymbol{x}^*(\boldsymbol{B}^*\boldsymbol{B} + \boldsymbol{C}^*\boldsymbol{C})\boldsymbol{x} \geq \boldsymbol{x}^*\boldsymbol{B}^*\boldsymbol{B}\boldsymbol{x} = \|\boldsymbol{B}\boldsymbol{x}\|_2^2.$$

But then $\|\boldsymbol{A}\boldsymbol{x}\|_2 \geq \|\boldsymbol{B}\boldsymbol{x}\|_2 \geq \sigma_{\boldsymbol{B}}$ and since $\boldsymbol{x}$ is arbitrary we obtain $\sigma_n \geq \sigma_{\boldsymbol{B}}$.

For $\boldsymbol{x} \in \mathbb{C}^m$, using (9.i)

$$\begin{aligned}
\|\boldsymbol{A}^\dagger \boldsymbol{x}\|_2 &= \|\mathbf{V}[\boldsymbol{\Sigma}^{-1}, \mathbf{0}]\mathbf{U}^*\boldsymbol{x}\|_2 \\
&= \|[\boldsymbol{\Sigma}^{-1}, \mathbf{0}]\mathbf{U}^*\boldsymbol{x}\|_2 && (\mathbf{V} \text{ is unitary}) \\
&\leq \|[\boldsymbol{\Sigma}^{-1}, \mathbf{0}]\|_2 \|\mathbf{U}^*\boldsymbol{x}\|_2 && (\|\cdot\|_2 \text{ subordinate}) \\
&= \|\boldsymbol{\Sigma}^{-1}\|_2 \|\boldsymbol{x}\|_2 && (\mathbf{U}^* \text{ unitary}) \\
&= \frac{1}{\sigma_n}\|\boldsymbol{x}\|_2 \leq \frac{1}{\sigma_{\boldsymbol{B}}}\|\boldsymbol{x}\|_2 = \|\boldsymbol{B}^{-1}\|_2 \|\boldsymbol{x}\|_2.
\end{aligned}$$

Taking the maximum over all $\boldsymbol{x}$ with $\|\boldsymbol{x}\|_2 = 1$ yields $\|\boldsymbol{A}^\dagger\|_2 \leq \|\boldsymbol{B}^{-1}\|_2$.

Exercises section 9.4

Exercise 9.19: Condition number

Let

$$\boldsymbol{A} = \begin{bmatrix} 1 & 2 \\ 1 & 1 \\ 1 & 1 \end{bmatrix}, \quad \boldsymbol{b} = \begin{bmatrix} b_1 \\ b_2 \\ b_3 \end{bmatrix}.$$

a) Determine the projections $\boldsymbol{b}_1$ and $\boldsymbol{b}_2$ of $\boldsymbol{b}$ on $\mathcal{R}(\boldsymbol{A})$ and $\mathcal{N}(\boldsymbol{A}^*)$.

Solution. By Exercise 9.7, the pseudoinverse of $\boldsymbol{A}$ is

$$\boldsymbol{A}^\dagger = (\boldsymbol{A}^{\mathrm{T}}\boldsymbol{A})^{-1}\boldsymbol{A}^{\mathrm{T}} = \begin{bmatrix} -1 & 1 & 1 \\ 1 & -\frac{1}{2} & -\frac{1}{2} \end{bmatrix}.$$

Theorem 9.6 tells us that the orthogonal projection of $\boldsymbol{b}$ into $\operatorname{span}(\boldsymbol{A})$ is

$$\boldsymbol{b}_1 := \boldsymbol{A}\boldsymbol{A}^\dagger \boldsymbol{b} = \begin{bmatrix} 1 & 0 & 0 \\ 0 & \frac{1}{2} & \frac{1}{2} \\ 0 & \frac{1}{2} & \frac{1}{2} \end{bmatrix} \begin{bmatrix} b_1 \\ b_2 \\ b_3 \end{bmatrix} = \frac{1}{2} \begin{bmatrix} 2b_1 \\ b_2 + b_3 \\ b_2 + b_3 \end{bmatrix},$$

so that the orthogonal projection of $\boldsymbol{b}$ into $\ker(\boldsymbol{A}^{\mathrm{T}})$ is

$$\boldsymbol{b}_2 := (\boldsymbol{I} - \boldsymbol{A}\boldsymbol{A}^\dagger)\boldsymbol{b} = \begin{bmatrix} 0 & 0 & 0 \\ 0 & \frac{1}{2} & -\frac{1}{2} \\ 0 & -\frac{1}{2} & \frac{1}{2} \end{bmatrix} \begin{bmatrix} b_1 \\ b_2 \\ b_3 \end{bmatrix} = \frac{1}{2} \begin{bmatrix} 0 \\ b_2 - b_3 \\ b_3 - b_2 \end{bmatrix},$$

where we used that $\boldsymbol{b} = \boldsymbol{b}_1 + \boldsymbol{b}_2$.

b) Compute $K(\boldsymbol{A}) = \|\boldsymbol{A}\|_2 \|\boldsymbol{A}^\dagger\|_2$.

Solution. By Theorem 8.3, the 2-norms $\|\boldsymbol{A}\|_2$ and $\|\boldsymbol{A}^\dagger\|_2$ can be found by computing the largest singular values of the matrices $\boldsymbol{A}$ and $\boldsymbol{A}^\dagger$. The largest singular value σ_1 of $\boldsymbol{A}$ is the square root of the largest eigenvalue λ_1 of $\boldsymbol{A}^{\mathrm{T}}\boldsymbol{A}$, which satisfies

$$0 = \det(\boldsymbol{A}^{\mathrm{T}}\boldsymbol{A} - \lambda_1 \boldsymbol{I}) = \det \begin{bmatrix} 3-\lambda_1 & 4 \\ 4 & 6-\lambda_1 \end{bmatrix} = \lambda_1^2 - 9\lambda_1 + 2.$$

It follows that $\sigma_1 = \frac{1}{2}\sqrt{2}\sqrt{9+\sqrt{73}}$. Similarly, the largest singular value σ_2 of $\boldsymbol{A}^\dagger$ is the square root of the largest eigenvalue λ_2 of $\boldsymbol{A}^{\dagger\mathrm{T}}\boldsymbol{A}^\dagger$, which satisfies

$$0 = \det(\boldsymbol{A}^{\dagger\mathrm{T}}\boldsymbol{A}^\dagger - \lambda_2 \boldsymbol{I}) = \det\left(\frac{1}{4}\begin{bmatrix} 8 & -6 & -6 \\ -6 & 5 & 5 \\ -6 & 5 & 5 \end{bmatrix} - \lambda_2 \boldsymbol{I}\right)$$
$$= -\frac{1}{2}\lambda_2\left(2\lambda_2^2 - 9\lambda_2 + 1\right).$$

Alternatively, we could have used that the largest singular value of $\boldsymbol{A}^\dagger$ is the inverse of the smallest singular value of $\boldsymbol{A}$ (this follows from the singular value factorization). It follows that $\sigma_2 = \frac{1}{2}\sqrt{9+\sqrt{73}} = \sqrt{2}/\sqrt{9-\sqrt{73}}$. We conclude

$$K(\boldsymbol{A}) = \|\boldsymbol{A}\|_2 \cdot \|\boldsymbol{A}^\dagger\|_2 = \sqrt{\frac{9+\sqrt{73}}{9-\sqrt{73}}} = \frac{1}{2\sqrt{2}}\left(9+\sqrt{73}\right) \approx 6.203.$$

Exercise 9.20: Equality in perturbation bound

Let $\boldsymbol{A} \in \mathbb{C}^{m\times n}$. Suppose $\boldsymbol{y}_{\boldsymbol{A}}$ and $\boldsymbol{y}_{\boldsymbol{A}^\dagger}$ are vectors with $\|\boldsymbol{y}_{\boldsymbol{A}}\| = \|\boldsymbol{y}_{\boldsymbol{A}^\dagger}\| = 1$ and $\|\boldsymbol{A}\| = \|\boldsymbol{A}\boldsymbol{y}_{\boldsymbol{A}}\|$ and $\|\boldsymbol{A}^\dagger\| = \|\boldsymbol{A}^\dagger\boldsymbol{y}_{\boldsymbol{A}^\dagger}\|$.

a) Show that we have equality to the right in (9.13) if $\boldsymbol{b} = \boldsymbol{A}\boldsymbol{y}_{\boldsymbol{A}}$, $\boldsymbol{e}_1 = \boldsymbol{y}_{\boldsymbol{A}^\dagger}$.

Solution. By assumption on $\boldsymbol{y}_{\boldsymbol{A}}$ and $\boldsymbol{y}_{\boldsymbol{A}^\dagger}$,

$$\|\boldsymbol{A}\boldsymbol{y}_{\boldsymbol{A}}\| = \|\boldsymbol{A}\| = \|\boldsymbol{A}\|\|\boldsymbol{y}_{\boldsymbol{A}}\|, \qquad \|\boldsymbol{A}^\dagger\boldsymbol{y}_{\boldsymbol{A}^\dagger}\| = \|\boldsymbol{A}^\dagger\| = \|\boldsymbol{A}^\dagger\|\|\boldsymbol{y}_{\boldsymbol{A}^\dagger}\|.$$

We have here that $\boldsymbol{e}_1 = \boldsymbol{y}_{\boldsymbol{A}^\dagger}$, and since $\boldsymbol{A}$ has linearly independent columns, $\boldsymbol{x} = \boldsymbol{y}_{\boldsymbol{A}}$.

We have equality in the right hand side if and only if

$$\|\boldsymbol{A}\boldsymbol{x}\| = \|\boldsymbol{A}\|\|\boldsymbol{x}\| \qquad \text{and} \qquad \|\boldsymbol{A}^\dagger\boldsymbol{e}_1\| = \|\boldsymbol{A}^\dagger\|\|\boldsymbol{e}_1\|.$$

Combining this with the observations above, the result follows.

b) Show that we have equality to the left if we switch $\boldsymbol{b}$ and $\boldsymbol{e}_1$ in **a)**.

Solution. We have equality in the left hand side if and only if $\|\boldsymbol{A}\|\|\boldsymbol{x}-\boldsymbol{y}\| = \|\boldsymbol{e}_1\|$, and $\|\boldsymbol{x}\| = \|\boldsymbol{A}^\dagger\|\|\boldsymbol{b}_1\|$. This is the same as $\|\boldsymbol{A}\|\|\boldsymbol{A}^\dagger\boldsymbol{e}_1\| = \|\boldsymbol{e}_1\|$, and $\|\boldsymbol{x}\| = \|\boldsymbol{A}^\dagger\|\|\boldsymbol{A}\boldsymbol{x}\|$. This is the same as

$$\|\boldsymbol{e}_1\| = \|\boldsymbol{A}\|\|\boldsymbol{A}^\dagger\boldsymbol{e}_1\| \qquad \text{and} \qquad \|\boldsymbol{A}^\dagger\boldsymbol{b}\| = \|\boldsymbol{A}^\dagger\|\|\boldsymbol{b}\|.$$

The first corresponding statement in **a)** can also be written as $\|\boldsymbol{b}\| = \|\boldsymbol{A}\|\|\boldsymbol{A}^\dagger\boldsymbol{b}\|$, so that the statements in **a)** can be written as

$$\|\boldsymbol{b}\| = \|\boldsymbol{A}\|\|\boldsymbol{A}^\dagger\boldsymbol{b}\| \qquad \text{and} \qquad \|\boldsymbol{A}^\dagger\boldsymbol{e}_1\| = \|\boldsymbol{A}^\dagger\|\|\boldsymbol{e}_1\|.$$

We now see that $\boldsymbol{b}$ and $\boldsymbol{e}_1$ simply have swapped roles.

> **c)** Let $\boldsymbol{A}$ be as in Example 9.7. Find extremal $\boldsymbol{b}$ and $\boldsymbol{e}$ when the l_∞ norm is used.
>
> This generalizes the sharpness results in Exercise 8.18. For if $m = n$ and $\boldsymbol{A}$ is nonsingular then $\boldsymbol{A}^\dagger = \boldsymbol{A}^{-1}$ and $\boldsymbol{e}_1 = \boldsymbol{e}$.

Solution. It is straightforward to check that $\boldsymbol{y}_{\boldsymbol{A}} = [1,1]^{\mathrm{T}}$ and $\boldsymbol{y}_{\boldsymbol{A}^\dagger} = [1,-1,0]^{\mathrm{T}}$ are vectors with infinity norm one, which satisfy the conditions in **a)** (they are not unique, however).

For right hand side equality, **a)** now gives

$$\boldsymbol{b} = \boldsymbol{A}\boldsymbol{y}_{\boldsymbol{A}} = \begin{bmatrix}1&1\\0&1\\0&0\end{bmatrix}\begin{bmatrix}1\\1\end{bmatrix} = \begin{bmatrix}2\\1\\0\end{bmatrix}.$$

One particular solution for $\boldsymbol{e}$ is

$$\boldsymbol{A}^\dagger\boldsymbol{e}_1 = \begin{bmatrix}1&-1&0\\0&1&0\end{bmatrix}\begin{bmatrix}1\\-1\\0\end{bmatrix} = \begin{bmatrix}2\\-1\end{bmatrix}.$$

For left hand side equality, **b)** gives $\boldsymbol{b} = \boldsymbol{y}_{\boldsymbol{A}^\dagger} = [1,-1,0]^{\mathrm{T}}$. One particular solution for $\boldsymbol{e}$ is

$$\boldsymbol{A}^\dagger\boldsymbol{e}_1 = \boldsymbol{A}^\dagger\boldsymbol{A}\begin{bmatrix}1\\1\end{bmatrix} = \begin{bmatrix}1&-1&0\\0&1&0\end{bmatrix}\begin{bmatrix}1&1\\0&1\\0&0\end{bmatrix}\begin{bmatrix}1\\1\end{bmatrix} = \begin{bmatrix}1\\1\end{bmatrix}.$$

Exercise 9.21: Problem using normal equations

Consider the least squares problems where

$$\boldsymbol{A} = \begin{bmatrix} 1 & 1 \\ 1 & 1 \\ 1 & 1+\epsilon \end{bmatrix}, \qquad \boldsymbol{b} = \begin{bmatrix} 2 \\ 3 \\ 2 \end{bmatrix}, \qquad \epsilon \in \mathbb{R}.$$

a) Find the normal equations and the exact least squares solution.

Solution. Let $\boldsymbol{A}$, $\boldsymbol{b}$, and ϵ be as in the exercise. The normal equations $\boldsymbol{A}^{\mathrm{T}}\boldsymbol{A}\boldsymbol{x} = \boldsymbol{A}^{\mathrm{T}}\boldsymbol{b}$ are then

$$\begin{bmatrix} 3 & 3+\epsilon \\ 3+\epsilon & (\epsilon+1)^2+2 \end{bmatrix} \begin{bmatrix} x_1 \\ x_2 \end{bmatrix} = \begin{bmatrix} 7 \\ 7+2\epsilon \end{bmatrix}.$$

If $\epsilon \neq 0$, inverting the matrix $\boldsymbol{A}^{\mathrm{T}}\boldsymbol{A}$ yields the unique solution

$$\begin{bmatrix} x_1 \\ x_2 \end{bmatrix} = \frac{1}{2\epsilon^2} \begin{bmatrix} (\epsilon+1)^2+2 & -3-\epsilon \\ -3-\epsilon & 3 \end{bmatrix} \begin{bmatrix} 7 \\ 7+2\epsilon \end{bmatrix} = \begin{bmatrix} \frac{5}{2}+\frac{1}{2\epsilon} \\ -\frac{1}{2\epsilon} \end{bmatrix}.$$

If $\epsilon = 0$, on the other hand, then any vector $\boldsymbol{x} = [x_1, x_2]^{\mathrm{T}}$ with $x_1 + x_2 = 7/3$ is a solution.

b) Suppose ϵ is small and we replace the $(2,2)$ entry $3 + 2\epsilon + \epsilon^2$ in $\boldsymbol{A}^*\boldsymbol{A}$ by $3 + 2\epsilon$. (This will be done in a computer if $\epsilon < \sqrt{u}$, u being the round-off unit). For example, if $u = 10^{-16}$ then $\sqrt{u} = 10^{-8}$. Solve $\boldsymbol{A}^*\boldsymbol{A}\boldsymbol{x} = \boldsymbol{A}^*\boldsymbol{b}$ for $\boldsymbol{x}$ and compare with the $\boldsymbol{x}$ found in **a)** (we will get a much more accurate result using the QR factorization or the singular value decomposition on this problem).

Solution. For $\epsilon = 0$, we get the same solution as in **a)**. For $\epsilon \neq 0$, however, the solution to the system

$$\begin{bmatrix} 3 & 3+\epsilon \\ 3+\epsilon & 3+2\epsilon \end{bmatrix} \begin{bmatrix} x_1 \\ x_2 \end{bmatrix} = \begin{bmatrix} 7 \\ 7+2\epsilon \end{bmatrix}$$

is

$$\begin{bmatrix} x_1' \\ x_2' \end{bmatrix} = -\frac{1}{\epsilon^2} \begin{bmatrix} 3+2\epsilon & -3-\epsilon \\ -3-\epsilon & 3 \end{bmatrix} \begin{bmatrix} 7 \\ 7+2\epsilon \end{bmatrix} = \begin{bmatrix} 2-\frac{1}{\epsilon} \\ \frac{1}{\epsilon} \end{bmatrix}.$$

We can compare this to the solution of **a)** by comparing the residuals,

$$\begin{aligned} \left\| \boldsymbol{A} \begin{bmatrix} \frac{5}{2}+\frac{1}{2\epsilon} \\ -\frac{1}{2\epsilon} \end{bmatrix} - \boldsymbol{b} \right\|_2 &= \left\| \begin{bmatrix} \frac{1}{2} \\ -\frac{1}{2} \\ 0 \end{bmatrix} \right\|_2 = \frac{1}{\sqrt{2}} \\ &\leq \sqrt{2} = \left\| \begin{bmatrix} 0 \\ -1 \\ 1 \end{bmatrix} \right\|_2 = \left\| \boldsymbol{A} \begin{bmatrix} 2-\frac{1}{\epsilon} \\ \frac{1}{\epsilon} \end{bmatrix} - \boldsymbol{b} \right\|_2, \end{aligned}$$

which shows that the solution from **a)** is more accurate.

Exercises section 9.5

Exercise 9.22: Singular values perturbation (Exam exercise 1980-2) ?

Let $\boldsymbol{A}(\epsilon) \in \mathbb{R}^{n\times n}$ be bidiagonal with $a_{i,j} = 0$ for $i, j = 1, \ldots, n$ and $j \neq i, i+1$. Moreover, for some $1 \leq k \leq n-1$ we have $a_{k,k+1} = \epsilon \in \mathbb{R}$. Show that

$$|\sigma_i(\epsilon) - \sigma_i(0)| \leq |\epsilon|, \qquad i = 1, \ldots, n,$$

where $\sigma_i(\epsilon)$, $i = 1, \ldots, n$ are the singular values of $\boldsymbol{A}(\epsilon)$.

Solution. Let $\boldsymbol{E} := \boldsymbol{A}(\epsilon) - \boldsymbol{A}(0)$ and $\boldsymbol{F} := \boldsymbol{E}^{\mathrm{T}}\boldsymbol{E}$. According to Theorem 9.11 we have

$$|\sigma_i(\epsilon) - \sigma_i(0)| \leq \|\boldsymbol{E}\|_2, \qquad i = 1, \ldots, n.$$

Now $f_{k+1,k+1} = \epsilon^2$, and all other elements of $\boldsymbol{F}$ are zero. It follows that the spectral radius of $\boldsymbol{F}$ is ϵ^2 and hence $\|\boldsymbol{E}\|_2 = |\epsilon|$.

As an alternative proof, we note that only one element of $\boldsymbol{E}$ is nonzero and therefore $\|\boldsymbol{E}\|_1 = \|\boldsymbol{E}\|_\infty = |\epsilon|$. But then $\|\boldsymbol{E}\|_2 = \sqrt{\|\boldsymbol{E}\|_1\|\boldsymbol{E}\|_\infty} \leq |\epsilon|$.

Chapter 10
The Kronecker Product

Exercises sections 10.1 and 10.2

Exercise 10.1: 4×4 **Poisson matrix**

Write down the Poisson matrix for $m = 2$ and show that it is strictly diagonally dominant.

Solution. For $m = 2$, the Poisson matrix $\boldsymbol{A}$ is the $2^2 \times 2^2$ matrix given by

$$\begin{bmatrix} 4 & -1 & -1 & 0 \\ -1 & 4 & 0 & -1 \\ -1 & 0 & 4 & -1 \\ 0 & -1 & -1 & 4 \end{bmatrix}.$$

In every row i, one has

$$|a_{ii}| = 4 > 2 = |-1| + |-1| + |0| = \sum_{j \neq i} |a_{ij}|.$$

In other words, $\boldsymbol{A}$ is strictly diagonally dominant.

Exercise 10.2: Properties of Kronecker products

Prove (10.13), i.e., that

$$\begin{aligned} (\lambda \boldsymbol{A}) \otimes (\mu \boldsymbol{B}) &= \lambda\mu(\boldsymbol{A} \otimes \boldsymbol{B}), \\ (\boldsymbol{A}_1 + \boldsymbol{A}_2) \otimes \boldsymbol{B} &= \boldsymbol{A}_1 \otimes \boldsymbol{B} + \boldsymbol{A}_2 \otimes \boldsymbol{B}, \\ \boldsymbol{A} \otimes (\boldsymbol{B}_1 + \boldsymbol{B}_2) &= \boldsymbol{A} \otimes \boldsymbol{B}_1 + \boldsymbol{A} \otimes \boldsymbol{B}_2, \\ (\boldsymbol{A} \otimes \boldsymbol{B}) \otimes \boldsymbol{C} &= \boldsymbol{A} \otimes (\boldsymbol{B} \otimes \boldsymbol{C}). \end{aligned}$$

T. Lyche et al., *Exercises in Numerical Linear Algebra and Matrix Factorizations*, Texts in Computational Science and Engineering 23, https://doi.org/10.1007/978-3-030-59789-4_10

Solution. Let be given matrices $\boldsymbol{A}, \boldsymbol{A}_1, \boldsymbol{A}_2 \in \mathbb{R}^{p \times q}$, $\boldsymbol{B}, \boldsymbol{B}_1, \boldsymbol{B}_2 \in \mathbb{R}^{r \times s}$, and $\boldsymbol{C} \in \mathbb{R}^{t \times u}$. Then $(\lambda \boldsymbol{A}) \otimes (\mu \boldsymbol{B}) = \lambda\mu(\boldsymbol{A} \otimes \boldsymbol{B})$ by definition of the Kronecker product and since

$$\begin{bmatrix} (\lambda\boldsymbol{A})\mu b_{11} & (\lambda\boldsymbol{A})\mu b_{12} & \cdots & (\lambda\boldsymbol{A})\mu b_{1s} \\ (\lambda\boldsymbol{A})\mu b_{21} & (\lambda\boldsymbol{A})\mu b_{22} & \cdots & (\lambda\boldsymbol{A})\mu b_{2s} \\ \vdots & \vdots & \ddots & \vdots \\ (\lambda\boldsymbol{A})\mu b_{r1} & (\lambda\boldsymbol{A})\mu b_{r2} & \cdots & (\lambda\boldsymbol{A})\mu b_{rs} \end{bmatrix} = \lambda\mu \begin{bmatrix} \boldsymbol{A}b_{11} & \boldsymbol{A}b_{12} & \cdots & \boldsymbol{A}b_{1s} \\ \boldsymbol{A}b_{21} & \boldsymbol{A}b_{22} & \cdots & \boldsymbol{A}b_{2s} \\ \vdots & \vdots & \ddots & \vdots \\ \boldsymbol{A}b_{r1} & \boldsymbol{A}b_{r2} & \cdots & \boldsymbol{A}b_{rs} \end{bmatrix}.$$

The identity $(\boldsymbol{A}_1 + \boldsymbol{A}_2) \otimes \boldsymbol{B} = (\boldsymbol{A}_1 \otimes \boldsymbol{B}) + (\boldsymbol{A}_2 \otimes \boldsymbol{B})$ follows from

$$\begin{aligned} &\begin{bmatrix} (\boldsymbol{A}_1 + \boldsymbol{A}_2)b_{11} & (\boldsymbol{A}_1 + \boldsymbol{A}_2)b_{12} & \cdots & (\boldsymbol{A}_1 + \boldsymbol{A}_2)b_{1s} \\ (\boldsymbol{A}_1 + \boldsymbol{A}_2)b_{21} & (\boldsymbol{A}_1 + \boldsymbol{A}_2)b_{22} & \cdots & (\boldsymbol{A}_1 + \boldsymbol{A}_2)b_{2s} \\ \vdots & \vdots & \ddots & \vdots \\ (\boldsymbol{A}_1 + \boldsymbol{A}_2)b_{r1} & (\boldsymbol{A}_1 + \boldsymbol{A}_2)b_{r2} & \cdots & (\boldsymbol{A}_1 + \boldsymbol{A}_2)b_{rs} \end{bmatrix} \\ = &\begin{bmatrix} \boldsymbol{A}_1 b_{11} + \boldsymbol{A}_2 b_{11} & \boldsymbol{A}_1 b_{12} + \boldsymbol{A}_2 b_{12} & \cdots & \boldsymbol{A}_1 b_{1s} + \boldsymbol{A}_2 b_{1s} \\ \boldsymbol{A}_1 b_{21} + \boldsymbol{A}_2 b_{21} & \boldsymbol{A}_1 b_{22} + \boldsymbol{A}_2 b_{22} & \cdots & \boldsymbol{A}_1 b_{2s} + \boldsymbol{A}_2 b_{2s} \\ \vdots & \vdots & \ddots & \vdots \\ \boldsymbol{A}_1 b_{r1} + \boldsymbol{A}_2 b_{r1} & \boldsymbol{A}_1 b_{r2} + \boldsymbol{A}_2 b_{r2} & \cdots & \boldsymbol{A}_1 b_{rs} + \boldsymbol{A}_2 b_{rs} \end{bmatrix} \\ = &\begin{bmatrix} \boldsymbol{A}_1 b_{11} & \boldsymbol{A}_1 b_{12} & \cdots & \boldsymbol{A}_1 b_{1s} \\ \boldsymbol{A}_1 b_{21} & \boldsymbol{A}_1 b_{22} & \cdots & \boldsymbol{A}_1 b_{2s} \\ \vdots & \vdots & \ddots & \vdots \\ \boldsymbol{A}_1 b_{r1} & \boldsymbol{A}_1 b_{r2} & \cdots & \boldsymbol{A}_1 b_{rs} \end{bmatrix} + \begin{bmatrix} \boldsymbol{A}_2 b_{11} & \boldsymbol{A}_2 b_{12} & \cdots & \boldsymbol{A}_2 b_{1s} \\ \boldsymbol{A}_2 b_{21} & \boldsymbol{A}_2 b_{22} & \cdots & \boldsymbol{A}_2 b_{2s} \\ \vdots & \vdots & \ddots & \vdots \\ \boldsymbol{A}_2 b_{r1} & \boldsymbol{A}_2 b_{r2} & \cdots & \boldsymbol{A}_2 b_{rs} \end{bmatrix}. \end{aligned}$$

A similar argument proves $\boldsymbol{A} \otimes (\boldsymbol{B}_1 + \boldsymbol{B}_2) = (\boldsymbol{A} \otimes \boldsymbol{B}_1) + (\boldsymbol{A} \otimes \boldsymbol{B}_2)$, and therefore the bilinearity of the Kronecker product. The associativity $(\boldsymbol{A} \otimes \boldsymbol{B}) \otimes \boldsymbol{C} = \boldsymbol{A} \otimes (\boldsymbol{B} \otimes \boldsymbol{C})$ follows from

$$\begin{bmatrix} \boldsymbol{A}b_{11} & \cdots & \boldsymbol{A}b_{1s} \\ \vdots & & \vdots \\ \boldsymbol{A}b_{r1} & \cdots & \boldsymbol{A}b_{rs} \end{bmatrix} \otimes \boldsymbol{C} = \left[\begin{array}{ccc|c|ccc} \boldsymbol{A}b_{11}c_{11} & \cdots & \boldsymbol{A}b_{1s}c_{11} & & \boldsymbol{A}b_{11}c_{1u} & \cdots & \boldsymbol{A}b_{1s}c_{1u} \\ \vdots & & \vdots & \cdots & \vdots & & \vdots \\ \boldsymbol{A}b_{r1}c_{11} & \cdots & \boldsymbol{A}b_{rs}c_{11} & & \boldsymbol{A}b_{r1}c_{1u} & \cdots & \boldsymbol{A}b_{rs}c_{1u} \\ \hline & \vdots & & & & \vdots & \\ \hline \boldsymbol{A}b_{11}c_{t1} & \cdots & \boldsymbol{A}b_{1s}c_{t1} & & \boldsymbol{A}b_{11}c_{tu} & \cdots & \boldsymbol{A}b_{1s}c_{tu} \\ \vdots & & \vdots & \cdots & \vdots & & \vdots \\ \boldsymbol{A}b_{r1}c_{t1} & \cdots & \boldsymbol{A}b_{rs}c_{t1} & & \boldsymbol{A}b_{r1}c_{tu} & \cdots & \boldsymbol{A}b_{rs}c_{tu} \end{array}\right]$$

$$= A \otimes \left[\begin{array}{ccc|c|ccc} b_{11}c_{11} & \cdots & b_{1s}c_{11} & & b_{11}c_{1u} & \cdots & b_{1s}c_{1u} \\ \vdots & & \vdots & \cdots & \vdots & & \vdots \\ b_{r1}c_{11} & \cdots & b_{rs}c_{11} & & b_{r1}c_{1u} & \cdots & b_{rs}c_{1u} \\ \hline & \vdots & & & & \vdots & \\ \hline b_{11}c_{t1} & \cdots & b_{1s}c_{t1} & & b_{11}c_{tu} & \cdots & b_{1s}c_{tu} \\ \vdots & & \vdots & \cdots & \vdots & & \vdots \\ b_{r1}c_{t1} & \cdots & b_{rs}c_{t1} & & b_{r1}c_{tu} & \cdots & b_{rs}c_{tu} \end{array}\right] = A \otimes \begin{bmatrix} Bc_{11} & \cdots & Bc_{1u} \\ \vdots & & \vdots \\ Bc_{t1} & \cdots & Bc_{tu} \end{bmatrix}.$$

Exercise 10.3: Eigenpairs of Kronecker products (Exam exercise 2008-3)

?

Let $A, B \in \mathbb{R}^{n \times n}$. Show that the eigenvalues of the Kronecker product $A \otimes B$ are products of the eigenvalues of A and B and that the eigenvectors of $A \otimes B$ are Kronecker products of the eigenvectors of A and B.

Solution. Suppose $Au = \lambda u$ and $Bv = \mu v$. Then

$$\begin{aligned} (A \otimes B)(u \otimes v) &= \begin{bmatrix} Ab_{11} & \cdots & Ab_{1n} \\ \vdots & & \vdots \\ Ab_{n1} & \cdots & Ab_{nn} \end{bmatrix} \begin{bmatrix} uv_1 \\ \vdots \\ uv_n \end{bmatrix} \\ &= \begin{bmatrix} Ab_{11}uv_1 + \cdots + Ab_{1n}uv_n \\ \vdots \\ Ab_{n1}uv_1 + \cdots + Ab_{nn}uv_n \end{bmatrix} = \begin{bmatrix} Au(Bv)_1 \\ \vdots \\ Au(Bv)_n \end{bmatrix} \\ &= (Au) \otimes (Bv) = (\lambda u) \otimes (\mu v) = (\lambda\mu)(u \otimes v). \end{aligned}$$

Exercises section 10.3

Exercise 10.4: 2. derivative matrix is positive definite ?

Write down the eigenvalues of $T = \mathrm{tridiag}(-1, 2, -1)$ using (10.15) and conclude that T is symmetric positive definite.

Solution. Applying Lemma 2.2 to the case that $a = -1$ and $d = 2$, one finds that the eigenvalues λ_j of the matrix $\mathrm{tridiag}(-1, 2, -1) \in \mathbb{R}^{m,m}$ are

$$\lambda_j = d + 2a\cos\left(\frac{j\pi}{m+1}\right) = 2\left(1 - \cos\left(\frac{j\pi}{m+1}\right)\right),$$

for $j = 1, \dots, m$. Moreover, as $|\cos(x)| < 1$ for any $x \in (0, \pi)$, it follows that $\lambda_j > 0$ for $j = 1, \dots, m$. Since, in addition, $\mathrm{tridiag}(-1, 2, -1)$ is symmetric, Lemma 4.5 implies that the matrix $\mathrm{tridiag}(-1, 2, -1)$ is symmetric positive definite.

Exercise 10.5: 1D test matrix is positive definite?

Show that the matrix $\boldsymbol{T}_1$ is symmetric positive definite if $d > 0$ and $d \geq 2|a|$.

Solution. The statement of this exercise is a generalization of the statement of Exercise 10.4. Consider a matrix $\boldsymbol{M} = \mathrm{tridiag}(a, d, a) \in \mathbb{R}^{m,m}$ for which $d > 0$ and $d \geq 2|a|$. By Lemma 2.2, the eigenvalues λ_j, with $j = 1, \dots, m$, of the matrix $\boldsymbol{M}$ are

$$\lambda_j = d + 2a \cos\left(\frac{j\pi}{m+1}\right).$$

If $a = 0$, then all these eigenvalues are equal to d and therefore positive. If $a \neq 0$, write $\mathrm{sgn}(a)$ for the sign of a. Then

$$\lambda_j \geq 2|a| \left[1 + \frac{a}{|a|} \cos\left(\frac{j\pi}{m+1}\right)\right] = 2|a| \left[1 + \mathrm{sgn}(a) \cos\left(\frac{j\pi}{m+1}\right)\right] > 0,$$

again because $|\cos(x)| < 1$ for any $x \in (0, \pi)$. Since, in addition, $\boldsymbol{M}$ is symmetric, Lemma 4.5 again implies that $\boldsymbol{M}$ is symmetric positive definite.

Exercise 10.6: Eigenvalues for 2D test matrix of order 4

For $m = 2$ the matrix (10.10) is given by

$$\boldsymbol{A} = \begin{bmatrix} 2d & a & a & 0 \\ a & 2d & 0 & a \\ a & 0 & 2d & a \\ 0 & a & a & 2d \end{bmatrix}.$$

Show that $\lambda = 2a + 2d$ is an eigenvalue corresponding to the eigenvector $\boldsymbol{x} = [1, 1, 1, 1]^{\mathrm{T}}$. Verify that apart from a scaling of the eigenvector this agrees with

$$\lambda_j = d + 2a\cos(j\pi h), \quad h := \frac{1}{m+1},$$
$$\boldsymbol{s}_j = [\sin(j\pi h), \sin(2j\pi h), \dots, \sin(mj\pi h)]^{\mathrm{T}}.$$

(i.e. (10.15) and (10.16)) for $j = k = 1$ and $m = 2$.

Solution. One has

$$\boldsymbol{A}\boldsymbol{x} = \begin{bmatrix} 2d & a & a & 0 \\ a & 2d & 0 & a \\ a & 0 & 2d & a \\ 0 & a & a & 2d \end{bmatrix} \begin{bmatrix} 1 \\ 1 \\ 1 \\ 1 \end{bmatrix} = \begin{bmatrix} 2d+2a \\ 2d+2a \\ 2d+2a \\ 2d+2a \end{bmatrix} = (2d+2a) \begin{bmatrix} 1 \\ 1 \\ 1 \\ 1 \end{bmatrix} = \lambda \boldsymbol{x},$$

which means that $(\lambda, \boldsymbol{x})$ is an eigenpair of $\boldsymbol{A}$. We also have for $m = 2$ that $\lambda_1 = d + 2a\cos(j\pi h) = d + 2a\cos(\pi/3) = d + a$, and $\boldsymbol{s}_1 = \begin{bmatrix} \sqrt{3}/2 \\ \sqrt{3}/2 \end{bmatrix}$. For $j = k = 1$ we now get

$$\lambda_j + \lambda_k = 2\lambda_1 = 2d + 2a$$

$$\boldsymbol{s}_j \otimes \boldsymbol{s}_k = \boldsymbol{s}_1 \otimes \boldsymbol{s}_1 = \begin{bmatrix} \sqrt{3}/2 \\ \sqrt{3}/2 \end{bmatrix} \otimes \begin{bmatrix} \sqrt{3}/2 \\ \sqrt{3}/2 \end{bmatrix} = \frac{3}{2} \begin{bmatrix} 1 \\ 1 \\ 1 \\ 1 \end{bmatrix} \frac{3}{2}\boldsymbol{x}.$$

Theorem 10.2 states that $\boldsymbol{A}(\boldsymbol{s}_j \otimes \boldsymbol{s}_k) = (\lambda_j + \lambda_k)(\boldsymbol{s}_j \otimes \boldsymbol{s}_k)$, i.e.,

$$\boldsymbol{A}\left(\frac{3}{2}\boldsymbol{x}\right) = (2d + 2a)\frac{3}{2}\boldsymbol{x}$$

This is thus just a scaling with $3/2$ of what we computed above.

Exercise 10.7: Nine point scheme for Poisson problem

Consider the following 9 point difference approximation to the Poisson problem $-\Delta u = f$, $u = 0$ on the boundary of the unit square (cf. (10.1))

$$\begin{aligned}
&\text{(a)}\ -(\Box_h v)_{j,k} = (\mu f)_{j,k} && j,k = 1,\ldots,m \\
&\text{(b)}\qquad\qquad 0 = v_{0,k} = v_{m+1,k} = v_{j,0} = v_{j,m+1}, && j,k = 0,1,\ldots,m+1, \\
&\text{(c)}\ -(\Box_h v)_{j,k} := [20v_{j,k} - 4v_{j-1,k} - 4v_{j,k-1} - 4v_{j+1,k} - 4v_{j,k+1} \\
&\qquad - v_{j-1,k-1} - v_{j+1,k-1} - v_{j-1,k+1} - v_{j+1,k+1}]/(6h^2), \\
&\text{(d)}\quad (\mu f)_{j,k} := [8f_{j,k} + f_{j-1,k} + f_{j,k-1} + f_{j+1,k} + f_{j,k+1}]/12.
\end{aligned} \tag{10.21}$$

a) Write down the 4-by-4 system we obtain for $m = 2$.

Solution. If $m = 2$, the boundary condition yields

$$\begin{bmatrix} v_{00} & v_{01} & v_{02} & v_{03} \\ v_{10} & & & v_{13} \\ v_{20} & & & v_{23} \\ v_{30} & v_{31} & v_{32} & v_{33} \end{bmatrix} = \begin{bmatrix} 0 & 0 & 0 & 0 \\ 0 & & & 0 \\ 0 & & & 0 \\ 0 & 0 & 0 & 0 \end{bmatrix},$$

leaving four equations to determine the interior points $v_{11}, v_{12}, v_{21}, v_{22}$. Since we have that $6h^2/12 = 1/(2(m+1)^2) = 1/18$ for $m = 2$, we obtain

$$20v_{11} - 4v_{01} - 4v_{10} - 4v_{21} - 4v_{12} - v_{00} - v_{20} - v_{02} - v_{22}$$

$$= \frac{1}{18}(8f_{11} + f_{01} + f_{10} + f_{21} + f_{12}),$$

$$20v_{21} - 4v_{11} - 4v_{20} - 4v_{31} - 4v_{22} - v_{10} - v_{30} - v_{12} - v_{32}$$
$$= \frac{1}{18}(8f_{21} + f_{11} + f_{20} + f_{31} + f_{22}),$$

$$20v_{12} - 4v_{02} - 4v_{11} - 4v_{22} - 4v_{13} - v_{01} - v_{21} - v_{03} - v_{23}$$
$$= \frac{1}{18}(8f_{12} + f_{02} + f_{11} + f_{22} + f_{13}),$$

$$20v_{22} - 4v_{12} - 4v_{21} - 4v_{32} - 4v_{23} - v_{11} - v_{31} - v_{13} - v_{33}$$
$$= \frac{1}{18}(8f_{22} + f_{12} + f_{21} + f_{32} + f_{23}).$$

Using the values known from the boundary condition, these equations can be simplified to

$$20v_{11} - 4v_{21} - 4v_{12} - v_{22} = \frac{1}{18}(8f_{11} + f_{01} + f_{10} + f_{21} + f_{12}),$$

$$20v_{21} - 4v_{11} - 4v_{22} - v_{12} = \frac{1}{18}(8f_{21} + f_{11} + f_{20} + f_{31} + f_{22}),$$

$$20v_{12} - 4v_{11} - 4v_{22} - v_{21} = \frac{1}{18}(8f_{12} + f_{02} + f_{11} + f_{22} + f_{13}),$$

$$20v_{22} - 4v_{12} - 4v_{21} - v_{11} = \frac{1}{18}(8f_{22} + f_{12} + f_{21} + f_{32} + f_{23}).$$

b) Find $v_{j,k}$ for $j, k = 1, 2$, if $f(x, y) = 2\pi^2 \sin(\pi x) \sin(\pi y)$ and $m = 2$.

Solution. For $f(x, y) = 2\pi^2 \sin(\pi x) \sin(\pi y)$, one finds

$$\begin{bmatrix} f_{00} & f_{01} & f_{02} & f_{03} \\ f_{10} & f_{11} & f_{12} & f_{13} \\ f_{20} & f_{21} & f_{22} & f_{23} \\ f_{30} & f_{31} & f_{32} & f_{33} \end{bmatrix} = \begin{bmatrix} 0 & 0 & 0 & 0 \\ 0 & 3\pi^2/2 & 3\pi^2/2 & 0 \\ 0 & 3\pi^2/2 & 3\pi^2/2 & 0 \\ 0 & 0 & 0 & 0 \end{bmatrix}.$$

Substituting these values in our linear system, we obtain

$$\begin{bmatrix} 20 & -4 & -4 & -1 \\ -4 & 20 & -1 & -4 \\ -4 & -1 & 20 & -4 \\ -1 & -4 & -4 & 20 \end{bmatrix} \begin{bmatrix} v_{11} \\ v_{21} \\ v_{12} \\ v_{22} \end{bmatrix} = \frac{8+1+1}{18}\frac{3\pi^2}{2} \begin{bmatrix} 1 \\ 1 \\ 1 \\ 1 \end{bmatrix} = \begin{bmatrix} 5\pi^2/6 \\ 5\pi^2/6 \\ 5\pi^2/6 \\ 5\pi^2/6 \end{bmatrix}.$$

Solving this system we find that $v_{11} = v_{12} = v_{21} = v_{22} = 5\pi^2/66$. It can be shown that (10.21) defines an $\mathcal{O}(h^4)$ approximation to (10.1).

Exercise 10.8: Matrix equation for nine point scheme

Consider the nine point difference approximation to (10.1) given by Equation (10.21) in Exercise 10.7.

a) Show that Equation (10.21) is equivalent to the matrix equation

$$\boldsymbol{T}\boldsymbol{V} + \boldsymbol{V}\boldsymbol{T} - \frac{1}{6}\boldsymbol{T}\boldsymbol{V}\boldsymbol{T} = h^2\mu\boldsymbol{F}. \tag{10.22}$$

Here $\mu\boldsymbol{F}$ has elements $(\mu f)_{j,k}$ given by (10.21**d**) and $\boldsymbol{T} = \text{tridiag}(-1, 2, -1)$.

Solution. Let

$$\boldsymbol{T} = \begin{bmatrix} 2 & -1 & 0 & & & \\ -1 & 2 & -1 & & & \\ 0 & \ddots & \ddots & \ddots & & \\ & & & & & 0 \\ & & & -1 & 2 & -1 \\ & & & 0 & -1 & 2 \end{bmatrix}, \qquad \boldsymbol{V} = \begin{bmatrix} v_{11} & \cdots & v_{1m} \\ \vdots & \ddots & \vdots \\ v_{m1} & \cdots & v_{mm} \end{bmatrix}$$

be of equal dimensions. Implicitly assuming the boundary condition

$$v_{0,k} = v_{m+1,k} = v_{j,0} = v_{j,m+1} = 0, \qquad \text{for } j, k = 0, \dots, m+1, \tag{10.i}$$

the (j, k)-th entry of $\boldsymbol{T}\boldsymbol{V} + \boldsymbol{V}\boldsymbol{T}$ can be written as

$$4v_{j,k} - v_{j-1,k} - v_{j+1,k} - v_{j,k-1} - v_{j,k+1}.$$

Similarly, writing out two matrix products, the (j, k)-th entry of $\boldsymbol{T}\boldsymbol{V}\boldsymbol{T} = \boldsymbol{T}(\boldsymbol{V}\boldsymbol{T})$ is found to be

$$\begin{array}{lll} -1(-1v_{j-1,k-1} & +2v_{j-1,k} & -1v_{j-1,k+1}) \\ +2(-1v_{j,k-1} & +2v_{j,k} & -1v_{j,k+1}) \\ -1(-1v_{j+1,k-1} & +2v_{j+1,k} & -1v_{j+1,k+1}) \end{array} = \begin{array}{lll} +v_{j-1,k-1} & -2v_{j-1,k} & +v_{j-1,k+1} \\ -2v_{j,k-1} & +4v_{j,k} & -2v_{j,k+1} \\ +v_{j+1,k-1} & -2v_{j+1,k} & +v_{j+1,k+1} \end{array}.$$

Together, these observations yield that the System (10.21) is equivalent to (10.i) and

$$\boldsymbol{T}\boldsymbol{V} + \boldsymbol{V}\boldsymbol{T} - \frac{1}{6}\boldsymbol{T}\boldsymbol{V}\boldsymbol{T} = h^2\mu\boldsymbol{F}.$$

b) Show that the standard form of the matrix equation (10.22) is $\boldsymbol{A}\boldsymbol{x} = \boldsymbol{b}$, where $\boldsymbol{A} = \boldsymbol{T} \otimes \boldsymbol{I} + \boldsymbol{I} \otimes \boldsymbol{T} - \frac{1}{6}\boldsymbol{T} \otimes \boldsymbol{T}$, $\boldsymbol{x} = \text{vec}(\boldsymbol{V})$, and $\boldsymbol{b} = h^2\text{vec}(\mu\boldsymbol{F})$.

Solution. It is a direct consequence of properties 7 and 8 of Theorem 10.1 that this equation can be rewritten as one of the form $\boldsymbol{A}\boldsymbol{x} = \boldsymbol{b}$, where

$$\boldsymbol{A} = \boldsymbol{T} \otimes \boldsymbol{I} + \boldsymbol{I} \otimes \boldsymbol{T} - \frac{1}{6}\boldsymbol{T} \otimes \boldsymbol{T}, \qquad \boldsymbol{x} = \text{vec}(\boldsymbol{V}), \qquad \boldsymbol{b} = h^2\text{vec}(\mu\boldsymbol{F}).$$

Exercise 10.9: Biharmonic equation

Consider the biharmonic equation

$$\begin{aligned} \Delta^2 u(s,t) &:= \Delta\big(\Delta u(s,t)\big) = f(s,t) & (s,t) &\in \Omega, \\ u(s,t) &= 0, \quad \Delta u(s,t) = 0 & (s,t) &\in \partial\Omega. \end{aligned} \tag{10.23}$$

Here Ω is the open unit square. The condition $\Delta u = 0$ is called the *Navier boundary condition*. Moreover, $\Delta^2 u = u_{xxxx} + 2u_{xxyy} + u_{yyyy}$.

a) Let $v := -\Delta u$. Show that (10.23) can be written as a system

$$\begin{aligned} -\Delta v(s,t) &= f(s,t) & (s,t) &\in \Omega \\ -\Delta u(s,t) &= v(s,t) & (s,t) &\in \Omega \\ u(s,t) &= v(s,t) = 0 & (s,t) &\in \partial\Omega. \end{aligned}$$

Solution. Writing $v = -\nabla u$, the second line in (10.23) is equivalent to

$$u(s,t) = v(s,t) = 0, \qquad \text{for } (s,t) \in \partial\Omega,$$

while the first line is equivalent to

$$f(s,t) = \Delta^2 u(s,t) = -\Delta\big(-\Delta u(s,t)\big) = -\Delta v(s,t), \qquad \text{for } (s,t) \in \Omega.$$

b) Discretizing, using (10.4), with $\boldsymbol{T} = \text{tridiag}(-1,2,-1) \in \mathbb{R}^{m\times m}$, $h = 1/(m+1)$, and $\boldsymbol{F} = \big(f(jh,kh)\big)_{j,k=1}^{m}$ we get two matrix equations

$$\boldsymbol{TV} + \boldsymbol{VT} = h^2\boldsymbol{F}, \qquad \boldsymbol{TU} + \boldsymbol{UT} = h^2\boldsymbol{V}.$$

Show that

$$(\boldsymbol{T}\otimes\boldsymbol{I}+\boldsymbol{I}\otimes\boldsymbol{T})\text{vec}(\boldsymbol{V}) = h^2\text{vec}(\boldsymbol{F}), \quad (\boldsymbol{T}\otimes\boldsymbol{I}+\boldsymbol{I}\otimes\boldsymbol{T})\text{vec}(\boldsymbol{U}) = h^2\text{vec}(\boldsymbol{V}).$$

and hence $\boldsymbol{A} = (\boldsymbol{T} \otimes \boldsymbol{I} + \boldsymbol{I} \otimes \boldsymbol{T})^2$ is the matrix for the standard form of the discrete biharmonic equation.

Solution. By property 8 of Theorem 10.1,

$$(\boldsymbol{A} \otimes \boldsymbol{I} + \boldsymbol{I} \otimes \boldsymbol{B})\text{vec}(\boldsymbol{V}) = \text{vec}(\boldsymbol{F}) \Longleftrightarrow \boldsymbol{AV} + \boldsymbol{VB}^{\mathrm{T}} = \boldsymbol{F},$$

whenever $\boldsymbol{A} \in \mathbb{R}^{r,r}, \boldsymbol{B} \in \mathbb{R}^{s,s}, \boldsymbol{F}, \boldsymbol{V} \in \mathbb{R}^{r,s}$ (the identity matrices are assumed to be of the appropriate dimensions). Using $\boldsymbol{T} = \boldsymbol{T}^{\mathrm{T}}$, this equation implies

$$\boldsymbol{TV} + \boldsymbol{VT} = h^2\boldsymbol{F} \Longleftrightarrow (\boldsymbol{T} \otimes \boldsymbol{I} + \boldsymbol{I} \otimes \boldsymbol{T})\text{vec}(\boldsymbol{V}) = h^2\text{vec}(\boldsymbol{F}),$$

$$TU + UT = h^2 V \iff (T \otimes I + I \otimes T)\mathrm{vec}(U) = h^2 \mathrm{vec}(V).$$

Substituting the equation for $\mathrm{vec}(V)$ into the equation for $\mathrm{vec}(F)$, one obtains the equation

$$A\mathrm{vec}(U) = h^4 \mathrm{vec}(F), \qquad \text{where } A := (T \otimes I + I \otimes T)^2,$$

which is a linear system of m^2 equations.

c) Show that with $n = m^2$ the vector form and standard form of the systems in **b)** can be written

$$T^2 U + 2TUT + UT^2 = h^4 F \quad \text{and} \quad Ax = b,$$

where $A = T^2 \otimes I + 2T \otimes T + I \otimes T^2 \in \mathbb{R}^{n \times n}$, $x = \mathrm{vec}(U)$, and $b = h^4 \mathrm{vec}(F)$.

Solution. The equations $h^2 V = TU + UT$ and $TV + VT = h^2 F$ together yield the normal form

$$T(TU + UT) + (TU + UT)T = T^2 U + 2TUT + UT^2 = h^4 F.$$

The vector form is given in **b)**. Using the distributive property of matrix multiplication and the mixed product rule of Lemma 10.1, the matrix $A = (T \otimes I + I \otimes T)^2$ can be rewritten as

$$\begin{aligned} A &= (T \otimes I)(T \otimes I) + (T \otimes I)(I \otimes T) + (I \otimes T)(T \otimes I) + (I \otimes T)(I \otimes T) \\ &= T^2 \otimes I + 2T \otimes T + I \otimes T^2. \end{aligned}$$

Writing $x := \mathrm{vec}(U)$ and $b := h^4 \mathrm{vec}(F)$, the linear system of **b)** can be written as $Ax = b$.

d) Determine the eigenvalues and eigenvectors of the matrix A in **c)** and show that it is positive definite. Also determine the bandwidth of A.

Solution. Property 6 of Theorem 10.1 says that $A \otimes I + I \otimes B$ is positive definite when of of A and B is positive definite, and the other is positive semidefinite. Since T is positive definite, it follows that $M := T \otimes I + I \otimes T$ is positive definite as well. The square of any symmetric positive definite matrix is symmetric positive definite as well, implying that $A = M^2$ is symmetric positive definite. Let us now show this more directly by calculating the eigenvalues of A.

By Lemma 2.2, we know the eigenpairs (λ_i, s_i), where $i = 1, \ldots, m$, of the matrix T. By property 5 of Theorem 10.1, it follows that the eigenpairs of M are $(\lambda_i + \lambda_j, s_i \otimes s_j)$, for $i, j = 1, \ldots, m$. If B is any matrix with eigenpairs (μ_i, v_i), where $i = 1, \ldots, m$, then B^2 has eigenpairs (μ_i^2, v_i), as

$$B^2 v_i = B(Bv_i) = B(\mu_i v_i) = \mu_i (Bv_i) = \mu_i^2 v_i, \qquad \text{for } i = 1, \ldots, m.$$

It follows that $\boldsymbol{A} = \boldsymbol{M}^2$ has eigenpairs $\big((\lambda_i + \lambda_j)^2, \boldsymbol{s}_i \otimes \boldsymbol{s}_j\big)$, for $i, j = 1, \ldots, m$. (Note that we can also verify this directly by multiplying $\boldsymbol{A}$ by $\boldsymbol{s}_i \otimes \boldsymbol{s}_j$ and using the mixed product rule.) Since the λ_i are positive, the eigenvalues of $\boldsymbol{A}$ are positive. We conclude that $\boldsymbol{A}$ is symmetric positive definite.

Writing $\boldsymbol{A} = \boldsymbol{T}^2 \otimes \boldsymbol{I} + 2\boldsymbol{T} \otimes \boldsymbol{T} + \boldsymbol{I} \otimes \boldsymbol{T}^2$ and computing the block structure of each of these terms, one finds that $\boldsymbol{A}$ has bandwidth $2m$, in the sense that any row has at most $4m + 1$ nonzero elements.

e) Suppose we want to solve the standard form equation $\boldsymbol{A}\boldsymbol{x} = \boldsymbol{b}$. We have two representations for the matrix $\boldsymbol{A}$, the product one in **b)** and the one in **c)**. Which one would you prefer for the basis of an algorithm? Why?

Solution. One can expect to solve the system of **b)** faster, as it is typically quicker to solve two simple systems instead of one complex system.

Chapter 11
Fast Direct Solution of a Large Linear System

Exercises section 11.3

Exercise 11.1: Fourier matrix

Show that the Fourier matrix $\boldsymbol{F}_4$ is symmetric, but not Hermitian.

Solution. The Fourier matrix $\boldsymbol{F}_N$ has entries

$$(\boldsymbol{F}_N)_{j,k} = \omega_N^{(j-1)(k-1)}, \qquad \omega_N := e^{-\frac{2\pi}{N}i} = \cos\left(\frac{2\pi}{N}\right) - i\sin\left(\frac{2\pi}{N}\right).$$

In particular for $N = 4$, this implies that $\omega_4 = -i$ and

$$\boldsymbol{F}_4 = \begin{bmatrix} 1 & 1 & 1 & 1 \\ 1 & -i & -1 & i \\ 1 & -1 & 1 & -1 \\ 1 & i & -1 & -i \end{bmatrix}.$$

Computing the transpose and Hermitian transpose gives

$$\boldsymbol{F}_4^{\mathrm{T}} = \begin{bmatrix} 1 & 1 & 1 & 1 \\ 1 & -i & -1 & i \\ 1 & -1 & 1 & -1 \\ 1 & i & -1 & -i \end{bmatrix} = \boldsymbol{F}_4, \qquad \boldsymbol{F}_4^H = \begin{bmatrix} 1 & 1 & 1 & 1 \\ 1 & i & -1 & -i \\ 1 & -1 & 1 & -1 \\ 1 & -i & -1 & i \end{bmatrix} \neq \boldsymbol{F}_4,$$

which is what needed to be shown.

Exercise 11.2: Sine transform as Fourier transform

Verify Lemma 11.1 directly when $m = 1$.

T. Lyche et al., *Exercises in Numerical Linear Algebra and Matrix Factorizations*, Texts in Computational Science and Engineering 23, https://doi.org/10.1007/978-3-030-59789-4_11

Solution. According to Lemma 11.1, the Discrete Sine Transform can be computed from the Discrete Fourier Transform by $(\boldsymbol{S}_m\boldsymbol{x})_k = \frac{i}{2}(\boldsymbol{F}_{2m+2}\boldsymbol{z})_{k+1}$, where

$$\boldsymbol{z} = [0, x_1, \ldots, x_m, 0, -x_m, \ldots, -x_1]^{\mathrm{T}}.$$

For $m = 1$ this means that

$$\boldsymbol{z} = [0, x_1, 0, -x_1]^{\mathrm{T}} \quad \text{and} \quad (\boldsymbol{S}_1\boldsymbol{x})_1 = \frac{i}{2}(\boldsymbol{F}_4\boldsymbol{z})_2.$$

Since $h = \frac{1}{m+1} = \frac{1}{2}$ for $m = 1$, computing the DST directly gives

$$(\boldsymbol{S}_1\boldsymbol{x})_1 = \sin(\pi h)x_1 = \sin\left(\frac{\pi}{2}\right)x_1 = x_1,$$

while computing the Fourier transform gives

$$\boldsymbol{F}_4\boldsymbol{z} = \begin{bmatrix} 1 & 1 & 1 & 1 \\ 1 & -i & -1 & i \\ 1 & -1 & 1 & -1 \\ 1 & i & -1 & -i \end{bmatrix} \begin{bmatrix} 0 \\ x_1 \\ 0 \\ -x_1 \end{bmatrix} = \begin{bmatrix} 0 \\ -2ix_1 \\ 0 \\ 2ix_1 \end{bmatrix} = -2i \begin{bmatrix} 0 \\ x_1 \\ 0 \\ -x_1 \end{bmatrix} = -2i\boldsymbol{z}.$$

Multiplying the Fourier transform with $\frac{i}{2}$, one finds $\frac{i}{2}\boldsymbol{F}_4\boldsymbol{z} = \boldsymbol{z}$, so that $\frac{i}{2}(\boldsymbol{F}_4\boldsymbol{z})_2 = x_1 = (\boldsymbol{S}_1\boldsymbol{x})_1$, which is what we needed to show.

Exercise 11.3: Explicit solution of the discrete Poisson equation

Show that the exact solution of the discrete Poisson equation (Equation (10.5)) can be written $\boldsymbol{V} = (v_{ij})_{i,j=1}^m$, where, with $h := 1/(m+1)$,

$$v_{ij} = h^4 \sum_{p=1}^m \sum_{r=1}^m \sum_{k=1}^m \sum_{l=1}^m \frac{\sin\left(\frac{ip\pi}{m+1}\right)\sin\left(\frac{jr\pi}{m+1}\right)\sin\left(\frac{kp\pi}{m+1}\right)\sin\left(\frac{lr\pi}{m+1}\right)}{\left[\sin\left(\frac{p\pi}{2(m+1)}\right)\right]^2 + \left[\sin\left(\frac{r\pi}{2(m+1)}\right)\right]^2} f_{kl}.$$

Solution. For $j = 1, \ldots, m$, let $\lambda_j = 4\sin^2(j\pi h/2)$, $\boldsymbol{D} = \operatorname{diag}(\lambda_1, \ldots, \lambda_m)$, and $\boldsymbol{S} = (s_{jk})_{jk} = \big(\sin(jk\pi h)\big)_{jk}$. By the results derived in Section 11.2, the solution to the discrete Poisson equation is $\boldsymbol{V} = \boldsymbol{S}\boldsymbol{X}\boldsymbol{S}$, where $\boldsymbol{X}$ is found by solving $\boldsymbol{D}\boldsymbol{X} + \boldsymbol{X}\boldsymbol{D} = 4h^4\boldsymbol{S}\boldsymbol{F}\boldsymbol{S}$. Since $\boldsymbol{D}$ is diagonal, one has $(\boldsymbol{D}\boldsymbol{X} + \boldsymbol{X}\boldsymbol{D})_{pr} = (\lambda_p + \lambda_r)x_{pr}$, so that

$$x_{pr} = 4h^4 \frac{(\boldsymbol{S}\boldsymbol{F}\boldsymbol{S})_{pr}}{\lambda_p + \lambda_r} = 4h^4 \sum_{k=1}^m \sum_{l=1}^m \frac{s_{pk} f_{kl} s_{lr}}{\lambda_p + \lambda_r},$$

and hence

$$v_{ij} = \sum_{p=1}^m \sum_{r=1}^m s_{ip} x_{pr} s_{rj} = 4h^4 \sum_{p=1}^m \sum_{r=1}^m \sum_{k=1}^m \sum_{l=1}^m \frac{s_{ip} s_{pk} s_{lr} s_{rj}}{\lambda_p + \lambda_r} f_{kl}$$

$$= h^4 \sum_{p=1}^{m}\sum_{r=1}^{m}\sum_{k=1}^{m}\sum_{l=1}^{m} \frac{\sin\left(\frac{ip\pi}{m+1}\right)\sin\left(\frac{pk\pi}{m+1}\right)\sin\left(\frac{lr\pi}{m+1}\right)\sin\left(\frac{rj\pi}{m+1}\right)}{\sin^2\left(\frac{p\pi}{2(m+1)}\right)+\sin^2\left(\frac{r\pi}{2(m+1)}\right)} f_{kl},$$

which is what needed to be shown.

Exercise 11.4: Improved version of Algorithm 11.1

Algorithm 11.1 involves multiplying a matrix by $\boldsymbol{S}$ four times. In this exercise we show that it is enough to multiply by $\boldsymbol{S}$ two times. We achieve this by diagonalizing only the second $\boldsymbol{T}$ in $\boldsymbol{TV}+\boldsymbol{VT}=h^2\boldsymbol{F}$. Let $\boldsymbol{D}=\operatorname{diag}(\lambda_1,\ldots,\lambda_m)$, where $\lambda_j = 4\sin^2(j\pi h/2)$, $j=1,\ldots,m$.

a) Show that

$$\boldsymbol{TX}+\boldsymbol{XD}=\boldsymbol{C}, \text{ where } \boldsymbol{X}=\boldsymbol{VS}, \text{ and } \boldsymbol{C}=h^2\boldsymbol{FS}.$$

Solution. Recall that

$$\boldsymbol{TV}+\boldsymbol{VT}=h^2\boldsymbol{F}, \tag{11.i}$$

with $\boldsymbol{T}=\boldsymbol{SDS}^{-1}$ the orthogonal diagonalization of $\boldsymbol{T}$ from Equation (11.4). We also write $\boldsymbol{X}=\boldsymbol{VS}$ and $\boldsymbol{C}=h^2\boldsymbol{FS}$.

Multiplying Equation (11.i) from the right by $\boldsymbol{S}$, one obtains

$$\boldsymbol{TX}+\boldsymbol{XD}=\boldsymbol{TVS}+\boldsymbol{VSD}=\boldsymbol{TVS}+\boldsymbol{VTS}=h^2\boldsymbol{FS}=\boldsymbol{C}.$$

b) Show that

$$(\boldsymbol{T}+\lambda_j\boldsymbol{I})\boldsymbol{x}_j=\boldsymbol{c}_j, \qquad j=1,\ldots,m$$

(System (11.9)), where $\boldsymbol{X}=[\boldsymbol{x}_1,\ldots,\boldsymbol{x}_m]$ and $\boldsymbol{C}=[\boldsymbol{c}_1,\ldots,\boldsymbol{c}_m]$. Thus we can find $\boldsymbol{X}$ by solving m linear systems, one for each of the columns of $\boldsymbol{X}$. Recall that a tridiagonal $m\times m$ system can be solved by Algorithms 2.1 and 2.2 in $8m-7$ arithmetic operations. Give an algorithm to find $\boldsymbol{X}$ which only requires $\mathcal{O}(\delta m^2)$ arithmetic operations for some constant δ independent of m.

Solution. Writing $\boldsymbol{C}=[\boldsymbol{c}_1,\ldots,\boldsymbol{c}_m]$, $\boldsymbol{X}=[\boldsymbol{x}_1,\ldots,\boldsymbol{x}_m]$ and applying the rules of block multiplication, we find

$$\begin{aligned}
[\boldsymbol{c}_1,\ldots,\boldsymbol{c}_m] &= \boldsymbol{C}\\
&= \boldsymbol{TX}+\boldsymbol{XD}\\
&= \boldsymbol{T}[\boldsymbol{x}_1,\ldots,\boldsymbol{x}_m]+\boldsymbol{X}[\lambda_1\boldsymbol{e}_1,\ldots,\lambda_m\boldsymbol{e}_m]\\
&= [\boldsymbol{Tx}_1+\lambda_1\boldsymbol{Xe}_1,\ldots,\boldsymbol{Tx}_m+\lambda_m\boldsymbol{Xe}_m]\\
&= [\boldsymbol{Tx}_1+\lambda_1\boldsymbol{x}_1,\ldots,\boldsymbol{Tx}_m+\lambda_m\boldsymbol{x}_m]\\
&= [(\boldsymbol{T}+\lambda_1\boldsymbol{I})\boldsymbol{x}_1,\ldots,(\boldsymbol{T}+\lambda_m\boldsymbol{I})\boldsymbol{x}_m],
\end{aligned}$$

which is equivalent to System (11.9). To find $\boldsymbol{X}$, we therefore need to solve the m tridiagonal linear systems of (11.9). Since the eigenvalues $\lambda_1, \ldots, \lambda_m$ are positive, each matrix $\boldsymbol{T} + \lambda_j \boldsymbol{I}$ is diagonally dominant. By Theorem 2.4, every such matrix is nonsingular and has a unique LU factorization. Algorithms 2.1 and 2.2 then solve the corresponding system $(\boldsymbol{T}+\lambda_j \boldsymbol{I})\boldsymbol{x}_j = \boldsymbol{c}_j$ in $\mathcal{O}(\delta m)$ operations for some constant δ. Doing this for all m columns $\boldsymbol{x}_1, \ldots, \boldsymbol{x}_m$, one finds the matrix $\boldsymbol{X}$ in $\mathcal{O}(\delta m^2)$ operations.

c) Describe a method to compute $\boldsymbol{V}$ which only requires $\mathcal{O}(4m^3) = \mathcal{O}(4n^{3/2})$ arithmetic operations.

Solution. To find $\boldsymbol{V}$, we first find $\boldsymbol{C} = h^2 \boldsymbol{F}\boldsymbol{S}$ by performing $\mathcal{O}(2m^3)$ operations. Next we find $\boldsymbol{X}$ as in step **b)** by performing $\mathcal{O}(\delta m^2)$ operations. Finally we compute $\boldsymbol{V} = 2h\boldsymbol{X}\boldsymbol{S}$ by performing $\mathcal{O}(2m^3)$ operations. In total, this amounts to $\mathcal{O}(4m^3)$ operations.

d) Describe a method based on the fast Fourier transform which requires $\mathcal{O}(2\gamma n \log_2 n)$ where γ is the same constant as mentioned at the end of the last section.

Solution. As explained in Section 11.3, multiplying by the matrix $\boldsymbol{S}$ can be done in $\mathcal{O}(2m^2 \log_2 m)$ operations by using the Fourier transform. The two matrix multiplications in **c)** can therefore be carried out in

$$\mathcal{O}(4\gamma m^2 \log_2 m) = \mathcal{O}(4\gamma n \log_2 n^{1/2}) = \mathcal{O}(2\gamma n \log_2 n)$$

operations.

Exercise 11.5: Fast solution of 9 point scheme

Consider the equation

$$\boldsymbol{TV} + \boldsymbol{VT} - \frac{1}{6}\boldsymbol{TVT} = h^2\mu\boldsymbol{F}, \tag{10.22}$$

that was derived in Exercise 10.8 for the 9-point scheme. Define the matrix $\boldsymbol{X} = (x_{j,k})$ by $\boldsymbol{V} = \boldsymbol{SXS}$, where $\boldsymbol{V}$ is the solution of (10.22). Show that

$$\boldsymbol{DX} + \boldsymbol{XD} - \frac{1}{6}\boldsymbol{DXD} = 4h^4\boldsymbol{G}, \qquad \text{where } \boldsymbol{G} = \boldsymbol{S}\mu\boldsymbol{FS},$$

where $\boldsymbol{D} = \operatorname{diag}(\lambda_1, \ldots, \lambda_m)$, with $\lambda_j = 4\sin^2(j\pi h/2)$, $j = 1, \ldots, m$, and that

$$x_{j,k} = \frac{h^4 g_{j,k}}{\sigma_j + \sigma_k - \frac{2}{3}\sigma_j\sigma_k}, \qquad \sigma_j := \sin^2\left(\frac{j\pi h}{2}\right), \qquad j, k = 1, 2, \ldots, m.$$

Show that $\sigma_j + \sigma_k - \frac{2}{3}\sigma_j\sigma_k > 0$ for $j, k = 1, 2, \ldots, m$. Conclude that the matrix $\boldsymbol{A}$ in Exercise 10.8**b)** is symmetric positive definite and that (10.21) always has a solution $\boldsymbol{V}$.

Solution. Analogously to Section 11.2, we use the relations between the matrices $\boldsymbol{T}, \boldsymbol{S}, \boldsymbol{X}, \boldsymbol{D}$ to rewrite Equation (10.22).

$$\begin{aligned}
& & h^2\mu\boldsymbol{F} &= \boldsymbol{TV} + \boldsymbol{VT} - \frac{1}{6}\boldsymbol{TVT} \\
&\iff & h^2\mu\boldsymbol{F} &= \boldsymbol{TSXS} + \boldsymbol{SXST} - \frac{1}{6}\boldsymbol{TSXST} \\
&\iff & h^2\mu\boldsymbol{SFS} &= \boldsymbol{STSXS}^2 + \boldsymbol{S}^2\boldsymbol{XSTS} - \frac{1}{6}\boldsymbol{STSXSTS} \\
&\iff & h^2\mu\boldsymbol{SFS} &= \boldsymbol{S}^2\boldsymbol{DXS}^2 + \boldsymbol{S}^2\boldsymbol{XS}^2\boldsymbol{D} - \frac{1}{6}\boldsymbol{S}^2\boldsymbol{DXS}^2\boldsymbol{D} \\
&\iff & 4h^4\boldsymbol{G} = 4h^4\mu\boldsymbol{SFS} &= \boldsymbol{DX} + \boldsymbol{XD} - \frac{1}{6}\boldsymbol{DXD}.
\end{aligned}$$

Writing $\boldsymbol{D} = \operatorname{diag}(\lambda_1, \ldots, \lambda_m)$, the (j,k)-th entry of $\boldsymbol{DX} + \boldsymbol{XD} - \frac{1}{6}\boldsymbol{DXD}$ is equal to $\lambda_j x_{jk} + x_{jk}\lambda_k - \frac{1}{6}\lambda_j x_{jk}\lambda_k$. Isolating x_{jk} and writing $\lambda_j = 4\sigma_j = 4\sin^2(j\pi h/2)$ then yields

$$x_{jk} = \frac{4h^4 g_{jk}}{\lambda_j + \lambda_k - \frac{1}{6}\lambda_j\lambda_k} = \frac{h^4 g_{jk}}{\sigma_j + \sigma_k - \frac{2}{3}\sigma_j\sigma_k}, \qquad \sigma_j = \sin^2\left(\frac{j\pi h}{2}\right).$$

Defining $\alpha := j\pi h/2$ and $\beta := k\pi h/2$, one has $0 < \alpha, \beta < \pi/2$. Note that

$$\sigma_j + \sigma_k - \frac{2}{3}\sigma_j\sigma_k > \sigma_j + \sigma_k - \sigma_j\sigma_k$$

$$
\begin{aligned}
&= 2 - \cos^2\alpha - \cos^2\beta - (1-\cos^2\alpha)(1-\cos^2\beta) \\
&= 1 - \cos^2\alpha\cos^2\beta \\
&\geq 1 - \cos^2\beta \\
&\geq 0.
\end{aligned}
$$

Let $\boldsymbol{A} = \boldsymbol{T}\otimes\boldsymbol{I} + \boldsymbol{I}\otimes\boldsymbol{T} - \frac{1}{6}\boldsymbol{T}\otimes\boldsymbol{T}$ be as in Exercise 10.8.**b)** and $\boldsymbol{s}_i$ as in Section 11.2. Applying the mixed-product rule, one obtains

$$
\begin{aligned}
\boldsymbol{A}(\boldsymbol{s}_i\otimes\boldsymbol{s}_j) &= (\boldsymbol{T}\otimes\boldsymbol{I} + \boldsymbol{I}\otimes\boldsymbol{T})(\boldsymbol{s}_i\otimes\boldsymbol{s}_j) - \frac{1}{6}(\boldsymbol{T}\otimes\boldsymbol{T})(\boldsymbol{s}_i\otimes\boldsymbol{s}_j) \\
&= (\lambda_i + \lambda_j)(\boldsymbol{s}_i\otimes\boldsymbol{s}_j) - \frac{1}{6}\lambda_i\lambda_j(\boldsymbol{s}_i\otimes\boldsymbol{s}_j) \\
&= (\lambda_i + \lambda_j - \frac{1}{6}\lambda_i\lambda_j)(\boldsymbol{s}_i\otimes\boldsymbol{s}_j).
\end{aligned}
$$

The matrix $\boldsymbol{A}$ therefore has eigenvectors $\boldsymbol{s}_i\otimes\boldsymbol{s}_j$, and counting them shows that these must be all of them. As shown above, the corresponding eigenvalues $\lambda_i + \lambda_j - \frac{1}{6}\lambda_i\lambda_j$ are positive, implying that the matrix $\boldsymbol{A}$ is positive definite. It follows that the System (10.21) always has a (unique) solution.

Exercise 11.6: Algorithm for fast solution of 9 point scheme

Derive an algorithm for solving System (10.21) which for large m requires essentially the same number of operations as in Algorithm 11.1. (We assume that $\mu\boldsymbol{F}$ already has been formed).

Solution. Algorithm 11.1 solves System (10.21).

code/ninepointscheme.m

```
function U = ninepointscheme(F)
   m = length(F);
   h = 1/(m+1);
   hv = pi*h*(1:m)';
   sigma = sin(hv/2).^2;
   S = sin(hv*(1:m));
   G = S*F*S;
   X = (h^4)*G./( sigma*ones(1,m)+ones(m,1)*sigma' - (2/3)*
      sigma.*sigma' );
   U = zeros(m+2,m+2);
   U(2:m+1,2:m+1) = S*X*S;
end
```

Listing 11.1: Solving the Poisson problem using a nine-point scheme.

Only two steps here are of order m^3: The ones which compute $\boldsymbol{SFS}$ and $\boldsymbol{SXS}$. Hence the overall complexity is determined by the four matrix multiplications and given by $\mathcal{O}(m^3)$.

Exercise 11.7: Fast solution of biharmonic equation

For the biharmonic problem we derived in Exercise 10.9 the equation

$$\boldsymbol{T}^2\boldsymbol{U} + 2\boldsymbol{TUT} + \boldsymbol{UT}^2 = h^4\boldsymbol{F}.$$

Define the matrix $\boldsymbol{X} = (x_{j,k})$ by $\boldsymbol{U} = \boldsymbol{SXS}$ where $\boldsymbol{U}$ is the solution of (10.25). Show that

$$\boldsymbol{D}^2\boldsymbol{X} + 2\boldsymbol{DXD} + \boldsymbol{XD}^2 = 4h^6\boldsymbol{G}, \text{ where } \boldsymbol{G} = \boldsymbol{SFS},$$

and that

$$x_{j,k} = \frac{h^6 g_{j,k}}{4(\sigma_j + \sigma_k)^2}, \qquad \sigma_j := \sin^2\left(\frac{j\pi h}{2}\right), \qquad j,k = 1,2,\ldots,m.$$

Solution. From Exercise 10.9 we know that $\boldsymbol{T} \in \mathbb{R}^{m\times m}$ is the second derivative matrix. According to Lemma 2.2, the eigenpairs $(\lambda_j, \boldsymbol{s}_j)$, with $j = 1,\ldots,m$, of $\boldsymbol{T}$ are given by

$$\begin{aligned}
\boldsymbol{s}_j &= [\sin(j\pi h), \sin(2j\pi h), \ldots, \sin(mj\pi h)]^{\mathrm{T}}, \\
\lambda_j &= 2 - 2\cos(j\pi h) = 4\sin^2(j\pi h/2),
\end{aligned}$$

and satisfy $\boldsymbol{s}_j^{\mathrm{T}}\boldsymbol{s}_k = \delta_{j,k}/(2h)$ for all j,k, where $h := 1/(m+1)$. Using, in order, that $\boldsymbol{U} = \boldsymbol{SXS}$, $\boldsymbol{TS} = \boldsymbol{SD}$, and $\boldsymbol{S}^2 = \boldsymbol{I}/(2h)$, one finds that

$$\begin{aligned}
& h^4\boldsymbol{F} = \boldsymbol{T}^2\boldsymbol{U} + 2\boldsymbol{TUT} + \boldsymbol{UT}^2 \\
\Longleftrightarrow\quad & h^4\boldsymbol{F} = \boldsymbol{T}^2\boldsymbol{SXS} + 2\boldsymbol{TSXST} + \boldsymbol{SXST}^2 \\
\Longleftrightarrow\quad & h^4\boldsymbol{SFS} = \boldsymbol{ST}^2\boldsymbol{SXS}^2 + 2\boldsymbol{STSXSTS} + \boldsymbol{S}^2\boldsymbol{XST}^2\boldsymbol{S} \\
\Longleftrightarrow\quad & h^4\boldsymbol{SFS} = \boldsymbol{S}^2\boldsymbol{D}^2\boldsymbol{XS}^2 + 2\boldsymbol{S}^2\boldsymbol{DXS}^2\boldsymbol{D} + \boldsymbol{S}^2\boldsymbol{XS}^2\boldsymbol{D}^2 \\
\Longleftrightarrow\quad & h^4\boldsymbol{SFS} = \boldsymbol{ID}^2\boldsymbol{XI}/(4h^2) + 2\boldsymbol{IDXID}/(4h^2) + \boldsymbol{IXID}^2/(4h^2) \\
\Longleftrightarrow\quad & 4h^6\boldsymbol{G} = \boldsymbol{D}^2\boldsymbol{X} + 2\boldsymbol{DXD} + \boldsymbol{XD}^2,
\end{aligned}$$

where $\boldsymbol{G} := \boldsymbol{SFS}$. The (j,k)-th entry of the latter matrix equation is

$$4h^6 g_{jk} = \lambda_j^2 x_{jk} + 2\lambda_j x_{jk}\lambda_k + x_{jk}\lambda_k^2 = x_{jk}(\lambda_j + \lambda_k)^2.$$

Writing $\sigma_j := \sin^2(j\pi h/2) = \lambda_j/4$, one obtains

$$x_{jk} = \frac{4h^6 g_{jk}}{(\lambda_j + \lambda_k)^2} = \frac{4h^6 g_{jk}}{\left(4\sin^2(j\pi h/2) + 4\sin^2(k\pi h/2)\right)^2} = \frac{h^6 g_{jk}}{4(\sigma_j + \sigma_k)^2}.$$

Exercise 11.8: Algorithm for fast solution of biharmonic equation

Use Exercise 11.7 to derive an algorithm `function U=simplefastbiharmonic(F)` which requires only $\mathcal{O}(\delta n^{3/2})$ operations to find $\boldsymbol{U}$ in Exercise 10.9. Here δ is some constant independent of n.

Solution. In order to derive an algorithm that computes $\boldsymbol{U}$ in Exercise 10.9, we can adjust Algorithm 11.1 by replacing the computation of the matrix $\boldsymbol{X}$ by the formula from Exercise 11.7. This adjustment does not change the complexity of Algorithm 11.1, which therefore remains $\mathcal{O}(\delta n^{3/2})$. The new algorithm can be implemented in MATLAB as in Listing 11.2.

code/simplefastbiharmonic.m

```
function U = simplefastbiharmonic(F)
   m = length(F);
   h = 1/(m+1);
   hv = pi*h*(1:m)';
   sigma = sin(hv/2).^2;
   S = sin(hv*(1:m));
   G = S*F*S;
   X = (h^6)*G./(4*(sigma*ones(1,m)+ones(m,1)*sigma').^2);
   U = zeros(m+2,m+2);
   U(2:m+1,2:m+1) = S*X*S;
end
```

Listing 11.2: A simple, fast solution to the biharmonic equation.

Exercise 11.9: Check algorithm for fast solution of biharmonic equation

In Exercise 11.8 compute the solution $\boldsymbol{U}$ corresponding to $\boldsymbol{F} =$ `ones(m,m)`. For some small m's check that you get the same solution obtained by solving the standard form $\boldsymbol{Ax} = \boldsymbol{b}$ in (10.25). You can use $\boldsymbol{x} = \boldsymbol{A}\backslash\boldsymbol{b}$ for solving $\boldsymbol{Ax} = \boldsymbol{b}$. Use `F(:)` to vectorize a matrix and `reshape(x,m,m)` to turn a vector $\boldsymbol{x} \in \mathbb{R}^{m^2}$ into an $m \times m$ matrix. Use the MATLAB command `surf(U)` for plotting $\boldsymbol{U}$ for, say, $m = 50$. Compare the result with Exercise 11.8 by plotting the difference between both matrices.

Solution. The MATLAB function from Listing 11.3 directly solves the standard form $\boldsymbol{Ax} = \boldsymbol{b}$ of Equation (10.25), making sure to return a matrix of the same dimension as the implementation from Listing 11.2.

code/standardbiharmonic.m

```
function V = standardbiharmonic(F)
```

```
    m = length(F);
    h = 1/(m+1);
    T = gallery('tridiag', m, -1, 2, -1);
    A = kron(T^2, eye(m)) + 2*kron(T,T) + kron(eye(m),T^2);
    b = h.^4*F(:);
    x = A\b;
    V = zeros(m+2, m+2);
    V(2:m+1,2:m+1) = reshape(x,m,m);
end
```

Listing 11.3: A direct solution to the biharmonic equation.

After specifying $m = 4$ by issuing the command `F = ones(4,4)`, the commands `simplefastbiharmonic(F)` and `standardbiharmonic(F)` both return the matrix

$$\begin{bmatrix} 0 & 0 & 0 & 0 & 0 & 0 \\ 0 & 0.0015 & 0.0024 & 0.0024 & 0.0015 & 0 \\ 0 & 0.0024 & 0.0037 & 0.0037 & 0.0024 & 0 \\ 0 & 0.0024 & 0.0037 & 0.0037 & 0.0024 & 0 \\ 0 & 0.0015 & 0.0024 & 0.0024 & 0.0015 & 0 \\ 0 & 0 & 0 & 0 & 0 & 0 \end{bmatrix}.$$

For large m, it is more insightful to plot the data returned by our MATLAB functions. For $m = 50$, we solve and plot our system with the commands in Listing 11.4, resulting in Figure 11.1.

code/biharmonic_compare.m

```
F = ones(50, 50);
U = simplefastbiharmonic(F);
V = standardbiharmonic(F);
surf(U);
surf(V);
```

Listing 11.4: Solving the biharmonic equation and plotting the result.

On the face of it, these plots seem to be virtually identical. But exactly how close are they? We investigate this by plotting the difference with the command `surf(U-V)`, which yields Figure 11.2. We conclude that their maximal difference is of the order of 10^{-14}, which makes them indeed very similar.

Exercise 11.10 : Fast solution of biharmonic equation using 9 point rule

Repeat Exercises 10.9, 11.8 and 11.9 using the nine point rule (10.21) to solve the system (10.24).

Solution. We here need to compute

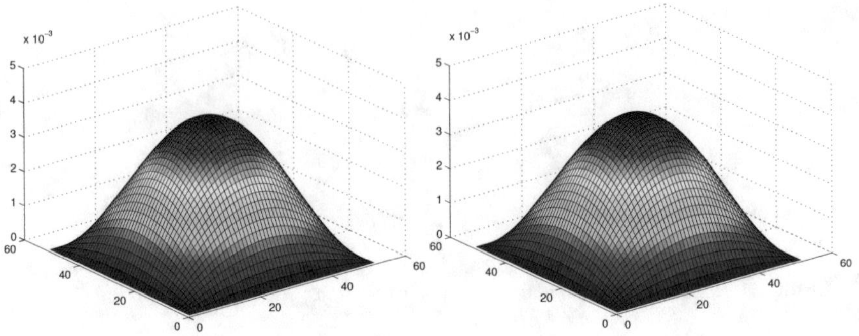

Figure 11.1: Solution of the biharmonic equation using the functions `simplefastbiharmonic` (left) and `standardbiharmonic` (right).

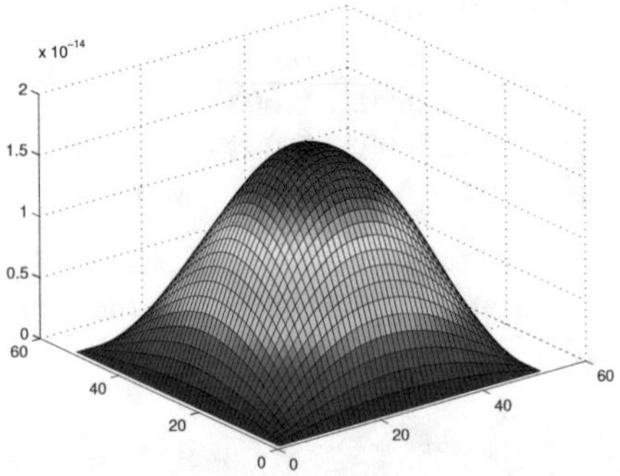

Figure 11.2: Difference of the solutions of the biharmonic equation using the functions `simplefastbiharmonic` and `standardbiharmonic`.

$$
\begin{aligned}
\boldsymbol{A} &= (\boldsymbol{T} \otimes \boldsymbol{I} + \boldsymbol{I} \otimes \boldsymbol{T} - \frac{1}{6}\boldsymbol{T} \otimes \boldsymbol{T})^2 \\
&= \boldsymbol{T}^2 \otimes \boldsymbol{I} + \boldsymbol{I} \otimes \boldsymbol{T}^2 + 2\boldsymbol{T} \otimes \boldsymbol{T} - \frac{1}{3}\boldsymbol{T} \otimes \boldsymbol{T}^2 - \frac{1}{3}\boldsymbol{T}^2 \otimes \boldsymbol{T} + \frac{1}{36}\boldsymbol{T}^2 \otimes \boldsymbol{T}^2.
\end{aligned}
$$

In vector form this can be written as $\boldsymbol{T}^2\boldsymbol{U} + 2\boldsymbol{T}\boldsymbol{U}\boldsymbol{T} + \boldsymbol{U}\boldsymbol{T}^2 - \frac{1}{3}\boldsymbol{T}^2\boldsymbol{U}\boldsymbol{T} - \frac{1}{3}\boldsymbol{T}\boldsymbol{U}\boldsymbol{T}^2 + \frac{1}{36}\boldsymbol{T}^2\boldsymbol{U}\boldsymbol{T}^2$. Following the deductions in Exercise 11.7 we write

$$
\begin{aligned}
h^4\boldsymbol{F} &= \boldsymbol{T}^2\boldsymbol{U} + 2\boldsymbol{T}\boldsymbol{U}\boldsymbol{T} + \boldsymbol{U}\boldsymbol{T}^2 - \frac{1}{3}\boldsymbol{T}^2\boldsymbol{U}\boldsymbol{T} - \frac{1}{3}\boldsymbol{T}\boldsymbol{U}\boldsymbol{T}^2 + \frac{1}{36}\boldsymbol{T}^2\boldsymbol{U}\boldsymbol{T}^2 \\
h^4\boldsymbol{F} &= \boldsymbol{T}^2\boldsymbol{S}\boldsymbol{X}\boldsymbol{S} + 2\boldsymbol{T}\boldsymbol{S}\boldsymbol{X}\boldsymbol{S}\boldsymbol{T} + \boldsymbol{S}\boldsymbol{X}\boldsymbol{S}\boldsymbol{T}^2 \\
&\quad - \frac{1}{3}\boldsymbol{T}^2\boldsymbol{S}\boldsymbol{X}\boldsymbol{S}\boldsymbol{T} - \frac{1}{3}\boldsymbol{T}\boldsymbol{S}\boldsymbol{X}\boldsymbol{S}\boldsymbol{T}^2 + \frac{1}{36}\boldsymbol{T}^2\boldsymbol{S}\boldsymbol{X}\boldsymbol{S}\boldsymbol{T}^2 \\
h^4\boldsymbol{S}\boldsymbol{F}\boldsymbol{S} &= \boldsymbol{S}\boldsymbol{T}^2\boldsymbol{S}\boldsymbol{X}\boldsymbol{S}^2 + 2\boldsymbol{S}\boldsymbol{T}\boldsymbol{S}\boldsymbol{X}\boldsymbol{S}\boldsymbol{T}\boldsymbol{S} + \boldsymbol{S}^2\boldsymbol{X}\boldsymbol{S}\boldsymbol{T}^2\boldsymbol{S} \\
&\quad - \frac{1}{3}\boldsymbol{S}\boldsymbol{T}^2\boldsymbol{S}\boldsymbol{X}\boldsymbol{S}\boldsymbol{T}\boldsymbol{S} - \frac{1}{3}\boldsymbol{S}\boldsymbol{T}\boldsymbol{S}\boldsymbol{X}\boldsymbol{S}\boldsymbol{T}^2\boldsymbol{S} + \frac{1}{36}\boldsymbol{S}\boldsymbol{T}^2\boldsymbol{S}\boldsymbol{X}\boldsymbol{S}\boldsymbol{T}^2\boldsymbol{S}
\end{aligned}
$$

$$
\begin{aligned}
h^4\boldsymbol{SFS} &= \boldsymbol{S}^2\boldsymbol{D}^2\boldsymbol{X}\boldsymbol{S}^2 + 2\boldsymbol{S}^2\boldsymbol{D}\boldsymbol{X}\boldsymbol{S}^2\boldsymbol{D} + \boldsymbol{S}^2\boldsymbol{X}\boldsymbol{S}^2\boldsymbol{D}^2 \\
&\quad - \frac{1}{3}\boldsymbol{S}^2\boldsymbol{D}^2\boldsymbol{X}\boldsymbol{S}^2\boldsymbol{D} - \frac{1}{3}\boldsymbol{S}^2\boldsymbol{D}\boldsymbol{X}\boldsymbol{S}^2\boldsymbol{D}^2 + \frac{1}{36}\boldsymbol{S}^2\boldsymbol{D}^2\boldsymbol{X}\boldsymbol{S}^2\boldsymbol{D}^2 \\
h^4\boldsymbol{SFS} &= (\boldsymbol{D}^2\boldsymbol{X} + 2\boldsymbol{D}\boldsymbol{X}\boldsymbol{D} + \boldsymbol{X}\boldsymbol{D}^2 \\
&\quad - \frac{1}{3}\boldsymbol{D}^2\boldsymbol{X}\boldsymbol{D} - \frac{1}{3}\boldsymbol{D}\boldsymbol{X}\boldsymbol{D}^2 + \frac{1}{36}\boldsymbol{D}^2\boldsymbol{X}\boldsymbol{D}^2)/(4h^2) \\
4h^6\boldsymbol{G} &= \boldsymbol{D}^2\boldsymbol{X} + 2\boldsymbol{D}\boldsymbol{X}\boldsymbol{D} + \boldsymbol{X}\boldsymbol{D}^2 - \frac{1}{3}\boldsymbol{D}^2\boldsymbol{X}\boldsymbol{D} - \frac{1}{3}\boldsymbol{D}\boldsymbol{X}\boldsymbol{D}^2 + \frac{1}{36}\boldsymbol{D}^2\boldsymbol{X}\boldsymbol{D}^2.
\end{aligned}
$$

Note that these deductions are very general: If we had an even more complex expression for $\boldsymbol{A}$, the same deductions apply so that $\boldsymbol{T}$ can be replaced in terms of $\boldsymbol{D}$ everywhere using that $\boldsymbol{TS} = \boldsymbol{SD}$. The (j,k)-th entry of the latter is

$$
4h^6 g_{jk} = \left(\lambda_j^2 + 2\lambda_j\lambda_k + \lambda_k^2 - \frac{1}{3}\lambda_j^2\lambda_k - \frac{1}{3}\lambda_j\lambda_k^2 + \frac{1}{36}\lambda_j^2\lambda_k^2\right) x_{jk}.
$$

Writing $\sigma_j := \sin^2(j\pi h/2) = \lambda_j/4$, one obtains

$$
\begin{aligned}
x_{jk} &= \frac{4h^6 g_{jk}}{(\lambda_j+\lambda_k)^2 - \frac{1}{3}\lambda_j^2\lambda_k - \frac{1}{3}\lambda_j\lambda_k^2 + \frac{1}{36}\lambda_j^2\lambda_k^2} \\
&= \frac{h^6 g_{jk}}{4\big(\sigma_j+\sigma_k\big)^2 - \frac{16}{3}\sigma_j^2\sigma_k - \frac{16}{3}\sigma_j\sigma_k^2 + \frac{16}{9}\sigma_j^2\sigma_k^2}.
\end{aligned}
$$

To replace the implementation of the fast solution to the biharmonic equation for the nine-point scheme, we simply replace a line in `simplefastbiharmonic.m` with the above. Otherwise the code is unchanged. This is shown in Listing 11.5 below.

code/simplefastbiharmonic_ninepointscheme.m

```
function U = simplefastbiharmonic_ninepointscheme(F)
    m = length(F);
    h = 1/(m+1);
    hv = pi*h*(1:m)';
    sigma = sin(hv/2).^2;
    S = sin(hv*(1:m));
    G = S*F*S;
    X = (h^6)*G./( 4*(sigma*ones(1,m)+ones(m,1)*sigma').^2 -
        (16/3)*sigma.*((sigma').^2) - (16/3)*(sigma.^2).*sigma
        ' + (16/9)*(sigma.^2).*((sigma').^2));
    U = zeros(m+2,m+2);
    U(2:m+1,2:m+1) = S*X*S;
end
```

Listing 11.5: A simple, fast solution to the biharmonic equation using the nine-point scheme.

In the code for solving the biharmonic equation directly, one simply replaces

$$T^2 \otimes I + I \otimes T^2 + 2T \otimes T$$

with

$$T^2 \otimes I + I \otimes T^2 + 2T \otimes T - \frac{1}{3}T \otimes T^2 - \frac{1}{3}T^2 \otimes T + \frac{1}{36}T^2 \otimes T^2,$$

so that only one line is changed also here. This is shown in Listing 11.6 below.

code/standardbiharmonic_ninepointscheme.m

```
function V = standardbiharmonic_ninepointscheme(F)
   m = length(F);
   h = 1/(m+1);
   T = gallery('tridiag', m, -1, 2, -1);
   A = kron(T^2, eye(m)) + 2*kron(T,T) + kron(eye(m),T^2) -
       (1/3)*kron(T^2,T) - 1/3*kron(T,T^2) + (1/36)*kron(T^2,
       T^2);
   b = h.^4*F(:);
   x = A\b;
   V = zeros(m+2, m+2);
   V(2:m+1,2:m+1) = reshape(x,m,m);
end
```

Listing 11.6: A direct solution to the biharmonic equation using the nine-point scheme.

The code where the two approaches for the nine-point scheme are compared is shown in Listing 11.7.

code/biharmonic_compare_ninepointscheme.m

```
F = ones(50, 50);
U = simplefastbiharmonic_ninepointscheme(F);
V = standardbiharmonic_ninepointscheme(F);
surf(U);
figure()
surf(V);
```

Listing 11.7: Solving the biharmonic equation and plotting the result using the nine-point scheme.

The plots look similar to those from the standard scheme.

Chapter 12
The Classical Iterative Methods

Exercises section 12.3

Exercise 12.1: Richardson and Jacobi

Show that if $a_{ii} = d \neq 0$ for all i then Richardson's method with $\alpha := 1/d$ is the same as Jacobi's method.

Solution. If $a_{11} = \cdots = a_{nn} = d \neq 0$ and $\alpha = 1/d$, Richardson's method (12.18) yields, for $i = 1, \ldots, n$,

$$\begin{aligned}
\boldsymbol{x}_{k+1}(i) &= \boldsymbol{x}_k(i) + \frac{1}{d}\left(b_i - \sum_{j=1}^{n} a_{ij}\boldsymbol{x}_k(j)\right) \\
&= \frac{1}{d}\left(d\boldsymbol{x}_k(i) - \sum_{j=1}^{n} a_{ij}\boldsymbol{x}_k(j) + b_i\right) \\
&= \frac{1}{a_{ii}}\left(a_{ii}\boldsymbol{x}_k(i) - \sum_{j=1}^{n} a_{ij}\boldsymbol{x}_k(j) + b_i\right) \\
&= \frac{1}{a_{ii}}\left(-\sum_{j=1}^{i-1} a_{ij}\boldsymbol{x}_k(j) - \sum_{j=i+1}^{n} a_{ij}\boldsymbol{x}_k(j) + b_i\right),
\end{aligned}$$

which is identical to Jacobi's method (12.2).

T. Lyche et al., *Exercises in Numerical Linear Algebra and Matrix Factorizations*, Texts in Computational Science and Engineering 23, https://doi.org/10.1007/978-3-030-59789-4_12

Exercise 12.2: R-method when eigenvalues have positive real part

Suppose all eigenvalues λ_j of $\boldsymbol{A}$ have positive real parts u_j for $j = 1, \ldots, n$ and that α is real. Show that the R-method converges if and only if $0 < \alpha < \min_j(2u_j/|\lambda_j|^2)$.

Solution. We can write Richardson's method as $\boldsymbol{x}_{k+1} = \boldsymbol{G}\boldsymbol{x}_k + \boldsymbol{c}$, with $\boldsymbol{G} = \boldsymbol{I} - \alpha\boldsymbol{A}$, $\boldsymbol{c} = \alpha\boldsymbol{b}$. We know from Theorem 12.4 that the method converges if and only if $\rho(\boldsymbol{G}) < 1$. The eigenvalues of $\boldsymbol{I} - \alpha\boldsymbol{A}$ are $1 - \alpha\lambda_j$, and we have that

$$|1 - \alpha\lambda_j|^2 = 1 + \alpha^2|\lambda_j|^2 - 2\alpha\Re(\lambda_j) = 1 + \alpha^2|\lambda_j|^2 - 2\alpha u_j.$$

This is less than 1 if and only if $\alpha^2|\lambda_j|^2 < 2\alpha u_j$. This can only hold if $\alpha > 0$, since $u_j > 0$. Dividing by α we get that $\alpha|\lambda_j|^2 < 2u_j$, so that $\alpha < 2u_j/|\lambda_j|^2$ (since $|\lambda_j| > 0$ since $u_j \neq 0$). We thus have that $\rho(G) < 1$ if and only if $\alpha < \min_j(2u_j/|\lambda_j|^2)$, and the result follows.

Exercise 12.3: Divergence example for J and GS

Show that both Jacobi's method and Gauss-Seidel's method diverge for $\boldsymbol{A} = \left[\begin{smallmatrix}1 & 2\\ 3 & 4\end{smallmatrix}\right]$.

Solution. We compute the matrices $\boldsymbol{G}_J$ and $\boldsymbol{G}_1$ from $\boldsymbol{A}$ and show that that the spectral radii $\rho(\boldsymbol{G}_J), \rho(\boldsymbol{G}_1) \geq 1$. Once this is shown, Theorem 12.4 implies that the Jacobi method and Gauss-Seidel's method diverge.

Write $\boldsymbol{A} = \boldsymbol{D} - \boldsymbol{A}_L - \boldsymbol{A}_R$, with $\boldsymbol{D}$ the diagonal part of $\boldsymbol{A}$, $\boldsymbol{A}_L$ lower triangular, and $\boldsymbol{A}_R$ upper triangular. From Equation (12.12), we find

$$\boldsymbol{G}_J = \boldsymbol{I} - \boldsymbol{M}_J^{-1}\boldsymbol{A} = \boldsymbol{I} - \boldsymbol{D}^{-1}\boldsymbol{A} = \begin{bmatrix}1 & 0\\ 0 & 1\end{bmatrix} - \begin{bmatrix}1 & 0\\ 0 & \frac{1}{4}\end{bmatrix}\begin{bmatrix}1 & 2\\ 3 & 4\end{bmatrix} = \begin{bmatrix}0 & -2\\ -\frac{3}{4} & 0\end{bmatrix},$$

$$\boldsymbol{G}_1 = \boldsymbol{I} - \boldsymbol{M}_1^{-1}\boldsymbol{A} = \boldsymbol{I} - (\boldsymbol{D} - \boldsymbol{A}_L)^{-1}\boldsymbol{A} = \begin{bmatrix}1 & 0\\ 0 & 1\end{bmatrix} - \begin{bmatrix}1 & 0\\ -\frac{3}{4} & \frac{1}{4}\end{bmatrix}\begin{bmatrix}1 & 2\\ 3 & 4\end{bmatrix} = \begin{bmatrix}0 & -2\\ 0 & \frac{3}{2}\end{bmatrix}.$$

From this, we find $\rho(\boldsymbol{G}_J) = \sqrt{3/2}$ and $\rho(\boldsymbol{G}_1) = 3/2$, both of which are greater than 1.

Exercise 12.4: 2 by 2 matrix

We want to show that Gauss-Seidel converges if and only if Jacobi converges for a 2 by 2 matrix $\boldsymbol{A} := \left[\begin{smallmatrix}a_{11} & a_{12}\\ a_{21} & a_{22}\end{smallmatrix}\right] \in \mathbb{R}^{2\times 2}$.

a) Show that the spectral radius for the Jacobi method is

$$\rho(\boldsymbol{G}_J) = \sqrt{|a_{21}a_{12}/a_{11}a_{22}|}.$$

Solution. The splitting matrix for the Jacobi method, and its inverse, are

$$\boldsymbol{M}_J = \boldsymbol{D} = \begin{bmatrix} a_{11} & 0 \\ 0 & a_{22} \end{bmatrix}, \qquad \boldsymbol{M}_J^{-1} = \begin{bmatrix} a_{11}^{-1} & 0 \\ 0 & a_{22}^{-1} \end{bmatrix}.$$

It follows that

$$\boldsymbol{G}_J = \boldsymbol{I} - \boldsymbol{M}_J^{-1}\boldsymbol{A} = \begin{bmatrix} 1 - a_{11}^{-1}a_{11} & -a_{11}^{-1}a_{12} \\ -a_{22}^{-1}a_{21} & 1 - a_{22}^{-1}a_{22} \end{bmatrix} = \begin{bmatrix} 0 & -a_{11}^{-1}a_{12} \\ -a_{22}^{-1}a_{21} & 0 \end{bmatrix},$$

which has eigenvalues $\pm\sqrt{a_{21}a_{12}/a_{11}a_{22}}$, and therefore spectral radius

$$\rho(\boldsymbol{G}_J) = \sqrt{|a_{21}a_{12}/a_{11}a_{22}|}.$$

b) Show that the spectral radius for the Gauss-Seidel method is

$$\rho(\boldsymbol{G}_1) = |a_{21}a_{12}/a_{11}a_{22}|.$$

Solution. The splitting matrix for the Gauss-Seidel method, and its inverse, are

$$\boldsymbol{M}_1 = \boldsymbol{D} - \boldsymbol{A}_L = \begin{bmatrix} a_{11} & 0 \\ a_{21} & a_{22} \end{bmatrix}, \qquad \boldsymbol{M}_1^{-1} = \frac{1}{a_{11}a_{22}} \begin{bmatrix} a_{22} & 0 \\ -a_{21} & a_{11} \end{bmatrix}.$$

It follows that

$$\boldsymbol{G}_1 = \boldsymbol{I} - \boldsymbol{M}_1^{-1}\boldsymbol{A} = \boldsymbol{I} - \frac{1}{a_{11}a_{22}} \begin{bmatrix} a_{22}a_{11} & a_{22}a_{12} \\ 0 & a_{11}a_{22} - a_{12}a_{21} \end{bmatrix} = \begin{bmatrix} 0 & -\frac{a_{12}}{a_{11}} \\ 0 & \frac{a_{12}a_{21}}{a_{11}a_{22}} \end{bmatrix},$$

which has eigenvalues 0 and $\frac{a_{12}a_{21}}{a_{11}a_{22}}$, and therefore spectral radius

$$\rho(\boldsymbol{G}_1) = |a_{21}a_{12}/a_{11}a_{22}|.$$

c) Conclude that Gauss-Seidel converges if and only if Jacobi converges.

Solution. From **a)** and **b)** it follows that $\rho(\boldsymbol{G}_J) = \sqrt{\rho(\boldsymbol{G}_1)}$, so that $\rho(\boldsymbol{G}_J) < 1$ if and only if $\rho(\boldsymbol{G}_1) < 1$. Since an iterative method converges if and only if $\rho(\boldsymbol{G}) < 1$, the result follows.

Exercise 12.5: Example: GS converges, J diverges

Show (by finding its eigenvalues) that the matrix

$$\begin{bmatrix} 1 & a & a \\ a & 1 & a \\ a & a & 1 \end{bmatrix}$$

is positive definite for $-1/2 < a < 1$. Thus, GS converges for these values of a. Show that the J method does not converge for $1/2 < a < 1$.

Solution. The eigenvalues of $\boldsymbol{A}$ are the zeros of $\det(\boldsymbol{A}-\lambda\boldsymbol{I}) = (-\lambda+2a+1)(\lambda+a-1)^2$. We find eigenvalues $\lambda_1 := 2a+1$ and $\lambda_2 := 1-a$, the latter having algebraic multiplicity two. Whenever $1/2 < a < 1$ these eigenvalues are positive, implying that $\boldsymbol{A}$ is positive definite for such a.

Let's compute the spectral radius of $\boldsymbol{G}_J = \boldsymbol{I} - \boldsymbol{D}^{-1}\boldsymbol{A}$, where $\boldsymbol{D}$ is the diagonal part of $\boldsymbol{A}$. The eigenvalues of $\boldsymbol{G}_J$ are the zeros of the characteristic polynomial

$$\det(\boldsymbol{G}_J - \lambda\boldsymbol{I}) = \begin{vmatrix} -\lambda & -a & -a \\ -a & -\lambda & -a \\ -a & -a & -\lambda \end{vmatrix} = (-\lambda-2a)(a-\lambda)^2,$$

and we find spectral radius $\rho(\boldsymbol{G}_J) = \max\{|a|, |2a|\}$. It follows that $\rho(\boldsymbol{G}_J) > 1$ whenever $1/2 < a < 1$, in which case Theorem 12.4 implies that the Jacobi method does not converge (even though $\boldsymbol{A}$ is symmetric positive definite).

Exercise 12.6: Example: GS diverges, J converges

Let $\boldsymbol{G}_J$ and $\boldsymbol{G}_1$ be the iteration matrices for the Jacobi and Gauss-Seidel methods applied to the matrix $\boldsymbol{A} := \begin{bmatrix} 1 & 0 & 1/2 \\ 1 & 1 & 0 \\ -1 & 1 & 1 \end{bmatrix}$[a].

a) Show that $\boldsymbol{G}_1 := \begin{bmatrix} 0 & 0 & -1/2 \\ 0 & 0 & 1/2 \\ 0 & 0 & -1 \end{bmatrix}$ and conclude that GS diverges.

[a] S. Venit, "The convergence of Jacobi and Gauss-Seidel iteration", Mathematics Magazine **48** (1975), 163–167.

Solution. The splitting matrix, and its inverse, are

$$\boldsymbol{M}_1 = \boldsymbol{D} - \boldsymbol{A}_L = \begin{bmatrix} 1 & 0 & 0 \\ 1 & 1 & 0 \\ -1 & 1 & 1 \end{bmatrix}, \qquad \boldsymbol{M}_1^{-1} = \begin{bmatrix} 1 & 0 & 0 \\ -1 & 1 & 0 \\ 2 & -1 & 1 \end{bmatrix},$$

so that

$$\boldsymbol{G}_1 = \boldsymbol{I} - \boldsymbol{M}_1^{-1}\boldsymbol{A} = \boldsymbol{I} - \begin{bmatrix} 1 & 0 & 1/2 \\ 0 & 1 & -1/2 \\ 0 & 0 & 2 \end{bmatrix} = \begin{bmatrix} 0 & 0 & -1/2 \\ 0 & 0 & 1/2 \\ 0 & 0 & -1 \end{bmatrix}.$$

This matrix has eigenvalues 0 and -1, and therefore $\rho(\boldsymbol{G}_1) = 1$, implying that Gauss-Seidel diverges for this matrix.

b) Show that $p(\lambda) := \det(\lambda\boldsymbol{I} - \boldsymbol{G}_J) = \lambda^3 + \frac{1}{2}\lambda + \frac{1}{2}$.

Solution. The splitting matrix $\boldsymbol{M}_J = \boldsymbol{D} = \boldsymbol{I}$, so that also $\boldsymbol{M}_J^{-1} = \boldsymbol{I}$, and

$$\boldsymbol{G}_J = \boldsymbol{D} - \boldsymbol{M}_J^{-1}\boldsymbol{A} = \boldsymbol{D} - \boldsymbol{A} = \begin{bmatrix} 0 & 0 & -1/2 \\ -1 & 0 & 0 \\ 1 & -1 & 0 \end{bmatrix},$$

so that

$$\det(\lambda \boldsymbol{I} - \boldsymbol{G}_J) = \begin{vmatrix} \lambda & 0 & 1/2 \\ 1 & \lambda & 0 \\ -1 & 1 & \lambda \end{vmatrix} = \lambda \begin{vmatrix} \lambda & 0 \\ 1 & \lambda \end{vmatrix} + \frac{1}{2} \begin{vmatrix} 1 & \lambda \\ -1 & 1 \end{vmatrix} = \lambda^3 + \frac{1}{2}\lambda + \frac{1}{2}.$$

c) Show that if $|\lambda| \geq 1$ then $p(\lambda) \neq 0$. Conclude that J converges.

Solution. Since $p'(\lambda) = 3\lambda^2 + \frac{1}{2} > 0$, p is increasing. Since $p(1) = 2$, $p(\lambda) > 0$ for $\lambda \geq 1$. Since $p(-1) = -1$, $p(\lambda) \leq -1$ for $\lambda \leq -1$. It follows that $p(\lambda)$ can be zero only if $-1 < \lambda < 1$, so that all eigenvalues are in $(-1, 1)$. It follows that $\rho(\boldsymbol{G}_J) < 1$, so that the Jacobi method converges.

Exercise 12.7: Strictly diagonally dominance; the J method

Show that the J method converges if $|a_{ii}| > \sum_{j \neq i} |a_{ij}|$ for $i = 1, \ldots, n$.

Solution. If, as assumed in the exercise, $\boldsymbol{A} = (a_{ij})_{ij}$ is strictly diagonally dominant, then it is nonsingular and $a_{11}, \ldots, a_{nn} \neq 0$. For the Jacobi method, one finds

$$\boldsymbol{G}_J = \boldsymbol{I} - \operatorname{diag}(a_{11}, \ldots, a_{nn})^{-1} \boldsymbol{A} = \begin{bmatrix} 0 & -\frac{a_{12}}{a_{11}} & -\frac{a_{13}}{a_{11}} & \cdots & -\frac{a_{1n}}{a_{11}} \\ -\frac{a_{21}}{a_{22}} & 0 & -\frac{a_{23}}{a_{22}} & \cdots & -\frac{a_{2n}}{a_{22}} \\ -\frac{a_{31}}{a_{33}} & -\frac{a_{32}}{a_{33}} & 0 & \cdots & -\frac{a_{3n}}{a_{33}} \\ \vdots & \vdots & \vdots & \ddots & \vdots \\ -\frac{a_{n1}}{a_{nn}} & -\frac{a_{n2}}{a_{nn}} & -\frac{a_{n3}}{a_{nn}} & \cdots & 0 \end{bmatrix}.$$

By Theorem 8.3, the ∞-norm can be expressed as the maximum, over all rows, of the sum of absolute values of the entries in a row. Using that $\boldsymbol{A}$ is strictly diagonally dominant, one finds

$$\|\boldsymbol{G}_J\|_\infty = \max_{1 \leq i \leq n} \sum_{j \neq i} \left| -\frac{a_{ij}}{a_{ii}} \right| = \max_{1 \leq i \leq n} \frac{1}{|a_{ii}|} \sum_{j \neq i} |a_{ij}| < 1.$$

As by Lemma 8.1 the ∞-norm is consistent, Corollary 12.3 implies that the Jacobi method converges for any strictly diagonally dominant matrix $\boldsymbol{A}$.

Exercise 12.8: Strictly diagonally dominance; the GS method

Consider the GS method. Let $\boldsymbol{x}$ be the exact solution to $\boldsymbol{A}\boldsymbol{x} = \boldsymbol{b}$, and let $\epsilon := \boldsymbol{x}_k - \boldsymbol{x}$ be its difference to the approximate solution $\boldsymbol{x}_k$ at iteration k. Suppose $r := \max_i r_i < 1$, where $r_i = \sum_{j \neq i} \frac{|a_{ij}|}{|a_{ii}|}$. Show using induction on i that $|\epsilon_{k+1}(j)| \leq r \|\epsilon_k\|_\infty$ for $j = 1, \ldots, i$. Conclude that Gauss-Seidel's method is convergent when $\boldsymbol{A}$ is strictly diagonally dominant.

Solution. Let $\boldsymbol{A} = -\boldsymbol{A}_L + \boldsymbol{D} - \boldsymbol{A}_R$ be decomposed as a sum of a lower triangular, a diagonal, and an upper triangular part. By Equation (12.3), the approximate solutions $\boldsymbol{x}_k$ are related by

$$\boldsymbol{D}\boldsymbol{x}_{k+1} = \boldsymbol{A}_L\boldsymbol{x}_{k+1} + \boldsymbol{A}_R\boldsymbol{x}_k + \boldsymbol{b}$$

in the Gauss Seidel method. Let $\boldsymbol{x}$ be the exact solution of $\boldsymbol{A}\boldsymbol{x} = \boldsymbol{b}$. It follows that the errors $\boldsymbol{\epsilon}_k := \boldsymbol{x}_k - \boldsymbol{x}$ are related by

$$\boldsymbol{D}\boldsymbol{\epsilon}_{k+1} = \boldsymbol{A}_L\boldsymbol{\epsilon}_{k+1} + \boldsymbol{A}_R\boldsymbol{\epsilon}_k.$$

Let r and r_i be as in the exercise. Let $k \geq 0$ be arbitrary. We show by induction that

$$|\boldsymbol{\epsilon}_{k+1}(j)| \leq r\|\boldsymbol{\epsilon}_k\|_\infty, \qquad \text{for } j = 1, 2\ldots, n. \tag{12.i}$$

For $j = 1$, the relation between the errors translates to

$$|\boldsymbol{\epsilon}_{k+1}(1)| = |a_{11}|^{-1}\,|-a_{12}\boldsymbol{\epsilon}_k(2) - \cdots - a_{1n}\boldsymbol{\epsilon}_k(n)| \leq r_1\|\boldsymbol{\epsilon}_k\|_\infty \leq r\|\boldsymbol{\epsilon}_k\|_\infty.$$

Assume that (12.i) holds for $1, \ldots, j-1$. The relation between the errors then yields the bound

$$\begin{aligned}|\boldsymbol{\epsilon}_{k+1}(j)| &\leq |a_{jj}|^{-1}\left|-\sum_{i=1}^{j-1} a_{j,i}\boldsymbol{\epsilon}_{k+1}(i) - \sum_{i=j+1}^{n} a_{j,i}\boldsymbol{\epsilon}_k(i)\right| \\ &\leq r_j \max\{r\|\boldsymbol{\epsilon}_k\|_\infty, \|\boldsymbol{\epsilon}_k\|_\infty\} = r_j\|\boldsymbol{\epsilon}_k\|_\infty \leq r\|\boldsymbol{\epsilon}_k\|_\infty.\end{aligned}$$

Equation (12.i) then follows by induction, and it also follows that $\|\boldsymbol{\epsilon}_{k+1}\|_\infty \leq r\|\boldsymbol{\epsilon}_k\|_\infty$. The matrix $\boldsymbol{A}$ is strictly diagonally dominant precisely when $r < 1$, implying

$$\lim_{k\to\infty} \|\boldsymbol{\epsilon}_k\|_\infty \leq \|\boldsymbol{\epsilon}_0\|_\infty \lim_{k\to\infty} r^k = 0.$$

We conclude that the Gauss Seidel method converges for strictly diagonally dominant matrices.

Exercise 12.9: Convergence example for fix point iteration

Consider for $a \in \mathbb{C}$

$$\boldsymbol{x} := \begin{bmatrix} x_1 \\ x_2 \end{bmatrix} = \begin{bmatrix} 0 & a \\ a & 0 \end{bmatrix} \begin{bmatrix} x_1 \\ x_2 \end{bmatrix} + \begin{bmatrix} 1-a \\ 1-a \end{bmatrix} =: \boldsymbol{G}\boldsymbol{x} + \boldsymbol{c}.$$

Starting with $\boldsymbol{x}_0 = \boldsymbol{0}$ show by induction

$$\boldsymbol{x}_k(1) = \boldsymbol{x}_k(2) = 1 - a^k, \quad k \geq 0,$$

and conclude that the iteration converges to the fixed-point $\boldsymbol{x} = [1,1]^{\mathrm{T}}$ for $|a| < 1$ and diverges for $|a| > 1$. Show that $\rho(\boldsymbol{G}) = 1 - \eta$ with $\eta = 1 - |a|$. Compute the estimate (12.31) for the rate of convergence for $a = 0.9$ and $s = 16$ and compare with the true number of iterations determined from $|a|^k \leq 10^{-16}$.

Solution. We show by induction that $\boldsymbol{x}_k(1) = \boldsymbol{x}_k(2) = 1 - a^k$ for every $k \geq 0$. Clearly the formula holds for $k = 0$. Assume the formula holds for some fixed k. Then

$$\boldsymbol{x}_{k+1} = \boldsymbol{G}\boldsymbol{x}_k + \boldsymbol{c} = \begin{bmatrix} 0 & a \\ a & 0 \end{bmatrix} \begin{bmatrix} 1-a^k \\ 1-a^k \end{bmatrix} + \begin{bmatrix} 1-a \\ 1-a \end{bmatrix} = \begin{bmatrix} 1-a^{k+1} \\ 1-a^{k+1} \end{bmatrix}.$$

It follows that the formula holds for any $k \geq 0$. When $|a| < 1$ we can evaluate the limit

$$\lim_{k\to\infty} \boldsymbol{x}_k(i) = \lim_{k\to\infty} 1 - a^k = 1 - \lim_{k\to\infty} a^k = 1, \qquad \text{for } i = 1, 2.$$

When $|a| > 1$, however, $|\boldsymbol{x}_k(1)| = |\boldsymbol{x}_k(2)| = |1 - a^k|$ becomes arbitrary large with k and $\lim_{k\to\infty} \boldsymbol{x}_k(i)$ diverges.

The eigenvalues of $\boldsymbol{G}$ are the zeros of the characteristic polynomial $\lambda^2 - a^2 = (\lambda - a)(\lambda + a)$, and we find that $\boldsymbol{G}$ has spectral radius $\rho(\boldsymbol{G}) = 1 - \eta$, where $\eta := 1 - |a|$. Equation (12.31) yields an estimate $\tilde{k} = \log(10)s/(1 - |a|)$ for the smallest number of iterations k so that $\rho(\boldsymbol{G})^k \leq 10^{-s}$. In particular, taking $a = 0.9$ and $s = 16$, one expects at least $\tilde{k} = 160\log(10) \approx 368$ iterations before $\rho(\boldsymbol{G})^k \leq 10^{-16}$. On the other hand, $0.9^k = |a|^k = 10^{-s} = 10^{-16}$ when $k \approx 350$, so in this case the estimate is fairly accurate.

Exercise 12.10: Estimate in Lemma 12.1 can be exact

Consider the iteration in Example 12.2. Show that $\rho(\boldsymbol{G}_J) = 1/2$. Then show that $\boldsymbol{x}_k(1) = \boldsymbol{x}_k(2) = 1 - 2^{-k}$ for $k \geq 0$. Thus the estimate in Lemma 12.1 is exact in this case.

Solution. As the eigenvalues of the matrix $\boldsymbol{G}_J$ are the zeros of $\lambda^2 - 1/4 = (\lambda - 1/2)(\lambda + 1/2) = 0$, one finds the spectral radius $\rho(\boldsymbol{G}_J) = 1/2$. In this example, the

Jacobi iteration process is described by

$$\boldsymbol{x}_{k+1} = \boldsymbol{G}_J \boldsymbol{x}_k + \boldsymbol{c}, \qquad \boldsymbol{G}_J = \begin{bmatrix} 0 & \frac{1}{2} \\ \frac{1}{2} & 0 \end{bmatrix}, \qquad \boldsymbol{c} = \begin{bmatrix} 2 & 0 \\ 0 & 2 \end{bmatrix}^{-1} \begin{bmatrix} 1 \\ 1 \end{bmatrix} = \begin{bmatrix} \frac{1}{2} \\ \frac{1}{2} \end{bmatrix}.$$

The initial guess

$$\boldsymbol{x}_0 = \begin{bmatrix} 0 \\ 0 \end{bmatrix}$$

satisfies the formula $\boldsymbol{x}_k(1) = \boldsymbol{x}_k(2) = 1 - 2^{-k}$ for $k = 0$. Moreover, if this formula holds for some $k \geq 0$, one finds

$$\boldsymbol{x}_{k+1} = \boldsymbol{G}_J \boldsymbol{x}_k + \boldsymbol{c} = \begin{bmatrix} 0 & \frac{1}{2} \\ \frac{1}{2} & 0 \end{bmatrix} \begin{bmatrix} 1 - 2^{-k} \\ 1 - 2^{-k} \end{bmatrix} + \begin{bmatrix} \frac{1}{2} \\ \frac{1}{2} \end{bmatrix} = \begin{bmatrix} 1 - 2^{-(k+1)} \\ 1 - 2^{-(k+1)} \end{bmatrix},$$

which means that it must then hold for $k+1$ as well. By induction we can conclude that the formula holds for all $k \geq 0$.

At iteration k, each entry of the approximation $\boldsymbol{x}_k$ differs by 2^{-k} from the fixed point, implying that $\|\boldsymbol{\epsilon}_k\|_\infty = 2^{-k}$. Therefore, for given s, the error $\|\boldsymbol{\epsilon}_k\|_\infty \leq 10^{-s}$ for the first time at $k = \lceil s \log(10)/\log(2) \rceil$. The bound $-s\log(10)/\log(\rho(\boldsymbol{G}))$ gives the same.

Exercise 12.11: Iterative method (Exam exercise 1991-3)

Let $\boldsymbol{A} \in \mathbb{R}^{n \times n}$ be a symmetric positive definite matrix with ones on the diagonal and let $\boldsymbol{b} \in \mathbb{R}^n$. We will consider an iterative method for the solution of $\boldsymbol{A}\boldsymbol{x} = \boldsymbol{b}$. Observe that $\boldsymbol{A}$ may be written $\boldsymbol{A} = \boldsymbol{I} - \boldsymbol{L} - \boldsymbol{L}^{\mathrm{T}}$, where $\boldsymbol{L}$ is lower triangular with zeros on the diagonal, $l_{i,j} = 0$, when $j \geq i$. The method is defined by

$$\boldsymbol{M}\boldsymbol{x}_{k+1} = \boldsymbol{N}\boldsymbol{x}_k + \boldsymbol{b}, \tag{12.45}$$

where $\boldsymbol{M}$ and $\boldsymbol{N}$ are given by the splitting

$$\boldsymbol{A} = \boldsymbol{M} - \boldsymbol{N}, \qquad \boldsymbol{M} = (\boldsymbol{I} - \boldsymbol{L})(\boldsymbol{I} - \boldsymbol{L}^{\mathrm{T}}), \qquad \boldsymbol{N} = \boldsymbol{L}\boldsymbol{L}^{\mathrm{T}}. \tag{12.46}$$

a) Let $\boldsymbol{x} \neq \boldsymbol{0}$ be an eigenvector of $\boldsymbol{M}^{-1}\boldsymbol{N}$ with eigenvalue λ. Show that

$$\lambda = \frac{\boldsymbol{x}^{\mathrm{T}}\boldsymbol{N}\boldsymbol{x}}{\boldsymbol{x}^{\mathrm{T}}\boldsymbol{A}\boldsymbol{x} + \boldsymbol{x}^{\mathrm{T}}\boldsymbol{N}\boldsymbol{x}}. \tag{12.47}$$

Solution. If $\boldsymbol{M}^{-1}\boldsymbol{N}\boldsymbol{x} = \lambda\boldsymbol{x}$ then $\boldsymbol{N}\boldsymbol{x} = \lambda\boldsymbol{M}\boldsymbol{x} = \lambda(\boldsymbol{A} + \boldsymbol{N})\boldsymbol{x}$. We multiply by $\boldsymbol{x}^{\mathrm{T}}$ and obtain $\boldsymbol{x}^{\mathrm{T}}\boldsymbol{N}\boldsymbol{x} = \lambda\boldsymbol{x}^{\mathrm{T}}(\boldsymbol{A} + \boldsymbol{N})\boldsymbol{x} = \lambda(\boldsymbol{x}^{\mathrm{T}}\boldsymbol{A}\boldsymbol{x} + \boldsymbol{x}^{\mathrm{T}}\boldsymbol{N}\boldsymbol{x})$ which after a division gives the result.

b) Show that the sequence $\{\boldsymbol{x}_k\}$ generated by (12.45) converges to the solution $\boldsymbol{x}$ of $\boldsymbol{A}\boldsymbol{x} = \boldsymbol{b}$ for any starting vector $\boldsymbol{x}_0$.

Solution. It is enough to show that all eigenvalues of $\boldsymbol{M}^{-1}\boldsymbol{N}$ are less than one in magnitude. Let $(\lambda, \boldsymbol{x})$ be an eigenpair of $\boldsymbol{M}^{-1}\boldsymbol{N}$. It satisfies (12.47). Since $\boldsymbol{A}$ is positive definite $\boldsymbol{x}^{\mathrm{T}}\boldsymbol{A}\boldsymbol{x} > 0$. Also, $\boldsymbol{x}^{\mathrm{T}}\boldsymbol{N}\boldsymbol{x} = \boldsymbol{x}^{\mathrm{T}}\boldsymbol{L}\boldsymbol{L}^{\mathrm{T}}\boldsymbol{x} = \|\boldsymbol{L}^{\mathrm{T}}\boldsymbol{x}\|_2^2 \geq 0$. Due to this (12.47) implies $\lambda \in [0, 1)$.

c) Consider the following algorithm

1. Choose $\boldsymbol{x} = [x(1), x(2), \ldots, x(n)]^{\mathrm{T}}$.
2. for $k = 1, 2, 3, \ldots$
   for $i = 1, 2, \ldots, n-1, n, n, n-1, n-2, \ldots, 1$
   $x(i) = b(i) - \sum_{j \neq i} a(i, j)x(j)$

(i.e., Algorithm (12.48) in the book). Is there a connection between this algorithm and the method of Gauss-Seidel? Show that the algorithm (12.48) leads up to the splitting (12.46).

Solution. The algorithm consists of two Gauss-Seidel iterations in each iteration. One in the order $1, 2, \ldots, n$ (forward) and then one in the order $n, n-1, \ldots, 1$ (backwards). We write the solution of $\boldsymbol{A}\boldsymbol{x} = \boldsymbol{b}$ as $\boldsymbol{x} = \boldsymbol{L}\boldsymbol{x} + \boldsymbol{L}^{\mathrm{T}}\boldsymbol{x} + b$. Forward and backward Gauss-Seidel can be written $\boldsymbol{y}_{k+1} = \boldsymbol{L}\boldsymbol{y}_{k+1} + \boldsymbol{L}^{\mathrm{T}}\boldsymbol{x}_k + \boldsymbol{b}$ and $\boldsymbol{x}_{k+1} = \boldsymbol{L}\boldsymbol{y}_{k+1} + \boldsymbol{L}^{\mathrm{T}}\boldsymbol{x}_{k+1} + \boldsymbol{b}$. We write these in the form

$$(\boldsymbol{I} - \boldsymbol{L})\boldsymbol{y}_{k+1} = \boldsymbol{L}^{\mathrm{T}}\boldsymbol{x}_k + \boldsymbol{b}, \tag{12.ii}$$

$$(\boldsymbol{I} - \boldsymbol{L}^{\mathrm{T}})\boldsymbol{x}_{k+1} = \boldsymbol{L}\boldsymbol{y}_{k+1} + \boldsymbol{b}. \tag{12.iii}$$

Multiplying (12.iii) by $\boldsymbol{I} - \boldsymbol{L}$ we obtain $\boldsymbol{M}\boldsymbol{x}_{k+1} = (\boldsymbol{I} - \boldsymbol{L})\boldsymbol{L}\boldsymbol{y}_{k+1} + (\boldsymbol{I} - \boldsymbol{L})\boldsymbol{b}$. Since $(\boldsymbol{I} - \boldsymbol{L})\boldsymbol{L} = \boldsymbol{L}(\boldsymbol{I} - \boldsymbol{L})$ we find

$$\begin{aligned}\boldsymbol{M}\boldsymbol{x}_{k+1} &= \boldsymbol{L}(\boldsymbol{I} - \boldsymbol{L})\boldsymbol{y}_{k+1} + (\boldsymbol{I} - \boldsymbol{L})\boldsymbol{b} \\ &\overset{(12.\mathrm{ii})}{=} \boldsymbol{L}\boldsymbol{L}^{\mathrm{T}}\boldsymbol{x}_k + \boldsymbol{L}\boldsymbol{b} + (\boldsymbol{I} - \boldsymbol{L})\boldsymbol{b} = \boldsymbol{N}\boldsymbol{x}_k + \boldsymbol{b}.\end{aligned}$$

This is the splitting (12.46).

Exercise 12.12: Gauss-Seidel method (Exam exercise 2008-1)

Consider the linear system $\boldsymbol{A}\boldsymbol{x} = \boldsymbol{b}$ in which

$$\boldsymbol{A} := \begin{bmatrix} 3 & 0 & 1 \\ 0 & 7 & 2 \\ 1 & 2 & 4 \end{bmatrix}$$

and $\boldsymbol{b} := [1, 9, -2]^{\mathrm{T}}$.

a) With $\boldsymbol{x}_0 = [1, 1, 1]^{\mathrm{T}}$, carry out one iteration of the Gauss-Seidel method to find $\boldsymbol{x}_1 \in \mathbb{R}^3$.

Solution. We obtain

$$\begin{aligned} x_1(1) &= \big(b(1) - x_0(3)\big)/3 = (1-1)/3 = 0 \\ x_1(2) &= \big(b(2) - 2x_0(3)\big)/7 = (9 - 2\cdot 1)/7 = 1 \\ x_1(3) &= \big(b(3) - x_1(1) - 2x_1(2)\big)/4 = (-2 - 0 - 2\cdot 1)/4 = -1. \end{aligned}$$

b) If we continue the iteration, will the method converge? Why?

Solution. Yes. Since $\boldsymbol{A}$ is strictly diagonally dominant, i.e.,

$$a_{ii} > \sum_{j\neq i} |a_{ij}|, \qquad i = 1, 2, 3,$$

all its eigenvalues are positive by Gershgorin's circle theorem. Therefore, since $\boldsymbol{A}$ is also symmetric it is symmetric positive definite, which guarantees convergence. That $\boldsymbol{A}$ is positive definite also follows since the three leading principal submatrices have positive determinants.

c) Write a MATLAB program for the Gauss-Seidel method applied to a matrix $\boldsymbol{A} \in \mathbb{R}^{n\times n}$ and right-hand side $\boldsymbol{b} \in \mathbb{R}^n$. Use the ratio of the current residual to the initial residual as the stopping criterion, as well as a maximum number of iterations.

Hint.
The function `C=tril(A)` extracts the lower part of `A` into a lower triangular matrix `C`.

Solution. The following code solves the system $\boldsymbol{A}\boldsymbol{x} = \boldsymbol{b}$ using the Gauss-Seidel method.

code/gs.m

```
function [x,it]=gs(A,b,x,tol,maxit)
nr=norm(b-A*x,2);
C = tril(A);
for k=1:maxit
r=b-A*x;
x=x+C\r;
if norm(r,2)/nr<tol
it=k; return;
end
end
it = maxit;
```

Listing 12.1: Solve $\boldsymbol{Ax} = \boldsymbol{b}$ using the Gauss-Seidel method.

Exercises section 12.4

Exercise 12.13: A special norm

Show that $\|\boldsymbol{B}\|_t := \|\boldsymbol{D}_t\boldsymbol{U}^*\boldsymbol{B}\boldsymbol{U}\boldsymbol{D}_t^{-1}\|_1$ defined in the proof of Theorem 12.12 is a consistent matrix norm on $\mathbb{C}^{n\times n}$.

Solution. We show that $\|\cdot\|_t$ inherits the three properties that define a norm from the operator norm $\|\cdot\|_1$. For arbitrary matrices $\boldsymbol{A}, \boldsymbol{B}$ and scalar a, we have

1. *Positivity*. One has $\|\boldsymbol{B}\|_t = \|\boldsymbol{D}_t\boldsymbol{U}^*\boldsymbol{B}\boldsymbol{U}\boldsymbol{D}_t^{-1}\|_1 \geq 0$, with equality holding precisely when $\boldsymbol{D}_t\boldsymbol{U}^*\boldsymbol{B}\boldsymbol{U}\boldsymbol{D}_t^{-1}$ is the zero matrix, which happens if and only if $\boldsymbol{B}$ is the zero matrix.
2. *Homogeneity*. For any scalar $a \in \mathbb{C}$,

$$\|a\boldsymbol{B}\|_t = \|a\boldsymbol{D}_t\boldsymbol{U}^*\boldsymbol{B}\boldsymbol{U}\boldsymbol{D}_t^{-1}\|_1 = |a| \cdot \|\boldsymbol{D}_t\boldsymbol{U}^*\boldsymbol{B}\boldsymbol{U}\boldsymbol{D}_t^{-1}\|_1 = |a| \cdot \|\boldsymbol{B}\|_t.$$

3. *Subadditivity*. One has

$$\begin{aligned}\|\boldsymbol{A}+\boldsymbol{B}\|_t &= \|\boldsymbol{D}_t\boldsymbol{U}^*(\boldsymbol{A}+\boldsymbol{B})\boldsymbol{U}\boldsymbol{D}_t^{-1}\|_1\\ &\leq \|\boldsymbol{D}_t\boldsymbol{U}^*\boldsymbol{A}\boldsymbol{U}\boldsymbol{D}_t^{-1}\|_1 + \|\boldsymbol{D}_t\boldsymbol{U}^*\boldsymbol{B}\boldsymbol{U}\boldsymbol{D}_t^{-1}\|_1\\ &= \|\boldsymbol{A}\|_t + \|\boldsymbol{B}\|_t.\end{aligned}$$

Since $\|\cdot\|_1$ is an operator norm, it is consistent. For any matrices $\boldsymbol{A}, \boldsymbol{B}$ for which the product $\boldsymbol{AB}$ is defined, therefore,

$$\begin{aligned}\|\boldsymbol{AB}\|_t &= \|\boldsymbol{D}_t\boldsymbol{U}^*\boldsymbol{A}\boldsymbol{B}\boldsymbol{U}\boldsymbol{D}_t^{-1}\|_1\\ &= \|\boldsymbol{D}_t\boldsymbol{U}^*\boldsymbol{A}\boldsymbol{U}\boldsymbol{D}_t^{-1}\boldsymbol{D}_t\boldsymbol{U}^*\boldsymbol{B}\boldsymbol{U}\boldsymbol{D}_t^{-1}\|_1\end{aligned}$$

$$\begin{aligned}&\le \|D_tU^*AUD_t^{-1}\|_1\|D_tU^*BUD_t^{-1}\|_1\\&= \|A\|_t\|B\|_t,\end{aligned}$$

proving that $\|\cdot\|_t$ is consistent.

Exercise 12.14: Is $A+E$ nonsingular?

Suppose $A \in \mathbb{C}^{n\times n}$ is nonsingular and $E \in \mathbb{C}^{n\times n}$. Show that $A+E$ is nonsingular if $\rho(A^{-1}E) < 1$.

Solution. Suppose $\rho(A^{-1}E) = \rho\big(A^{-1}(-E)\big) < 1$. By part 2 of Theorem 12.14, $I + A^{-1}E$ is nonsingular and therefore so is the product $A(I + A^{-1}E) = A + E$.

Exercise 12.15: Slow spectral radius convergence

The convergence $\lim_{k\to\infty}\|A^k\|^{1/k} = \rho(A)$ can be quite slow. Consider

$$A := \begin{bmatrix} \lambda & a & 0 & \cdots & 0 & 0\\ 0 & \lambda & a & \cdots & 0 & 0\\ 0 & 0 & \lambda & \cdots & 0 & 0\\ \vdots & \vdots & \vdots & \ddots & \vdots & \vdots\\ 0 & 0 & 0 & \cdots & \lambda & a\\ 0 & 0 & 0 & \cdots & 0 & \lambda \end{bmatrix} \in \mathbb{R}^{n\times n}.$$

If $|\lambda| = \rho(A) < 1$ then $\lim_{k\to\infty} A^k = 0$ for any $a \in \mathbb{R}$. We show below that the $(1, n)$ element of A^k is given by $f(k) := \binom{k}{n-1}a^{n-1}\lambda^{k-n+1}$ for $k \ge n-1$.

a) Pick an n, e.g. $n = 5$, and make a plot of $f(k)$ for $\lambda = 0.9$, $a = 10$, and $n-1 \le k \le 200$. Your program should also compute $\max_k f(k)$. Use your program to determine how large k must be before $f(k) < 10^{-8}$.

Solution. The MATLAB code

code/exsgst1.m

```
n = 5
a = 10
l = 0.9

for k = n-1:200
   L(k) = nchoosek(k,n-1)*a^(n-1)*l^(k-n+1);
end

```

```
stairs(L)
```

Listing 12.2: For $n = 5$, $\lambda = 0.9$ and $a = 10$, plot the $(1, n)$ element $f(k)$ of the matrix $\boldsymbol{A}^k$ for $n - 1 \le k \le 200$.

yields the following stairstep graph of f:

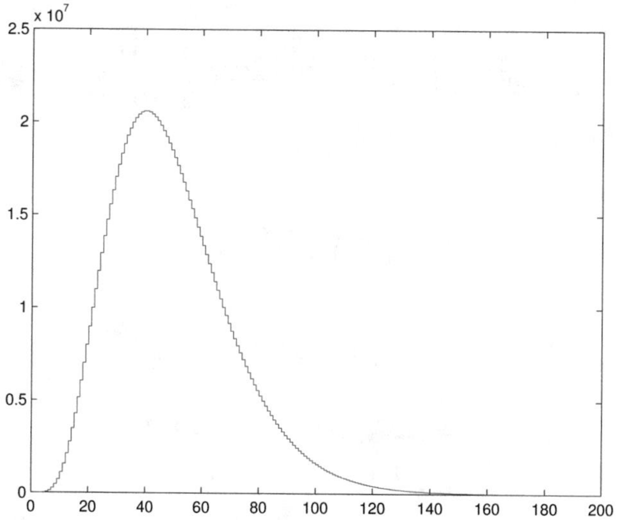

The command `max(L)` returns a maximum of $\approx 2.0589 \cdot 10^7$ of f on the interval $n - 1 \le k \le 200$. Moreover, the code

code/exsgst2.m

```
k = n-1;

while nchoosek(k,n-1)*a^(n-1)*1^(k-n+1) >= 10^(-8)
   k = k + 1;
end

k
```

Listing 12.3: Find the first integer k for which $f(k) < 10^{-8}$.

finds that $f(k)$ dives for the first time below 10^{-8} at $k = 470$. We conclude that the matrix $\boldsymbol{A}^k$ is close to zero only for a very high power k.

b) $\boldsymbol{A}^k$ can be found explicitly for any k. Let $\boldsymbol{E} := (\boldsymbol{A} - \lambda \boldsymbol{I})/a$. Show by induction that $\boldsymbol{E}^k = \left[\begin{smallmatrix} \mathbf{0} & \boldsymbol{I}_{n-k} \\ \mathbf{0} & \mathbf{0} \end{smallmatrix}\right]$ for $1 \le k \le n - 1$ and that $\boldsymbol{E}^n = \mathbf{0}$.

Solution. Let $\boldsymbol{E} = \boldsymbol{E}_1 := (\boldsymbol{A} - \lambda \boldsymbol{I})/a$ be the $n \times n$ matrix in the exercise, and write

$$\boldsymbol{E}_k := \begin{bmatrix} \boldsymbol{0} & \boldsymbol{I}_{n-k} \\ \boldsymbol{0} & \boldsymbol{0} \end{bmatrix} \in \mathbb{R}^{n,n}.$$

Clearly $\boldsymbol{E}^k = \boldsymbol{E}_k$ for $k = 1$. Suppose that $\boldsymbol{E}^k = \boldsymbol{E}_k$ for some k satisfying $1 \le k \le n-1$. Using the rules of block multiplication,

$$\begin{aligned} \boldsymbol{E}^{k+1} &= \boldsymbol{E}^k \boldsymbol{E}^1 \\ &= \left[\begin{array}{c|c} \boldsymbol{0}_{n-k,k} & \boldsymbol{I}_{n-k} \\ \hline \boldsymbol{0}_{k,k} & \boldsymbol{0}_{k,n-k} \end{array}\right] \left[\begin{array}{c|c} \boldsymbol{0}_{k,1}\ ,\ \boldsymbol{I}_k & \boldsymbol{0}_{k,n-k-1} \\ \hline \boldsymbol{0}_{n-k,k+1} & \begin{matrix} \boldsymbol{I}_{n-k-1} \\ \boldsymbol{0}_{1,n-k-1} \end{matrix} \end{array}\right] = \left[\begin{array}{c|c} \boldsymbol{0}_{n-k,k+1} & \begin{matrix} \boldsymbol{I}_{n-k-1} \\ \boldsymbol{0}_{1,n-k-1} \end{matrix} \\ \hline \boldsymbol{0}_{k,k+1} & \boldsymbol{0}_{k,n-k-1} \end{array}\right] \\ &= \boldsymbol{E}_{k+1}. \end{aligned}$$

Alternatively, since

$$(\boldsymbol{E}^k)_{ij} = \begin{cases} 1, \text{ if } j = i+k, \\ 0, \text{ otherwise,} \end{cases}$$

one has

$$\begin{aligned} (\boldsymbol{E}^{k+1})_{ij} &= (\boldsymbol{E}^k \boldsymbol{E}^1)_{ij} = \sum_{\ell} (\boldsymbol{E}^k)_{i\ell} (\boldsymbol{E}^1)_{\ell j} \\ &= (\boldsymbol{E}^k)_{i,i+k} (\boldsymbol{E}^1)_{i+k,j} = 1 \cdot (\boldsymbol{E}^1)_{i+k,j} \\ &= \begin{cases} 1 \text{ if } j = i+k+1, \\ 0 \text{ otherwise,} \end{cases} \end{aligned}$$

By induction we conclude that $\boldsymbol{E}^k = \boldsymbol{E}_k$ for any k satisfying $1 \le k \le n$, with the convention that $\boldsymbol{E}^n = \boldsymbol{E}_n = \boldsymbol{0}_{n,n}$. We summarize that the matrix $\boldsymbol{E}$ is *nilpotent* of degree n.

c) We have $\boldsymbol{A}^k = (a\boldsymbol{E} + \lambda \boldsymbol{I})^k = \sum_{j=0}^{\min\{k,n-1\}} \binom{k}{j} a^j \lambda^{k-j} \boldsymbol{E}^j$ and conclude that the $(1,n)$ element is given by $f(k)$ for $k \ge n-1$.

Solution. Since the matrices $\boldsymbol{E}$ and $\boldsymbol{I}$ commute, the binomial theorem and **b)** yield

$$\boldsymbol{A}^k = (a\boldsymbol{E} + \lambda \boldsymbol{I})^k = \sum_{j=0}^{\min\{k,n-1\}} \binom{k}{j} \lambda^{k-j} a^j \boldsymbol{E}^j.$$

Since $(\boldsymbol{E}^j)_{1,n} = 0$ for $1 \le j \le n-2$ and $(\boldsymbol{E}^{n-1})_{1,n} = 1$, it follows that

$$(\boldsymbol{A}^k)_{1,n} = \sum_{j=0}^{\min\{k,n-1\}} \binom{k}{j} \lambda^{k-j} a^j (\boldsymbol{E}^j)_{1,n} = \binom{k}{n-1} \lambda^{k-n+1} a^{n-1} = f(k),$$

which is what needed to be shown.

Chapter 13
The Conjugate Gradient Method

Exercises section 13.1

Exercise 13.1: A-norm

One can show that the $\boldsymbol{A}$-norm is a vector norm on $\mathbb{R}^n$ without using the fact that it is an inner product norm. Show this with the help of the Cholesky factorization of $\boldsymbol{A}$.

Solution. Let $\boldsymbol{A} = \boldsymbol{L}\boldsymbol{L}^*$ be a Cholesky factorization of $\boldsymbol{A}$, i.e., $\boldsymbol{L}$ is lower triangular with positive diagonal elements. The $\boldsymbol{A}$-norm then takes the form $\|\boldsymbol{x}\|_{\boldsymbol{A}} = \sqrt{\boldsymbol{x}^{\mathrm{T}}\boldsymbol{L}\boldsymbol{L}^*\boldsymbol{x}} = \|\boldsymbol{L}^*\boldsymbol{x}\|$. Let us verify the three properties of a vector norm:

1. *Positivity*: Clearly $\|\boldsymbol{x}\|_{\boldsymbol{A}} = \|\boldsymbol{L}^*\boldsymbol{x}\| \geq 0$. Since $\boldsymbol{L}^*$ is nonsingular, $\|\boldsymbol{x}\|_{\boldsymbol{A}} = \|\boldsymbol{L}^*\boldsymbol{x}\| = 0$ if and only if $\boldsymbol{L}^*\boldsymbol{x} = 0$ if and only if $\boldsymbol{x} = 0$.
2. *Homogeneity*: $\|a\boldsymbol{x}\|_{\boldsymbol{A}} = \|\boldsymbol{L}^*(a\boldsymbol{x})\| = \|a\boldsymbol{L}^*\boldsymbol{x}\| = |a|\|\boldsymbol{L}^*(\boldsymbol{x})\| = |a|\|\boldsymbol{x}\|_{\boldsymbol{A}}$.
3. *Subadditivity*:

$$\|\boldsymbol{x}+\boldsymbol{y}\|_{\boldsymbol{A}} = \|\boldsymbol{L}^*(\boldsymbol{x}+\boldsymbol{y})\| = \|\boldsymbol{L}^*\boldsymbol{x}+\boldsymbol{L}^*\boldsymbol{y}\| \leq \|\boldsymbol{L}^*\boldsymbol{x}\|+\|\boldsymbol{L}^*\boldsymbol{y}\| = \|\boldsymbol{x}\|_{\boldsymbol{A}}+\|\boldsymbol{y}\|_{\boldsymbol{A}}.$$

Exercise 13.2: Paraboloid

Let $\boldsymbol{A} = \boldsymbol{U}\boldsymbol{D}\boldsymbol{U}^{\mathrm{T}}$ be the spectral decomposition of $\boldsymbol{A}$, i.e., $\boldsymbol{U}$ is orthonormal and $\boldsymbol{D} = \mathrm{diag}(\lambda_1, \ldots, \lambda_n)$ is diagonal. Define new variables $\boldsymbol{v} = [v_1, \ldots, v_n]^{\mathrm{T}} := \boldsymbol{U}^{\mathrm{T}}\boldsymbol{y}$, and set $\boldsymbol{c} := \boldsymbol{U}^{\mathrm{T}}\boldsymbol{b} = [c_1, \ldots, c_n]^{\mathrm{T}}$. Show that

$$Q(\boldsymbol{y}) := \frac{1}{2}\boldsymbol{y}^{\mathrm{T}}\boldsymbol{A}\boldsymbol{y} - \boldsymbol{b}^{\mathrm{T}}\boldsymbol{y} = \frac{1}{2}\sum_{j=1}^{n}\lambda_j v_j^2 - \sum_{j=1}^{n} c_j v_j.$$

Solution. One has

T. Lyche et al., *Exercises in Numerical Linear Algebra and Matrix Factorizations*, Texts in Computational Science and Engineering 23, https://doi.org/10.1007/978-3-030-59789-4_13

$$Q(\boldsymbol{y}) = \frac{1}{2}\boldsymbol{y}^{\mathrm{T}}\boldsymbol{U}\boldsymbol{D}\boldsymbol{U}^{\mathrm{T}}\boldsymbol{y} - \boldsymbol{b}^{\mathrm{T}}\boldsymbol{y} = \frac{1}{2}\boldsymbol{v}^{\mathrm{T}}\boldsymbol{D}\boldsymbol{v} - \boldsymbol{c}^{\mathrm{T}}\boldsymbol{v} = \frac{1}{2}\sum_{j=1}^{n}\lambda_j v_j^2 - \sum_{j=1}^{n} c_j v_j.$$

Exercise 13.3: Steepest descent iteration

Verify the numbers in Example 13.1, i.e., show that

$$\boldsymbol{t}_{2k-2} = 3 \cdot 4^{1-k}\begin{bmatrix}1\\-1/2\end{bmatrix}, \ \boldsymbol{x}_{2k-1} = -4^{-k}\begin{bmatrix}1\\2\end{bmatrix}, \ \boldsymbol{r}_{2k-1} = 3 \cdot 4^{-k}\begin{bmatrix}0\\1\end{bmatrix}, \tag{13.i}$$

$$\boldsymbol{t}_{2k-1} = 3 \cdot 4^{-k}\begin{bmatrix}-1\\2\end{bmatrix}, \ \boldsymbol{x}_{2k} = -4^{-k}\begin{bmatrix}1\\1/2\end{bmatrix}, \ \boldsymbol{r}_{2k} = 3 \cdot 4^{-k}\begin{bmatrix}1/2\\0\end{bmatrix}. \tag{13.ii}$$

Solution. In the steepest descent method we choose, at the kth iteration, the search direction $\boldsymbol{p}_k = \boldsymbol{r}_k = \boldsymbol{b} - \boldsymbol{A}\boldsymbol{x}_k$ and optimal step length

$$\alpha_k := \frac{\boldsymbol{r}_k^{\mathrm{T}}\boldsymbol{r}_k}{\boldsymbol{r}_k^{\mathrm{T}}\boldsymbol{A}\boldsymbol{r}_k}.$$

Given is a quadratic function

$$Q(x,y) = \frac{1}{2}\,[x\ y]\,\boldsymbol{A}\begin{bmatrix}x\\y\end{bmatrix} - \boldsymbol{b}^{\mathrm{T}}\begin{bmatrix}x\\y\end{bmatrix}, \quad \boldsymbol{A} = \begin{bmatrix}2 & -1\\-1 & 2\end{bmatrix}, \quad \boldsymbol{b} = \begin{bmatrix}0\\0\end{bmatrix},$$

and an initial guess $\boldsymbol{x}_0 = [-1, -1/2]^{\mathrm{T}}$ of its minimum. The corresponding residual is

$$\boldsymbol{r}_0 = \boldsymbol{b} - \boldsymbol{A}\boldsymbol{x}_0 = \begin{bmatrix}0\\0\end{bmatrix} - \begin{bmatrix}2 & -1\\-1 & 2\end{bmatrix}\begin{bmatrix}-1\\-1/2\end{bmatrix} = \begin{bmatrix}3/2\\0\end{bmatrix}.$$

Performing the steps in Equation (13.8) twice yields

$$\boldsymbol{t}_0 = \boldsymbol{A}\boldsymbol{r}_0 = \begin{bmatrix}2 & -1\\-1 & 2\end{bmatrix}\begin{bmatrix}3/2\\0\end{bmatrix} = \begin{bmatrix}3\\-3/2\end{bmatrix}, \qquad \alpha_0 = \frac{\boldsymbol{r}_0^{\mathrm{T}}\boldsymbol{r}_0}{\boldsymbol{r}_0^{\mathrm{T}}\boldsymbol{t}_0} = \frac{9/4}{9/2} = \frac{1}{2},$$

$$\boldsymbol{x}_1 = \begin{bmatrix}-1\\-1/2\end{bmatrix} + \frac{1}{2}\begin{bmatrix}3/2\\0\end{bmatrix} = \begin{bmatrix}-1/4\\-1/2\end{bmatrix}, \qquad \boldsymbol{r}_1 = \begin{bmatrix}3/2\\0\end{bmatrix} - \frac{1}{2}\begin{bmatrix}3\\-3/2\end{bmatrix} = \begin{bmatrix}0\\3/4\end{bmatrix},$$

$$\boldsymbol{t}_1 = \boldsymbol{A}\boldsymbol{r}_1 = \begin{bmatrix}2 & -1\\-1 & 2\end{bmatrix}\begin{bmatrix}0\\3/4\end{bmatrix} = \begin{bmatrix}-3/4\\3/2\end{bmatrix}, \qquad \alpha_1 = \frac{\boldsymbol{r}_1^{\mathrm{T}}\boldsymbol{r}_1}{\boldsymbol{r}_1^{\mathrm{T}}\boldsymbol{t}_1} = \frac{9/16}{9/8} = \frac{1}{2},$$

$$\boldsymbol{x}_2 = \begin{bmatrix}-1/4\\-1/2\end{bmatrix} + \frac{1}{2}\begin{bmatrix}0\\3/4\end{bmatrix} = \begin{bmatrix}-1/4\\-1/8\end{bmatrix}, \qquad \boldsymbol{r}_2 = \begin{bmatrix}0\\3/4\end{bmatrix} - \frac{1}{2}\begin{bmatrix}-3/4\\3/2\end{bmatrix} = \begin{bmatrix}3/8\\0\end{bmatrix}.$$

Moreover, assume that for some $k \geq 1$, (13.i) and (13.ii) hold. Then

$$
\begin{aligned}
\boldsymbol{t}_{2k} &= 3\cdot 4^{-k}\begin{bmatrix}2 & -1\\ -1 & 2\end{bmatrix}\begin{bmatrix}1/2\\ 0\end{bmatrix} = 3\cdot 4^{1-(k+1)}\begin{bmatrix}1\\ -1/2\end{bmatrix},\\
\alpha_{2k} &= \frac{\boldsymbol{r}_{2k}^{\mathrm{T}}\boldsymbol{r}_{2k}}{\boldsymbol{r}_{2k}^{\mathrm{T}}\boldsymbol{t}_{2k}} = \frac{9\cdot 4^{-2k}\cdot(\frac{1}{2})^2}{9\cdot 4^{-2k}\cdot\frac{1}{2}} = \frac{1}{2},\\
\boldsymbol{x}_{2k+1} &= -4^{-k}\begin{bmatrix}1\\ 1/2\end{bmatrix} + \frac{1}{2}\cdot 3\cdot 4^{-k}\begin{bmatrix}1/2\\ 0\end{bmatrix} = -4^{-(k+1)}\begin{bmatrix}1\\ 2\end{bmatrix},\\
\boldsymbol{r}_{2k+1} &= 3\cdot 4^{-k}\begin{bmatrix}1/2\\ 0\end{bmatrix} - \frac{1}{2}\cdot 3\cdot 4^{1-(k+1)}\begin{bmatrix}1\\ -1/2\end{bmatrix} = 3\cdot 4^{-(k+1)}\begin{bmatrix}0\\ 1\end{bmatrix},\\
\boldsymbol{t}_{2k+1} &= 3\cdot 4^{-(k+1)}\begin{bmatrix}2 & -1\\ -1 & 2\end{bmatrix}\begin{bmatrix}0\\ 1\end{bmatrix} = 3\cdot 4^{-(k+1)}\begin{bmatrix}-1\\ 2\end{bmatrix},\\
\alpha_{2k+1} &= \frac{\boldsymbol{r}_{2k+1}^{\mathrm{T}}\boldsymbol{r}_{2k+1}}{\boldsymbol{r}_{2k+1}^{\mathrm{T}}\boldsymbol{t}_{2k+1}} = \frac{9\cdot 4^{-2(k+1)}}{9\cdot 4^{-2(k+1)}\cdot 2} = \frac{1}{2},\\
\boldsymbol{x}_{2k+2} &= -4^{-(k+1)}\begin{bmatrix}1\\ 2\end{bmatrix} + \frac{1}{2}\cdot 3\cdot 4^{-(k+1)}\begin{bmatrix}0\\ 1\end{bmatrix} = -4^{-(k+1)}\begin{bmatrix}1\\ 1/2\end{bmatrix},\\
\boldsymbol{r}_{2k+2} &= 3\cdot 4^{-(k+1)}\begin{bmatrix}0\\ 1\end{bmatrix} - \frac{1}{2}\cdot 3\cdot 4^{-(k+1)}\begin{bmatrix}-1\\ 2\end{bmatrix} = 3\cdot 4^{-(k+1)}\begin{bmatrix}1/2\\ 0\end{bmatrix}.
\end{aligned}
$$

Using the method of induction, we conclude that (13.i), (13.ii), and $\alpha_k = 1/2$ hold for any $k \geq 1$.

Exercise 13.4: Steepest descent (Exam exercise 2011-1)

The steepest descent method can be used to solve a linear system $\boldsymbol{A}\boldsymbol{x} = \boldsymbol{b}$ for $\boldsymbol{x} \in \mathbb{R}^n$, where $\boldsymbol{A} \in \mathbb{R}^{n,n}$ is symmetric and positive definite, and $\boldsymbol{b} \in \mathbb{R}^n$. With $\boldsymbol{x}_0 \in \mathbb{R}^n$ an initial guess, the iteration is $\boldsymbol{x}_{k+1} = \boldsymbol{x}_k + \alpha_k \boldsymbol{r}_k$, where $\boldsymbol{r}_k$ is the residual, $\boldsymbol{r}_k = \boldsymbol{b} - \boldsymbol{A}\boldsymbol{x}_k$, and $\alpha_k = \frac{\boldsymbol{r}_k^{\mathrm{T}}\boldsymbol{r}_k}{\boldsymbol{r}_k^{\mathrm{T}}\boldsymbol{A}\boldsymbol{r}_k}$.

a) Compute $\boldsymbol{x}_1$ if $\boldsymbol{A} = \begin{bmatrix}2 & -1\\ -1 & 2\end{bmatrix}$, $\boldsymbol{b} = [1\ 1]^{\mathrm{T}}$ and $\boldsymbol{x}_0 = \boldsymbol{0}$.

Solution. We find $\boldsymbol{r}_0 = [1\ 1]^{\mathrm{T}}$, $\alpha_0 = 1$, and $\boldsymbol{x}_1 = [1\ 1]^{\mathrm{T}}$, the exact solution.

b) If the k-th error, $\boldsymbol{e}_k = \boldsymbol{x}_k - \boldsymbol{x}$, is an eigenvector of $\boldsymbol{A}$, what can you say about $\boldsymbol{x}_{k+1}$?

Solution. If $\boldsymbol{A}\boldsymbol{e}_k = \lambda\boldsymbol{e}_k$ for some $\lambda \in \mathbb{R}$ then $\boldsymbol{r}_k = -\lambda\boldsymbol{e}_k$, and $\boldsymbol{A}\boldsymbol{r}_k = -\lambda\boldsymbol{A}\boldsymbol{e}_k = -\lambda^2\boldsymbol{e}_k$. We therefore have $\boldsymbol{r}_k^{\mathrm{T}}\boldsymbol{r}_k = \lambda^2\boldsymbol{e}_k^{\mathrm{T}}\boldsymbol{e}_k$ and $\boldsymbol{r}_k^{\mathrm{T}}\boldsymbol{A}\boldsymbol{r}_k = \lambda^3\boldsymbol{e}_k^{\mathrm{T}}\boldsymbol{e}_k$, so that $\alpha_k = 1/\lambda$, and therefore, $\boldsymbol{x}_{k+1} = \boldsymbol{x}_k + \alpha_k\boldsymbol{r}_k = \boldsymbol{x}_k - \boldsymbol{e}_k = \boldsymbol{x}$, which is the solution.

Exercises section 13.2

Exercise 13.5: Conjugate gradient iteration, II

Do one iteration with the conjugate gradient method when $\boldsymbol{x}_0 = \boldsymbol{0}$.

Solution. Using $\boldsymbol{x}_0 = \boldsymbol{0}$, one finds

$$\boldsymbol{x}_1 = \boldsymbol{x}_0 + \frac{(\boldsymbol{b} - \boldsymbol{A}\boldsymbol{x}_0)^{\mathrm{T}}(\boldsymbol{b} - \boldsymbol{A}\boldsymbol{x}_0)}{(\boldsymbol{b} - \boldsymbol{A}\boldsymbol{x}_0)^{\mathrm{T}}\boldsymbol{A}(\boldsymbol{b} - \boldsymbol{A}^2\boldsymbol{x}_0)}(\boldsymbol{b} - \boldsymbol{A}\boldsymbol{x}_0) = \frac{\boldsymbol{b}^{\mathrm{T}}\boldsymbol{b}}{\boldsymbol{b}^{\mathrm{T}}\boldsymbol{A}\boldsymbol{b}}\boldsymbol{b}.$$

Exercise 13.6: Conjugate gradient iteration, III

Do two conjugate gradient iterations for the system

$$\begin{bmatrix} 2 & -1 \\ -1 & 2 \end{bmatrix}\begin{bmatrix} x_1 \\ x_2 \end{bmatrix} = \begin{bmatrix} 0 \\ 3 \end{bmatrix},$$

starting with $\boldsymbol{x}_0 = \boldsymbol{0}$.

Solution. By Exercise 13.5,

$$\boldsymbol{x}_1 = \frac{\boldsymbol{b}^{\mathrm{T}}\boldsymbol{b}}{\boldsymbol{b}^{\mathrm{T}}\boldsymbol{A}\boldsymbol{b}}\boldsymbol{b} = \frac{9}{18}\begin{bmatrix} 0 \\ 3 \end{bmatrix} = \begin{bmatrix} 0 \\ 3/2 \end{bmatrix}.$$

We find, in order,

$$\boldsymbol{p}_0 = \boldsymbol{r}_0 = \begin{bmatrix} 0 \\ 3 \end{bmatrix}, \qquad \alpha_0 = \frac{1}{2}, \ \boldsymbol{r}_1 = \begin{bmatrix} \frac{3}{2} \\ 0 \end{bmatrix},$$

$$\beta_0 = \frac{1}{4}, \qquad \boldsymbol{p}_1 = \begin{bmatrix} \frac{3}{2} \\ \frac{3}{4} \end{bmatrix}, \qquad \alpha_1 = \frac{2}{3}, \ \boldsymbol{x}_2 = \begin{bmatrix} 1 \\ 2 \end{bmatrix}.$$

Since the residual vectors $\boldsymbol{r}_0, \boldsymbol{r}_1, \boldsymbol{r}_2$ must be orthogonal, it follows that $\boldsymbol{r}_2 = \boldsymbol{0}$ and $\boldsymbol{x}_2$ must be an exact solution. This can be verified directly by hand.

Exercise 13.7: The cg step length is optimal

Show that the step length α_k in the conjugate gradient method is optimal.

Hint.

Use induction on k to show that $\boldsymbol{p}_k = \boldsymbol{r}_k + \sum_{j=0}^{k-1} a_{k,j}\boldsymbol{r}_j$ for some constants $a_{k,j}$.

Solution. For any fixed search direction $\boldsymbol{p}_k$, the step length α_k is optimal if $Q(\boldsymbol{x}_{k+1})$ is as small as possible, that is

$$Q(\boldsymbol{x}_{k+1}) = Q(\boldsymbol{x}_k + \alpha_k \boldsymbol{p}_k) = \min_{\alpha \in \mathbb{R}} f(\alpha),$$

where, by (13.5),

$$f(\alpha) := Q(\boldsymbol{x}_k + \alpha \boldsymbol{p}_k) = Q(\boldsymbol{x}_k) - \alpha \boldsymbol{p}_k^{\mathrm{T}} \boldsymbol{r}_k + \frac{1}{2}\alpha^2 \boldsymbol{p}_k^{\mathrm{T}} \boldsymbol{A} \boldsymbol{p}_k$$

is a quadratic polynomial in α. Since $\boldsymbol{A}$ is assumed to be positive definite, necessarily $\boldsymbol{p}_k^{\mathrm{T}} \boldsymbol{A} \boldsymbol{p}_k > 0$. Therefore f has a minimum, which it attains at

$$\alpha = \frac{\boldsymbol{p}_k^{\mathrm{T}} \boldsymbol{r}_k}{\boldsymbol{p}_k^{\mathrm{T}} \boldsymbol{A} \boldsymbol{p}_k}.$$

Applying (13.17) repeatedly, one finds that the search direction $\boldsymbol{p}_k$ for the conjugate gradient method satisfies

$$\boldsymbol{p}_k = \boldsymbol{r}_k + \frac{\boldsymbol{r}_k^{\mathrm{T}} \boldsymbol{r}_k}{\boldsymbol{r}_{k-1}^{\mathrm{T}} \boldsymbol{r}_{k-1}} \boldsymbol{p}_{k-1} = \boldsymbol{r}_k + \frac{\boldsymbol{r}_k^{\mathrm{T}} \boldsymbol{r}_k}{\boldsymbol{r}_{k-1}^{\mathrm{T}} \boldsymbol{r}_{k-1}} \left(\boldsymbol{r}_{k-1} + \frac{\boldsymbol{r}_{k-1}^{\mathrm{T}} \boldsymbol{r}_{k-1}}{\boldsymbol{r}_{k-2}^{\mathrm{T}} \boldsymbol{r}_{k-2}} \boldsymbol{p}_{k-2} \right) = \cdots$$

As $\boldsymbol{p}_0 = \boldsymbol{r}_0$, the difference $\boldsymbol{p}_k - \boldsymbol{r}_k$ is a linear combination of the vectors $\boldsymbol{r}_{k-1}, \ldots, \boldsymbol{r}_0$, each of which is orthogonal to $\boldsymbol{r}_k$. It follows that $\boldsymbol{p}_k^{\mathrm{T}} \boldsymbol{r}_k = \boldsymbol{r}_k^{\mathrm{T}} \boldsymbol{r}_k$ and that the step length α is optimal for

$$\alpha = \frac{\boldsymbol{r}_k^{\mathrm{T}} \boldsymbol{r}_k}{\boldsymbol{p}_k^{\mathrm{T}} \boldsymbol{A} \boldsymbol{p}_k} = \alpha_k.$$

Exercise 13.8: Starting value in cg

Show that the conjugate gradient method (13.18) for $\boldsymbol{A}\boldsymbol{x} = \boldsymbol{b}$ starting with $\boldsymbol{x}_0$ is the same as applying the method to the system $\boldsymbol{A}\boldsymbol{y} = \boldsymbol{r}_0 := \boldsymbol{b} - \boldsymbol{A}\boldsymbol{x}_0$ starting with $\boldsymbol{y}_0 = \boldsymbol{0}$.

Hint.

The conjugate gradient method for $\boldsymbol{A}\boldsymbol{y} = \boldsymbol{r}_0$ can be written $\boldsymbol{y}_{k+1} := \boldsymbol{y}_k + \gamma_k \boldsymbol{q}_k$, $\gamma_k := \frac{\boldsymbol{s}_k^{\mathrm{T}} \boldsymbol{s}_k}{\boldsymbol{q}_k^{\mathrm{T}} \boldsymbol{A} \boldsymbol{q}_k}$, $\boldsymbol{s}_{k+1} := \boldsymbol{s}_k - \gamma_k \boldsymbol{A} \boldsymbol{q}_k$, $\boldsymbol{q}_{k+1} := \boldsymbol{s}_{k+1} + \delta_k \boldsymbol{q}_k$, $\delta_k := \frac{\boldsymbol{s}_{k+1}^{\mathrm{T}} \boldsymbol{s}_{k+1}}{\boldsymbol{s}_k^{\mathrm{T}} \boldsymbol{s}_k}$. Show that $\boldsymbol{y}_k = \boldsymbol{x}_k - \boldsymbol{x}_0$, $\boldsymbol{s}_k = \boldsymbol{r}_k$, and $\boldsymbol{q}_k = \boldsymbol{p}_k$, for $k = 0, 1, 2 \ldots$.

Solution. As in the exercise, we consider the conjugate gradient method for $\boldsymbol{A}\boldsymbol{y} = \boldsymbol{r}_0$, with $\boldsymbol{r}_0 = \boldsymbol{b} - \boldsymbol{A}\boldsymbol{x}_0$. Starting with

$$\boldsymbol{y}_0 = \boldsymbol{0}, \qquad \boldsymbol{s}_0 = \boldsymbol{r}_0 - \boldsymbol{A}\boldsymbol{y}_0 = \boldsymbol{r}_0, \qquad \boldsymbol{q}_0 = \boldsymbol{s}_0 = \boldsymbol{r}_0,$$

one computes, for any $k \geq 0$,

$$\gamma_k := \frac{\boldsymbol{s}_k^{\mathrm{T}}\boldsymbol{s}_k}{\boldsymbol{q}_k^{\mathrm{T}}\boldsymbol{A}\boldsymbol{q}_k}, \qquad \boldsymbol{y}_{k+1} = \boldsymbol{y}_k + \gamma_k \boldsymbol{q}_k, \qquad \boldsymbol{s}_{k+1} = \boldsymbol{s}_k - \gamma_k \boldsymbol{A}\boldsymbol{q}_k,$$

$$\delta_k := \frac{\boldsymbol{s}_{k+1}^{\mathrm{T}}\boldsymbol{s}_{k+1}}{\boldsymbol{s}_k^{\mathrm{T}}\boldsymbol{s}_k}, \qquad \boldsymbol{q}_{k+1} = \boldsymbol{s}_{k+1} + \delta_k \boldsymbol{q}_k.$$

How are the iterates $\boldsymbol{y}_k$ and $\boldsymbol{x}_k$ related? As remarked above, $\boldsymbol{s}_0 = \boldsymbol{r}_0$ and $\boldsymbol{q}_0 = \boldsymbol{r}_0 = \boldsymbol{p}_0$. Suppose $\boldsymbol{s}_k = \boldsymbol{r}_k$ and $\boldsymbol{q}_k = \boldsymbol{p}_k$ for some $k \geq 0$. Then

$$\boldsymbol{s}_{k+1} = \boldsymbol{s}_k - \gamma_k \boldsymbol{A}\boldsymbol{q}_k = \boldsymbol{r}_k - \frac{\boldsymbol{r}_k^{\mathrm{T}}\boldsymbol{r}_k}{\boldsymbol{p}_k^{\mathrm{T}}\boldsymbol{A}\boldsymbol{p}_k}\boldsymbol{A}\boldsymbol{p}_k = \boldsymbol{r}_k - \alpha_k \boldsymbol{A}\boldsymbol{p}_k = \boldsymbol{r}_{k+1},$$

$$\boldsymbol{q}_{k+1} = \boldsymbol{s}_{k+1} + \delta_k \boldsymbol{q}_k = \boldsymbol{r}_{k+1} + \frac{\boldsymbol{r}_{k+1}^{\mathrm{T}}\boldsymbol{r}_{k+1}}{\boldsymbol{r}_k^{\mathrm{T}}\boldsymbol{r}_k}\boldsymbol{p}_k = \boldsymbol{p}_{k+1}.$$

It follows by induction that $\boldsymbol{s}_k = \boldsymbol{r}_k$ and $\boldsymbol{q}_k = \boldsymbol{p}_k$ for all $k \geq 0$. In addition,

$$\boldsymbol{y}_{k+1} - \boldsymbol{y}_k = \gamma_k \boldsymbol{q}_k = \frac{\boldsymbol{r}_k^{\mathrm{T}}\boldsymbol{r}_k}{\boldsymbol{p}_k^{\mathrm{T}}\boldsymbol{A}\boldsymbol{p}_k}\boldsymbol{p}_k = \boldsymbol{x}_{k+1} - \boldsymbol{x}_k, \qquad \text{for any } k \geq 0,$$

so that $\boldsymbol{y}_k = \boldsymbol{x}_k - \boldsymbol{x}_0$.

Exercise 13.9: Program code for testing steepest descent

Write a function `K=sdtest(m,a,d,tol,itmax)` to test the steepest descent method on the matrix $\boldsymbol{T}_2$. Make the analogues of Table 13.1 and Table 13.2. For Table 13.2 it is enough to test for say $n = 100, 400, 1600, 2500$, and tabulate K/n instead of $K/\sqrt{n}$ in the last row. Conclude that the upper bound (13.19) is realistic. Compare also with the number of iterations for the J and GS method in Table 12.1.

Solution. Replacing the steps in (13.18) by those in (13.8), Algorithm 13.2 changes into the following algorithm for testing the steepest descent method.

code/sdtest.m

```
function [V,K] = sdtest(m, a, d, tol, itmax)
R = ones(m)/(m+1)^2; rho = sum(sum(R.*R)); rho0 = rho;
V = zeros(m,m);
T1=sparse(toeplitz([d, a, zeros(1,m-2)]));
for k=1:itmax
   if sqrt(rho/rho0) <= tol
      K = k; return
   end
   T = T1*R + R*T1;
   a = rho/sum(sum(R.*T)); V = V + a*R; R = R - a*T;
   rhos = rho; rho = sum(sum(R.*R));
end
```

```
K = itmax + 1;
```

Listing 13.1: Testing the steepest descent method.

To check that this program is correct, we compare its output with that of `cgtest`.

code/exsdexample.m

```
[V1, K] = sdtest(50, -1, 2, 10^(-8), 1000000);
[V2, K] = cgtest(50, -1, 2, 10^(-8), 1000000);
surf(V2 - V1);
```

Listing 13.2: Compare the steepest descent and conjugate gradient methods.

Running these commands yields Figure 13.1, which shows that the difference between both tests is of the order of 10^{-9}, well within the specified tolerance.

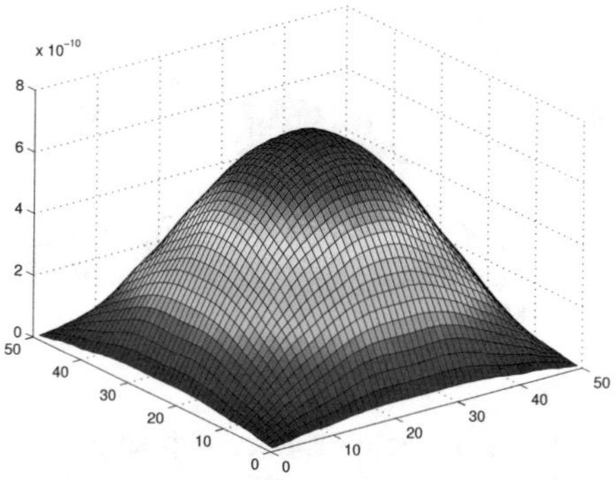

Figure 13.1: For a 50×50 Poisson matrix and a tolerance of 10^{-8}, the figure shows the difference of the outputs of `cgtest` and `sdtest`.

As in Tables 13.1 and 13.2, we let the tolerance be $\text{tol} = 10^{-8}$ and run `sdtest` for the $m \times m$ grid for various m, to find the number of iterations K_{sd} required before $||\boldsymbol{r}_{K_{\text{sd}}}||_2 \le \text{tol} \cdot ||\boldsymbol{r}_0||_2$. Choosing $a = 1/9$ and $d = 5/18$ yields the averaging matrix, and we obtain Table 13.1.

Choosing $a = -1$ and $d = 2$ yields the Poisson matrix, and we obtain Table 13.2. Here the number of iterations K_{J}, K_{GS}, and K_{SOR} of the Jacobi, Gauss-Seidel and SOR methods are taken from Table 12.1, and K_{cg} is the number of iterations in the conjugate gradient method.

Since K_{sd}/n seems to tend towards a constant, it seems that the steepest descent method requires $\mathcal{O}(n)$ iterations for solving the Poisson problem for some given accuracy, as opposed to the $\mathcal{O}(\sqrt{n})$ iterations required by the conjugate gradient method. The number of iterations in the steepest descent method is comparable to

n	2 500	10 000	40 000	1 000 000	4 000 000
K_{sd}	37	35	32	26	24

Table 13.1: For tol $= 10^{-8}$ and averaging matrices ($a = 1/9, d = 5/18$) of various sizes n, the table shows the number of iterations K_{sd} required before the residuals satisfy $||\boldsymbol{r}_{K_{sd}}||_2 \le \text{tol} \cdot ||\boldsymbol{r}_0||_2$.

n	100	400	1 600	2 500	10 000	40 000
K_J	385			8 386		
K_{GS}	194			4 194		
K_{SOR}	35			164	324	645
K_{cg}	16	37	75	94	188	370
K_{sd}	419	1 613	6 258	9 708	38 235	151 451
K_{sd}/n	4.1900	4.0325	3.9112	3.8832	3.8235	3.7863

Table 13.2: For tol $= 10^{-8}$ and Poisson matrices ($a = -1, d = 2$) of various sizes n and various iterative methods, the table shows the number of iterations K required before the residuals satisfy $||\boldsymbol{r}_K||_2 \le \text{tol} \cdot ||\boldsymbol{r}_0||_2$.

the number of iterations in the Jacobi method, while the number of iterations in the conjugate gradient method is of the same order as in the SOR method.

The spectral condition number of the $m \times m$ Poisson matrix is $\kappa = \big(1 + \cos(\pi h)\big)/\big(1 - \cos(\pi h)\big)$. Theorem 13.3 therefore states that

$$\frac{||\boldsymbol{x} - \boldsymbol{x}_k||_{\boldsymbol{A}}}{||\boldsymbol{x} - \boldsymbol{x}_0||_{\boldsymbol{A}}} \le \left(\frac{\kappa - 1}{\kappa + 1}\right)^k = \cos^k\left(\frac{\pi}{m+1}\right). \tag{13.iii}$$

How can we relate this to the tolerance in the algorithm, which is specified in terms of the Euclidean norm? Since

$$\frac{||\boldsymbol{x}||_{\boldsymbol{A}}^2}{||\boldsymbol{x}||_2^2} = \frac{\boldsymbol{x}^{\mathrm{T}}\boldsymbol{A}\boldsymbol{x}}{\boldsymbol{x}^{\mathrm{T}}\boldsymbol{x}}$$

is the Rayleigh quotient of $\boldsymbol{x}$, Lemma 6.8 implies the bound

$$\lambda_{\min}||\boldsymbol{x}||_2^2 \le ||\boldsymbol{x}||_{\boldsymbol{A}}^2 \le \lambda_{\max}||\boldsymbol{x}||_2^2,$$

with $\lambda_{\min} = 4\big(1 - \cos(\pi h)\big)$ the smallest and $\lambda_{\max} = 4\big(1 + \cos(\pi h)\big)$ the largest eigenvalue of $\boldsymbol{A}$. Combining these bounds with Equation (13.iii) yields

$$\frac{\|\boldsymbol{x} - \boldsymbol{x}_k\|_2}{\|\boldsymbol{x} - \boldsymbol{x}_0\|_2} \leq \sqrt{\kappa}\left(\frac{\kappa - 1}{\kappa + 1}\right)^k = \sqrt{\frac{1 + \cos\left(\frac{\pi}{m+1}\right)}{1 - \cos\left(\frac{\pi}{m+1}\right)}} \cos^k\left(\frac{\pi}{m+1}\right).$$

Replacing k by the number of iterations K_{sd} for the various values of m shows that this estimate holds for the tolerance of 10^{-8}.

Exercise 13.10: Using cg to solve normal equations

Consider solving the linear system $\boldsymbol{A}^{\mathrm{T}}\boldsymbol{A}\boldsymbol{x} = \boldsymbol{A}^{\mathrm{T}}\boldsymbol{b}$ by using the conjugate gradient method. Here $\boldsymbol{A} \in \mathbb{R}^{m,n}$, $\boldsymbol{b} \in \mathbb{R}^m$ and $\boldsymbol{A}^{\mathrm{T}}\boldsymbol{A}$ is positive definite. Explain why only the following modifications in Algorithm 13.1 are necessary

1. `r=A'(b-A*x); p=r;`
2. `a=rho/(t'*t);`
3. `r=r-a*A'*t;`

Note that the condition number of the normal equations is $\mathrm{cond}_2(\boldsymbol{A})^2$, the square of the condition number of $\boldsymbol{A}$.

Hint.
This system, known as the *normal equations*, appears in linear least squares problems and was considered in this context in Chapter 9.

Solution. We need to perform Algorithm 13.1 with $\boldsymbol{A}^{\mathrm{T}}\boldsymbol{A}$ replacing $\boldsymbol{A}$ and $\boldsymbol{A}^{\mathrm{T}}\boldsymbol{b}$ replacing $\boldsymbol{b}$. For the system $\boldsymbol{A}^{\mathrm{T}}\boldsymbol{A}\boldsymbol{x} = \boldsymbol{A}^{\mathrm{T}}\boldsymbol{b}$, Equations (13.15), (13.16), and (13.17) become

$$\boldsymbol{x}_{k+1} = \boldsymbol{x}_k + \alpha_k \boldsymbol{p}_k, \qquad \alpha_k = \frac{\boldsymbol{r}_k^{\mathrm{T}}\boldsymbol{r}_k}{\boldsymbol{p}_k^{\mathrm{T}}\boldsymbol{A}^{\mathrm{T}}\boldsymbol{A}\boldsymbol{p}_k} = \frac{\boldsymbol{r}_k^{\mathrm{T}}\boldsymbol{r}_k}{(\boldsymbol{A}\boldsymbol{p}_k)^{\mathrm{T}}\boldsymbol{A}\boldsymbol{p}_k},$$

$$\boldsymbol{r}_{k+1} = \boldsymbol{r}_k - \alpha_k \boldsymbol{A}^{\mathrm{T}}\boldsymbol{A}\boldsymbol{p}_k,$$

$$\boldsymbol{p}_{k+1} = \boldsymbol{r}_{k+1} + \beta_k \boldsymbol{p}_k, \qquad \beta_k = \frac{\boldsymbol{r}_{k+1}^{\mathrm{T}}\boldsymbol{r}_{k+1}}{\boldsymbol{r}_k^{\mathrm{T}}\boldsymbol{r}_k},$$

with $\boldsymbol{p}_0 = \boldsymbol{r}_0 = \boldsymbol{b} - \boldsymbol{A}^{\mathrm{T}}\boldsymbol{A}\boldsymbol{x}_0$. Hence we only need to change the computation of $\boldsymbol{r}_0$, α_k, and $\boldsymbol{r}_{k+1}$ in Algorithm 13.1, which yields the implementation in Listing 13.3.

code/cg_leastSquares.m

```
function [x,K]=cg_leastSquares(A,b,x,tol,itmax)
r=b-A'*A*x; p=r;
rho=r'*r; rho0=rho;
```

```
for k=0:itmax
  if sqrt(rho/rho0)<= tol
    K=k;
    return
  end
  t=A*p; a=rho /(t'*t);
  x=x+a*p; r=r-a*A'*t;
  rhos=rho; rho=r'*r;
  p=r+(rho/rhos)*p;
end
K=itmax+1;
```

Listing 13.3: Conjugate gradient method for least squares

Exercise 13.11: $A^{\mathrm{T}}A$ inner product (Exam exercise 2018-3)

In this exercise we consider linear systems of the form $\boldsymbol{A}\boldsymbol{x} = \boldsymbol{b}$, where $\boldsymbol{A} \in \mathbb{R}^{n\times n}$ and $\boldsymbol{b} \in \mathbb{R}^n$ are given, and $\boldsymbol{x} \in \mathbb{R}^n$ is the unknown vector. We assume throughout that $\boldsymbol{A}$ is nonsingular.

a) Let $\{\boldsymbol{v}_i\}_{i=1}^k$ be a set of linearly independent vectors in $\mathbb{R}^n$, and let $\langle\cdot,\cdot\rangle$ be an inner product in $\mathbb{R}^n$. Explain that the $k \times k$-matrix $\boldsymbol{N}$ with entries $n_{ij} = \langle \boldsymbol{v}_i, \boldsymbol{v}_j\rangle$ is symmetric positive definite.

Solution. The matrix is symmetric by symmetry of the inner product. For any $\boldsymbol{x} \neq 0$, one has

$$\boldsymbol{x}^{\mathrm{T}}\boldsymbol{N}\boldsymbol{x} = \sum_{i,j=1}^{k} x_i\langle \boldsymbol{v}_i, \boldsymbol{v}_j\rangle x_j = \left\langle \sum_{i=1}^{k} x_i\boldsymbol{v}_i, \sum_{j=1}^{k} x_j\boldsymbol{v}_j \right\rangle = \left\| \sum_{i=1}^{k} x_i\boldsymbol{v}_i \right\|^2 > 0,$$

where $\|\cdot\|$ is the norm associated with $\langle\cdot,\cdot\rangle$. This inequality is strict due to the linear independence of the $\boldsymbol{v}_i$. It follows that the matrix is positive definite.

b) Let $\mathbb{W} \subset \mathbb{R}^n$ be any linear subspace. Show that there is one and only one vector $\hat{\boldsymbol{x}} \in \mathbb{W}$ so that

$$\boldsymbol{w}^{\mathrm{T}}\boldsymbol{A}^{\mathrm{T}}\boldsymbol{A}\hat{\boldsymbol{x}} = \boldsymbol{w}^{\mathrm{T}}\boldsymbol{A}^{\mathrm{T}}\boldsymbol{b}, \qquad \text{for all } \boldsymbol{w} \in \mathbb{W},$$

and that $\hat{\boldsymbol{x}}$ satisfies

$$\|\boldsymbol{b} - \boldsymbol{A}\hat{\boldsymbol{x}}\|_2 \leq \|\boldsymbol{b} - \boldsymbol{A}\boldsymbol{w}\|_2, \qquad \text{for all } \boldsymbol{w} \in \mathbb{W}.$$

Solution. Let $\{\boldsymbol{w}_i\}_i$ be a basis for $\mathbb{W}$. The first equation is satisfied if and only if

$$\boldsymbol{w}_i^{\mathrm{T}}\boldsymbol{A}^{\mathrm{T}}\boldsymbol{A}\hat{\boldsymbol{x}} = \boldsymbol{w}_i^{\mathrm{T}}\boldsymbol{A}^{\mathrm{T}}\boldsymbol{b}$$

for all i. Any $\hat{\boldsymbol{x}} \in \mathbb{W}$ can be written uniquely as $\sum_j c_j \boldsymbol{w}_j$, and inserting this gives $\sum_j \boldsymbol{w}_i^{\mathrm{T}} \boldsymbol{A}^{\mathrm{T}} \boldsymbol{A} \boldsymbol{w}_j c_j = \boldsymbol{w}_i^{\mathrm{T}} \boldsymbol{A}^{\mathrm{T}} \boldsymbol{b}$. With $\boldsymbol{W}$ the matrix with $\boldsymbol{w}_i$ as columns, this can also be written as $\boldsymbol{W}^{\mathrm{T}} \boldsymbol{A}^{\mathrm{T}} \boldsymbol{A} \boldsymbol{W} \boldsymbol{c} = \boldsymbol{W}^{\mathrm{T}} \boldsymbol{A} \boldsymbol{b}$, or

$$(\boldsymbol{A}\boldsymbol{W})^{\mathrm{T}} \boldsymbol{A} \boldsymbol{W} \boldsymbol{c} = (\boldsymbol{A}^{\mathrm{T}} \boldsymbol{W})^{\mathrm{T}} \boldsymbol{b}. \tag{13.iv}$$

The matrix on the left hand side has entries $\langle \boldsymbol{A}\boldsymbol{w}_i, \boldsymbol{A}\boldsymbol{w}_j \rangle$, which we proved in **a)** to be positive definite (the $\boldsymbol{A}\boldsymbol{w}_i$ are linearly independent whenever the $\boldsymbol{w}_i$ are, since $\boldsymbol{A}$ is nonsingular), and thus nonsingular. It follows that the c_j, and hence $\hat{\boldsymbol{x}}$, is unique.

Equation (13.iv) can also be viewed as the normal equations for the least squares problem $\min \|\boldsymbol{A}\boldsymbol{W}\boldsymbol{c} - \boldsymbol{b}\|$. Thus, $\boldsymbol{c}$ is the least squares solutions to $\boldsymbol{A}\boldsymbol{W}\boldsymbol{c} = \boldsymbol{b}$, so that $\hat{\boldsymbol{x}} = \boldsymbol{W}\boldsymbol{c}$ is the least squares solution to $\boldsymbol{A}\hat{\boldsymbol{x}} = \boldsymbol{b}$ in the subspace $\mathbb{W}$. This is equivalent to the second statement. The least squares view also gives an alternative proof for the first statement: $\boldsymbol{A}\boldsymbol{W}$ has linearly independent columns, which ensures uniqueness of the least squares solution.

c) In the rest of this exercise we consider the situation above, but where the vector space $\mathbb{W}$ is taken to be the Krylov space

$$\mathbb{W}_k := \operatorname{span}(\boldsymbol{b}, \boldsymbol{A}\boldsymbol{b}, \ldots, \boldsymbol{A}^{k-1}\boldsymbol{b}).$$

We use the inner product in $\mathbb{R}^n$ given by

$$\langle \boldsymbol{v}, \boldsymbol{w} \rangle_{\boldsymbol{A}} := \boldsymbol{v}^{\mathrm{T}} \boldsymbol{A}^{\mathrm{T}} \boldsymbol{A} \boldsymbol{w}, \qquad \boldsymbol{v}, \boldsymbol{w} \in \mathbb{R}^n.$$

The associated approximations of $\boldsymbol{x}$, corresponding to $\hat{\boldsymbol{x}}$ in $\mathbb{W}_k$, are then denoted $\boldsymbol{x}_k$. Assume that $\boldsymbol{x}_k \in \mathbb{W}_k$ is already determined. In addition, assume that we already have computed a "search direction" $\boldsymbol{p}_k \in \mathbb{W}_{k+1}$ such that $\|\boldsymbol{A}\boldsymbol{p}_k\|_2 = \|\boldsymbol{p}_k\|_{\boldsymbol{A}} = 1$, and such that

$$\langle \boldsymbol{p}_k, \boldsymbol{w} \rangle_{\boldsymbol{A}} = 0, \quad \text{for all } \boldsymbol{w} \in \mathbb{W}_k.$$

Show that $\boldsymbol{x}_{k+1} = \boldsymbol{x}_k + \alpha_k \boldsymbol{p}_k$ for a suitable $\alpha_k \in \mathbb{R}$, and express α_k in terms of the residual $\boldsymbol{r}_k := \boldsymbol{b} - \boldsymbol{A}\boldsymbol{x}_k$, and $\boldsymbol{p}_k$.

Solution. From **b)** it follows that, for all $\boldsymbol{w} \in \mathbb{W}_k$, $\boldsymbol{w}^{\mathrm{T}} \boldsymbol{A}^{\mathrm{T}} \boldsymbol{A}(\boldsymbol{x}_{k+1} - \boldsymbol{x}_k) = 0$, so that $\boldsymbol{x}_{k+1} - \boldsymbol{x}_k$ is orthogonal to $\mathbb{W}_k$ with respect to $\langle \cdot, \cdot \rangle_{\boldsymbol{A}}$. Since the orthogonal complement of $\mathbb{W}_k$ in $\mathbb{W}_{k+1}$ is one-dimensional, the stated vector $\boldsymbol{p}_k$ spans this orthogonal complement, so that we can write $\boldsymbol{x}_{k+1} - \boldsymbol{x}_k = \alpha_k \boldsymbol{p}_k$ for some scalar α_k, i.e., $\boldsymbol{x}_{k+1} = \boldsymbol{x}_k + \alpha_k \boldsymbol{p}_k$. From **b)** it follows that

$$\boldsymbol{w}^{\mathrm{T}} \boldsymbol{A}^{\mathrm{T}} \boldsymbol{A} \boldsymbol{x}_{k+1} = \boldsymbol{w}^{\mathrm{T}} \boldsymbol{A}^{\mathrm{T}} \boldsymbol{b}, \qquad \text{for all } \boldsymbol{w} \in \mathbb{W}_{k+1}.$$

To determine α_k, insert $\boldsymbol{w} = \boldsymbol{p}_k$ and $\boldsymbol{x}_{k+1} = \boldsymbol{x}_k + \alpha_k \boldsymbol{p}_k$ here to obtain

$$\boldsymbol{p}_k^{\mathrm{T}} \boldsymbol{A}^{\mathrm{T}} \boldsymbol{A}(\boldsymbol{x}_k + \alpha_k \boldsymbol{p}_k) = \boldsymbol{p}_k^{\mathrm{T}} \boldsymbol{A}^{\mathrm{T}} \boldsymbol{b},$$

so that

$$\alpha_k \boldsymbol{p}_k^{\mathrm{T}} \boldsymbol{A}^{\mathrm{T}} \boldsymbol{A} \boldsymbol{p}_k = \boldsymbol{p}_k^{\mathrm{T}} \boldsymbol{A}^{\mathrm{T}} (\boldsymbol{b} - \boldsymbol{A} \boldsymbol{x}_k) = \boldsymbol{p}_k^{\mathrm{T}} \boldsymbol{A}^{\mathrm{T}} \boldsymbol{r}_k.$$

By assumption, the left side computes to α_k, so that $\alpha_k = \boldsymbol{p}_k^{\mathrm{T}} \boldsymbol{A}^{\mathrm{T}} \boldsymbol{r}_k$.

d) Assume that $\boldsymbol{A}$ is symmetric, but not necessarily positive definite. Assume further that the vectors $\boldsymbol{p}_{k-2}, \boldsymbol{p}_{k-1}$, and $\boldsymbol{p}_k$ are already known with properties as above. Show that

$$\boldsymbol{A}\boldsymbol{p}_{k-1} \in \operatorname{span}(\boldsymbol{p}_{k-2}, \boldsymbol{p}_{k-1}, \boldsymbol{p}_k).$$

Use this to suggest how the search vectors $\boldsymbol{p}_k$ can be computed recursively.

Solution. It follows by the definitions of the spaces $\mathbb{W}_k$ and the property $p_{k-1} \in \mathbb{W}_k$ stated in the exercise that $\boldsymbol{A}\boldsymbol{p}_{k-1} \in \mathbb{W}_{k+1}$. As a consequence, since $\{\boldsymbol{p}_j\}_{j=0}^k$ is an orthonormal basis for $\mathbb{W}_{k+1}$ with respect to $\langle \cdot, \cdot \rangle_{\boldsymbol{A}}$, we must have that

$$\boldsymbol{A}\boldsymbol{p}_{k-1} = \sum_{j=0}^{k} \langle \boldsymbol{A}\boldsymbol{p}_{k-1}, \boldsymbol{p}_j \rangle_{\boldsymbol{A}} \boldsymbol{p}_j = \sum_{j=0}^{k} \langle \boldsymbol{p}_{k-1}, \boldsymbol{A}\boldsymbol{p}_j \rangle_{\boldsymbol{A}} \boldsymbol{p}_j,$$

where we have applied the symmetry of $\boldsymbol{A}$ to the definition of $\langle \cdot, \cdot \rangle_{\boldsymbol{A}}$. However, for $j < k-2$, $\boldsymbol{A}\boldsymbol{p}_j \in \mathbb{W}_{k-1}$, and as a consequence $\langle \boldsymbol{p}_{k-1}, \boldsymbol{A}\boldsymbol{p}_j \rangle_{\boldsymbol{A}} = 0$. We therefore obtain

$$\boldsymbol{A}\boldsymbol{p}_{k-1} = \sum_{j=k-2}^{k} \langle \boldsymbol{p}_{k-1}, \boldsymbol{A}\boldsymbol{p}_j \rangle_{\boldsymbol{A}} \boldsymbol{p}_j,$$

so that

$$\begin{aligned}
&\langle \boldsymbol{p}_{k-1}, \boldsymbol{A}\boldsymbol{p}_k \rangle_{\boldsymbol{A}} \boldsymbol{p}_k \\
&= \boldsymbol{A}\boldsymbol{p}_{k-1} - \langle \boldsymbol{A}\boldsymbol{p}_{k-1}, \boldsymbol{p}_{k-1} \rangle_{\boldsymbol{A}} \boldsymbol{p}_{k-1} - \langle \boldsymbol{A}\boldsymbol{p}_{k-1}, \boldsymbol{p}_{k-2} \rangle_{\boldsymbol{A}} \boldsymbol{p}_{k-2}.
\end{aligned}$$

The right hand side now gives a vector $\boldsymbol{q}$ with the same direction as $\boldsymbol{p}_k$. Finally we compute $\|\boldsymbol{q}\|_{\boldsymbol{A}}$, and define $\boldsymbol{p}_k := \boldsymbol{q}/\|\boldsymbol{q}\|_{\boldsymbol{A}}$.

Exercises section 13.3

Exercise 13.12: Krylov space and cg iterations

Consider the linear system $\boldsymbol{A}\boldsymbol{x} = \boldsymbol{b}$ where

$$\boldsymbol{A} = \begin{bmatrix} 2 & -1 & 0 \\ -1 & 2 & -1 \\ 0 & -1 & 2 \end{bmatrix}, \quad \text{and} \quad \boldsymbol{b} = \begin{bmatrix} 4 \\ 0 \\ 0 \end{bmatrix}.$$

a) Determine the vectors defining the Krylov spaces for $k \leq 3$, taking as initial approximation $\boldsymbol{x} = \boldsymbol{0}$.

Hint.

Answer: $[\boldsymbol{b}, \boldsymbol{A}\boldsymbol{b}, \boldsymbol{A}^2\boldsymbol{b}] = \begin{bmatrix} 4 & 8 & 20 \\ 0 & -4 & -16 \\ 0 & 0 & 4 \end{bmatrix}$.

Solution. The Krylov spaces $\mathbb{W}_k$ are defined as

$$\mathbb{W}_k := \operatorname{span}\left\{\boldsymbol{r}_0, \boldsymbol{A}\boldsymbol{r}_0, \ldots, \boldsymbol{A}^{k-1}\boldsymbol{r}_0\right\}.$$

Taking $\boldsymbol{A}, \boldsymbol{b}, \boldsymbol{x} = \boldsymbol{0}$, and $\boldsymbol{r}_0 = \boldsymbol{b} - \boldsymbol{A}\boldsymbol{x} = \boldsymbol{b}$ as in the exercise, these vectors can be expressed as

$$\left[\boldsymbol{r}_0, \boldsymbol{A}\boldsymbol{r}_0, \boldsymbol{A}^2\boldsymbol{r}_0\right] = \left[\boldsymbol{b}, \boldsymbol{A}\boldsymbol{b}, \boldsymbol{A}^2\boldsymbol{b}\right] = \left[\begin{bmatrix} 4 \\ 0 \\ 0 \end{bmatrix}, \begin{bmatrix} 8 \\ -4 \\ 0 \end{bmatrix}, \begin{bmatrix} 20 \\ -16 \\ 4 \end{bmatrix}\right].$$

b) Carry out three CG-iterations on $\boldsymbol{A}\boldsymbol{x} = \boldsymbol{b}$.

Solution. As $\boldsymbol{x}_0 = \boldsymbol{0}$ we have $\boldsymbol{p}_0 = \boldsymbol{r}_0 = \boldsymbol{b}$. We have for $k = 0, 1, 2, \ldots$ Equations (13.15), (13.16), and (13.17),

$$\begin{aligned} \boldsymbol{x}_{k+1} &= \boldsymbol{x}_k + \alpha_k \boldsymbol{p}_k, & \alpha_k &= \frac{\boldsymbol{r}_k^{\mathrm{T}}\boldsymbol{r}_k}{\boldsymbol{p}_k^{\mathrm{T}}\boldsymbol{A}\boldsymbol{p}_k}, \\ \boldsymbol{r}_{k+1} &= \boldsymbol{r}_k - \alpha_k \boldsymbol{A}\boldsymbol{p}_k, \\ \boldsymbol{p}_{k+1} &= \boldsymbol{r}_{k+1} + \beta_k \boldsymbol{p}_k, & \beta_k &= \frac{\boldsymbol{r}_{k+1}^{\mathrm{T}}\boldsymbol{r}_{k+1}}{\boldsymbol{r}_k^{\mathrm{T}}\boldsymbol{r}_k}, \end{aligned}$$

which determine the approximations $\boldsymbol{x}_k$. For $k = 0, 1, 2$ these give

$$\alpha_0 = \frac{1}{2}, \quad \boldsymbol{x}_1 = \begin{bmatrix} 2 \\ 0 \\ 0 \end{bmatrix}, \quad \boldsymbol{r}_1 = \begin{bmatrix} 0 \\ 2 \\ 0 \end{bmatrix}, \quad \beta_0 = \frac{1}{4}, \quad \boldsymbol{p}_1 = \begin{bmatrix} 1 \\ 2 \\ 0 \end{bmatrix},$$

$$\alpha_1 = \frac{2}{3}, \quad \boldsymbol{x}_2 = \frac{1}{3}\begin{bmatrix} 8 \\ 4 \\ 0 \end{bmatrix}, \quad \boldsymbol{r}_2 = \frac{1}{3}\begin{bmatrix} 0 \\ 0 \\ 4 \end{bmatrix}, \quad \beta_1 = \frac{4}{9}, \quad \boldsymbol{p}_2 = \frac{1}{9}\begin{bmatrix} 4 \\ 8 \\ 12 \end{bmatrix},$$

$$\alpha_2 = \frac{3}{4}, \quad \boldsymbol{x}_3 = \begin{bmatrix} 3 \\ 2 \\ 1 \end{bmatrix}, \quad \boldsymbol{r}_3 = \begin{bmatrix} 0 \\ 0 \\ 0 \end{bmatrix}, \quad \beta_2 = 0, \quad \boldsymbol{p}_3 = \begin{bmatrix} 0 \\ 0 \\ 0 \end{bmatrix}.$$

c) Verify that

- $\dim(\mathbb{W}_k) = k$ for $k = 0, 1, 2, 3$.
- $\boldsymbol{x} = \boldsymbol{x}_3$ is the exact solution of $\boldsymbol{A}\boldsymbol{x} = \boldsymbol{b}$.
- $\boldsymbol{r}_0, \ldots, \boldsymbol{r}_{k-1}$ is an orthogonal basis for $\mathbb{W}_k$ for $k = 1, 2, 3$.
- $\boldsymbol{p}_0, \ldots, \boldsymbol{p}_{k-1}$ is an $\boldsymbol{A}$-orthogonal basis for $\mathbb{W}_k$ for $k = 1, 2, 3$.
- $(\|\boldsymbol{r}_k\|)_k$ is monotonically decreasing.
- $(\|\boldsymbol{x}_k - \boldsymbol{x}\|)_k$ is monotonically decreasing.

Solution. By definition we have $\mathbb{W}_0 = \{\boldsymbol{0}\}$. From the solution of part **a)** we know that $\mathbb{W}_k = \operatorname{span}(\boldsymbol{b}_0, \boldsymbol{A}\boldsymbol{b}_0, \ldots, \boldsymbol{A}^{k-1}\boldsymbol{b}_0)$, where the vectors $\boldsymbol{b}$, $\boldsymbol{A}\boldsymbol{b}$ and $\boldsymbol{A}^2\boldsymbol{b}$ are linearly independent. Hence we have $\dim \mathbb{W}_k = k$ for $k = 0, 1, 2, 3$.

From **b)** we know that the residual $\boldsymbol{r}_3 = \boldsymbol{b} - \boldsymbol{A}\boldsymbol{x}_3 = \boldsymbol{0}$. Hence $\boldsymbol{x}_3$ is the exact solution to $\boldsymbol{A}\boldsymbol{x} = \boldsymbol{b}$.

We observe that $\boldsymbol{r}_0 = 4\boldsymbol{e}_1$, $\boldsymbol{r}_1 = 2\boldsymbol{e}_2$ and $\boldsymbol{r}_2 = (4/3)\boldsymbol{e}_3$ and hence the $\boldsymbol{r}_k$ for $k = 0, 1, 2$ are linear independent and orthogonal to each other. Thus we are only left to show that $\mathbb{W}_k$ is the span of $\boldsymbol{r}_0, \ldots, \boldsymbol{r}_{k-1}$. We observe that $\boldsymbol{b} = \boldsymbol{r}_0$, $\boldsymbol{A}\boldsymbol{b} = 2\boldsymbol{r}_0 - 2\boldsymbol{r}_1$ and $\boldsymbol{A}^2\boldsymbol{b} = 5\boldsymbol{r}_0 - 8\boldsymbol{r}_1 + 3\boldsymbol{r}_2$. Hence $\operatorname{span}(\boldsymbol{b}, \boldsymbol{A}\boldsymbol{b}, \ldots, \boldsymbol{A}^{k-1})\boldsymbol{b} = \operatorname{span}(\boldsymbol{r}_0, \ldots, \boldsymbol{r}_{k-1})$ for $k = 1, 2, 3$. We conclude that, for $k = 1, 2, 3$, the vectors $\boldsymbol{r}_0, \ldots, \boldsymbol{r}_{k-1}$ form an orthogonal basis for $\mathbb{W}_k$.

One can verify directly that $\boldsymbol{p}_0, \boldsymbol{p}_1$, and $\boldsymbol{p}_2$ are $\boldsymbol{A}$-orthogonal. Moreover, observing that $\boldsymbol{b} = \boldsymbol{p}_0$, $\boldsymbol{A}\boldsymbol{b} = (5/2)\boldsymbol{p}_0 - 2\boldsymbol{p}_1$, and $\boldsymbol{A}^2\boldsymbol{b} = 7\boldsymbol{p}_0 - (28/3)\boldsymbol{p}_1 + 3\boldsymbol{p}_2$, it follows that

$$\operatorname{span}(\boldsymbol{b}, \boldsymbol{A}\boldsymbol{b}, \ldots, \boldsymbol{A}^{k-1})\boldsymbol{b} = \operatorname{span}(\boldsymbol{p}_0, \ldots, \boldsymbol{p}_{k-1}), \quad \text{for } k = 1, 2, 3.$$

We conclude that, for $k = 1, 2, 3$, the vectors $\boldsymbol{p}_0, \ldots, \boldsymbol{p}_{k-1}$ form an $\boldsymbol{A}$-orthogonal basis for $\mathbb{W}_k$.

By computing the Euclidean norms of $\boldsymbol{r}_0, \boldsymbol{r}_1, \boldsymbol{r}_2, \boldsymbol{r}_3$, we get

$$\|\boldsymbol{r}_0\|_2 = 4, \qquad \|\boldsymbol{r}_1\|_2 = 2, \qquad \|\boldsymbol{r}_2\|_2 = 4/3, \qquad \|\boldsymbol{r}_3\|_2 = 0.$$

It follows that the sequence $(\|\boldsymbol{r}_k\|)_k$ is monotonically decreasing. Similarly, one finds

$$\left(\|\boldsymbol{x}_k - \boldsymbol{x}\|_2\right)_{k=0}^{3} = \left(\sqrt{10}, \sqrt{6}, \sqrt{14/9}, 0\right),$$

which is clearly monotonically decreasing.

Exercise 13.13: Antisymmetric system (Exam exercise 1983-3)

In this and the next exercise $\langle \boldsymbol{x}, \boldsymbol{y} \rangle = \boldsymbol{x}^{\mathrm{T}}\boldsymbol{y}$ is the usual inner product in $\mathbb{R}^n$. We note that

$$\langle \boldsymbol{x}, \boldsymbol{y} \rangle = \langle \boldsymbol{y}, \boldsymbol{x} \rangle, \quad \boldsymbol{x}, \boldsymbol{y} \in \mathbb{R}^n, \tag{13.61}$$

$$\langle \boldsymbol{C}\boldsymbol{x}, \boldsymbol{y} \rangle = \langle \boldsymbol{x}, \boldsymbol{C}^{\mathrm{T}}\boldsymbol{y} \rangle = \langle \boldsymbol{C}^{\mathrm{T}}\boldsymbol{y}, \boldsymbol{x} \rangle, \quad \boldsymbol{x}, \boldsymbol{y} \in \mathbb{R}^n, \boldsymbol{C} \in \mathbb{R}^{n\times n}. \tag{13.62}$$

Let $\boldsymbol{B} \in \mathbb{R}^{n\times n}$ be an antisymmetric matrix, i.e., $\boldsymbol{B}^{\mathrm{T}} = -\boldsymbol{B}$, and let $\boldsymbol{A} := \boldsymbol{I} - \boldsymbol{B}$, where $\boldsymbol{I}$ is the unit matrix in $\mathbb{R}^n$.

a) Show that

$$\langle \boldsymbol{B}\boldsymbol{x}, \boldsymbol{x} \rangle = 0, \quad \boldsymbol{x} \in \mathbb{R}^n, \tag{13.63}$$

$$\langle \boldsymbol{A}\boldsymbol{x}, \boldsymbol{x} \rangle = \langle \boldsymbol{x}, \boldsymbol{x} \rangle = \|\boldsymbol{x}\|_2^2. \tag{13.64}$$

Solution. We have

$$\langle \boldsymbol{B}\boldsymbol{x}, \boldsymbol{x} \rangle \overset{(13.62)}{=} \langle \boldsymbol{B}^{\mathrm{T}}\boldsymbol{x}, \boldsymbol{x} \rangle = -\langle \boldsymbol{B}\boldsymbol{x}, \boldsymbol{x} \rangle,$$

which implies that $\langle \boldsymbol{B}\boldsymbol{x}, \boldsymbol{x} \rangle = 0$ for all $\boldsymbol{x} \in \mathbb{R}^n$. Moreover,

$$\langle \boldsymbol{A}\boldsymbol{x}, \boldsymbol{x} \rangle = \langle \boldsymbol{x}, \boldsymbol{x} \rangle - \langle \boldsymbol{B}\boldsymbol{x}, \boldsymbol{x} \rangle \overset{(13.63)}{=} \langle \boldsymbol{x}, \boldsymbol{x} \rangle = \|\boldsymbol{x}\|_2^2.$$

b) Show that $\|\boldsymbol{A}\boldsymbol{x}\|_2^2 = \|\boldsymbol{x}\|_2^2 + \|\boldsymbol{B}\boldsymbol{x}\|_2^2$ and that $\|\boldsymbol{A}\|_2 = \sqrt{1 + \|\boldsymbol{B}\|_2^2}$.

Solution. Since $\boldsymbol{B}^{\mathrm{T}} + \boldsymbol{B} = \boldsymbol{0}$ we have

$$\begin{aligned}
\|\boldsymbol{A}\boldsymbol{x}\|_2^2 &= \boldsymbol{x}^{\mathrm{T}}(\boldsymbol{I} - \boldsymbol{B})^{\mathrm{T}}(\boldsymbol{I} - \boldsymbol{B})\boldsymbol{x} \\
&= \boldsymbol{x}^{\mathrm{T}}(\boldsymbol{I} - \boldsymbol{B}^{\mathrm{T}} - \boldsymbol{B} + \boldsymbol{B}^{\mathrm{T}}\boldsymbol{B})\boldsymbol{x} \\
&= \boldsymbol{x}^{\mathrm{T}}\boldsymbol{x} + \boldsymbol{x}^{\mathrm{T}}\boldsymbol{B}^{\mathrm{T}}\boldsymbol{B}\boldsymbol{x} = \|\boldsymbol{x}\|_2^2 + \|\boldsymbol{B}\boldsymbol{x}\|_2^2.
\end{aligned}$$

But then

$$\frac{\|\boldsymbol{A}\boldsymbol{x}\|_2^2}{\|\boldsymbol{x}\|_2^2} = 1 + \frac{\|\boldsymbol{B}\boldsymbol{x}\|_2^2}{\|\boldsymbol{x}\|_2^2} \le 1 + \|\boldsymbol{B}\|_2^2$$

Taking max on the left side we see that $\|\boldsymbol{A}\|_2^2 \le 1 + \|\boldsymbol{B}\|_2^2$. Similarly, maximizing

$$\frac{\|\boldsymbol{B}\boldsymbol{x}\|_2^2}{\|\boldsymbol{x}\|_2^2} = \frac{\|\boldsymbol{A}\boldsymbol{x}\|_2^2}{\|\boldsymbol{x}\|_2^2} - 1$$

we obtain $\|\boldsymbol{B}\|_2^2 \leq \|\boldsymbol{A}\|_2^2 - 1$ and hence $\|\boldsymbol{A}\|_2^2 = 1 + \|\boldsymbol{B}\|_2^2$. Taking square roots the result follows.

c) Show that $\boldsymbol{A}$ is nonsingular,

$$\|\boldsymbol{A}^{-1}\|_2 = \max_{\boldsymbol{x}\neq\boldsymbol{0}} \frac{\|\boldsymbol{x}\|_2}{\|\boldsymbol{A}\boldsymbol{x}\|_2},$$

and $\|\boldsymbol{A}\|_2 \leq 1$.

Solution. If $\boldsymbol{A}\boldsymbol{x} = \boldsymbol{0}$ then $0 = \|\boldsymbol{A}\boldsymbol{x}\|_2^2 = \|\boldsymbol{x}\|_2^2 + \|\boldsymbol{B}\boldsymbol{x}\|_2^2$, which implies that $\boldsymbol{x} = \boldsymbol{0}$ and $\boldsymbol{A}$ is nonsingular. Let $\boldsymbol{y} \in \mathbb{R}^n$ and $\boldsymbol{x} := \boldsymbol{A}^{-1}\boldsymbol{y}$. Then

$$\frac{\|\boldsymbol{A}^{-1}\boldsymbol{y}\|_2}{\|\boldsymbol{y}\|_2} = \frac{\|\boldsymbol{x}\|_2}{\|\boldsymbol{A}\boldsymbol{x}\|_2}.$$

Taking the maximum on the right hand side and then on the left hand side we obtain

$$\|\boldsymbol{A}^{-1}\|_2 \leq \max_{\boldsymbol{x}\neq\boldsymbol{0}} \frac{\|\boldsymbol{x}\|_2}{\|\boldsymbol{A}\boldsymbol{x}\|_2}.$$

Taking the maximum on the left hand side and then on the right hand side we obtain the inequality the other way. From **b)** we have $\|\boldsymbol{A}\boldsymbol{x}\|_2 \geq \|\boldsymbol{x}\|_2$ and then

$$\|\boldsymbol{A}^{-1}\|_2 = \max_{\boldsymbol{x}\neq\boldsymbol{0}} \frac{\|\boldsymbol{x}\|_2}{\|\boldsymbol{A}\boldsymbol{x}\|_2} \leq 1.$$

d) Let $1 \leq k \leq n$, $\mathcal{W} = \text{span}(\boldsymbol{w}_1, \ldots, \boldsymbol{w}_k)$ a k-dimensional subspace of $\mathbb{R}^n$ and $\boldsymbol{b} \in \mathbb{R}^n$. Show that if $\boldsymbol{x} \in \mathcal{W}$ is such that

$$\langle \boldsymbol{A}\boldsymbol{x}, \boldsymbol{w} \rangle = \langle \boldsymbol{b}, \boldsymbol{w} \rangle \text{ for all } \boldsymbol{w} \in \mathcal{W}, \tag{13.65}$$

then $\|\boldsymbol{x}\|_2 \leq \|\boldsymbol{b}\|_2$.

With $\boldsymbol{x} := \sum_{j=1}^k x_j \boldsymbol{w}_j$ the problem (13.65) is equivalent to finding real numbers $x_1, \ldots, x_k$ solving the linear system

$$\sum_{j=1}^k x_j \langle \boldsymbol{A}\boldsymbol{w}_j, \boldsymbol{w}_i \rangle = \langle \boldsymbol{b}, \boldsymbol{w}_i \rangle, \quad i = 1, \ldots, k. \tag{13.66}$$

Show that (13.65) has a unique solution $\boldsymbol{x} \in \mathcal{W}$.

Solution. Taking $\boldsymbol{w} = \boldsymbol{x}$ in (13.65), and using the Cauchy-Schwarz inequality we find

$$\|\boldsymbol{x}\|_2^2 \overset{(13.64)}{=} \langle \boldsymbol{A}\boldsymbol{x}, \boldsymbol{x}\rangle \overset{(13.65)}{=} \langle \boldsymbol{b}, \boldsymbol{x}\rangle \le \|\boldsymbol{b}\|_2 \|\boldsymbol{x}\|_2.$$

Dividing by $\|\boldsymbol{x}\|_2$ we find $\|\boldsymbol{x}\|_2 \le \|\boldsymbol{b}\|_2$. To show that the linear system (13.66) has a unique solution, it is enough to show that the homogeneous system only has the zero solution. But if $\boldsymbol{b} = \boldsymbol{0}$ then $\boldsymbol{x} = \boldsymbol{0}$ is the only solution since $\|\boldsymbol{x}\|_2 \le \|\boldsymbol{b}\|_2 = 0$.

e) Let $\boldsymbol{x}^* := \boldsymbol{A}^{-1}\boldsymbol{b}$. Show that

$$\|\boldsymbol{x}^* - \boldsymbol{x}\|_2 \le \|\boldsymbol{A}\|_2 \min_{\boldsymbol{w}\in\mathcal{W}} \|\boldsymbol{x}^* - \boldsymbol{w}\|_2. \tag{13.67}$$

Solution. For any $\boldsymbol{w} \in \mathcal{W}$

$$\begin{aligned}
\|\boldsymbol{x}^* - \boldsymbol{x}\|_2^2 &\overset{(13.64)}{=} \langle \boldsymbol{A}(\boldsymbol{x}^* - \boldsymbol{x}), \boldsymbol{x}^* - \boldsymbol{x}\rangle \\
&= \langle \boldsymbol{A}(\boldsymbol{x}^* - \boldsymbol{x}), \boldsymbol{x}^* - \boldsymbol{w}\rangle + \langle \boldsymbol{A}(\boldsymbol{x}^* - \boldsymbol{x}), \boldsymbol{w} - \boldsymbol{x}\rangle \\
&= \langle \boldsymbol{A}(\boldsymbol{x}^* - \boldsymbol{x}), \boldsymbol{x}^* - \boldsymbol{w}\rangle + \langle \boldsymbol{b} - \boldsymbol{A}\boldsymbol{x}, \boldsymbol{w} - \boldsymbol{x}\rangle.
\end{aligned}$$

By (13.65) the last term is zero since $\boldsymbol{w} - \boldsymbol{x} \in \mathcal{W}$, and we obtain

$$\|\boldsymbol{x}^* - \boldsymbol{x}\|_2^2 = \langle \boldsymbol{A}(\boldsymbol{x}^* - \boldsymbol{x}), \boldsymbol{x}^* - \boldsymbol{w}\rangle \le \|\boldsymbol{A}\|_2 \|\boldsymbol{x}^* - \boldsymbol{x}\|_2 \|\boldsymbol{x}^* - \boldsymbol{w}\|_2.$$

by the Cauchy-Schwarz inequality. Dividing by $\|\boldsymbol{x}^* - \boldsymbol{x}\|_2$ the estimate follows.

Exercise 13.14: cg antisymmetric system (Exam exercise 1983-4)

(It is recommended to study Exercise 13.13 before starting this exercise.) As in Exercise 13.13 let $\boldsymbol{B} \in \mathbb{R}^{n\times n}$ be an antisymmetric matrix, i.e., $\boldsymbol{B}^{\mathrm{T}} = -\boldsymbol{B}$, let $\langle \boldsymbol{x}, \boldsymbol{y}\rangle = \boldsymbol{x}^{\mathrm{T}}\boldsymbol{y}$ be the usual inner product in $\mathbb{R}^n$, let $\boldsymbol{A} := \boldsymbol{I} - \boldsymbol{B}$, where $\boldsymbol{I}$ is the unit matrix in $\mathbb{R}^n$ and $\boldsymbol{b} \in \mathbb{R}^n$. The purpose of this exercise is to develop an iterative algorithm for the linear system $\boldsymbol{A}\boldsymbol{x} = \boldsymbol{b}$. The algorithm is partly built on the same idea as for the conjugate gradient method for positive definite systems.

Let $\boldsymbol{x}_0 = 0$ be the initial approximation to the exact solution $\boldsymbol{x}^* := \boldsymbol{A}^{-1}\boldsymbol{b}$. For $k = 1, 2, \ldots, n$ we let

$$\mathcal{W}_k := \operatorname{span}(\boldsymbol{b}, \boldsymbol{B}\boldsymbol{b}, \ldots, \boldsymbol{B}^{k-1}\boldsymbol{b}).$$

For $k = 1, 2, \ldots, n$ we define $\boldsymbol{x}_k \in \mathcal{W}_k$ by

$$\langle \boldsymbol{A}\boldsymbol{x}_k, \boldsymbol{w}\rangle = \langle \boldsymbol{b}, \boldsymbol{w}\rangle, \text{ for all } \boldsymbol{w} \in \mathcal{W}_k.$$

The vector $\boldsymbol{x}_k$ is uniquely determined as shown in Exercise 13.13**d)** and that it is a "good" approximation to $\boldsymbol{x}^*$ follows from (13.67). In this exercise we will derive a recursive algorithm to determine $\boldsymbol{x}_k$.

For $k = 0, \ldots, n$ we set

$$\boldsymbol{r}_k := \boldsymbol{b} - \boldsymbol{A}\boldsymbol{x}_k, \text{ and } \rho_k := \|\boldsymbol{r}_k\|_2^2.$$

Let $m \in \mathbb{N}$ be such that

$$\rho_k \neq 0, \quad k = 0, \ldots, m.$$

Let $\omega_0, \omega_1, \ldots, \omega_m$ be real numbers defined recursively for $k = 1, 2, \ldots, m$ by

$$\omega_k := \begin{cases} 1, & \text{if } k = 0, \\ (1 + \omega_{k-1}^{-1}\rho_k/\rho_{k-1})^{-1}, & \text{otherwise.} \end{cases} \tag{13.68}$$

We will show below that $\boldsymbol{x}_k$ and $\boldsymbol{r}_k$ satisfy the following recurrence relations for $k = 0, 1, \ldots, m-1$

$$\boldsymbol{x}_{k+1} = (1 - \omega_k)\boldsymbol{x}_{k-1} + \omega_k(\boldsymbol{x}_k + \boldsymbol{r}_k), \tag{13.69}$$

$$\boldsymbol{r}_{k+1} = (1 - \omega_k)\boldsymbol{r}_{k-1} + \omega_k\boldsymbol{B}\boldsymbol{r}_k, \tag{13.70}$$

starting with $\boldsymbol{x}_0 = \boldsymbol{x}_{-1} = 0$ and $\boldsymbol{r}_0 = \boldsymbol{r}_{-1} = \boldsymbol{b}$.

a) Show that $0 < \omega_k < 1$ for $k = 1, 2, \ldots, m$.

Solution. Since $\omega_0 = 1$ and $\rho_k > 0$ for $k = 0, \ldots, m$ we clearly have $0 < \omega_1 < 1$. Suppose $0 < \omega_{k-1} < 1$ for some $k \geq 2$. Now $\omega_{k-1}^{-1}\rho_k/\rho_{k-1} > 0$ and therefore $0 < \omega_k < 1$. The result follows by induction.

b) Explain briefly how to define an iterative algorithm for determining $\boldsymbol{x}_k$ using the formulas (13.68), (13.69), (13.70) and estimate the number of arithmetic operations in each iteration.

Solution. The following code runs the algorithm a fixed number of iterations, for a given matrix $\boldsymbol{B}$ and vector $\boldsymbol{b}$.

code/cgantisymm.m

```
function x=cgantisymm(B,b)
   m = 10; % Number of iterations
   n = size(B,1);
   x = zeros(n,m+2);
   r = zeros(n,m+2); r(:,1) = b; r(:,2) = b;
   rho = zeros(m+2,1); rho(2) = norm(b)^2;
   omega = 1;
   for k=2:(m+1)
      x(:,k+1) = (1-omega)*x(:,k-1) + omega*(x(:,k)+r(:,k));
      r(:,k+1) = (1-omega)*r(:,k-1) + omega*B*r(:,k);
      rho(k+1) = norm(r(:,k+1))^2;
      omega=1/(1+rho(k+1)/(rho(k)*omega));
   end
   x = x(:,m+2);
end
```

Listing 13.4: A conjugate gradient-like method for antisymmetric systems.

The number of operations if $\boldsymbol{B}$ is a full matrix are $\mathcal{O}(n)$ for the first and third lines in the `for`-loop, $\mathcal{O}(n^2)$ for the second line and $\mathcal{O}(1)$ for the fourth line. The following code tests the algorithm.

code/test_cgantisymm.m

```
n=8;
C=tril(rand(n));
B = C-C';
b=rand(n,1);

x=cgantisymm(B,b);
norm( (eye(n)-B)*x-b )
```

Listing 13.5: Testing the conjugate gradient-like method for antisymmetric systems.

The last line here tests if the solution found by `cgantisymm` solves the system. Ideally it would return zero, but due to roundoff errors a small number is returned. If space is at a premium we note that it is enough to use two arrays for $\boldsymbol{x}$, two arrays for $\boldsymbol{r}$ and two real numbers for ρ. The code above, however, stores all iterations of these.

c) Show that $\langle \boldsymbol{r}_k, \boldsymbol{r}_j \rangle = 0$ for $j = 0, 1, \ldots, k-1$.

Solution. We first note that $\boldsymbol{r}_j \in \mathcal{W}_{j+1}$ since $\boldsymbol{x}_j \in \mathcal{W}_j$ and $\boldsymbol{r}_j = \boldsymbol{b} - \boldsymbol{A}\boldsymbol{x}_j = \boldsymbol{b} - \boldsymbol{x}_j + \boldsymbol{B}\boldsymbol{x}_j$. Since the exercise imposes that $\langle \boldsymbol{A}\boldsymbol{x}_k, \boldsymbol{w} \rangle = \langle \boldsymbol{b}, \boldsymbol{w} \rangle$ for all $\boldsymbol{w} \in \mathcal{W}_k$, it follows that $\langle \boldsymbol{r}_k, \boldsymbol{w} \rangle = 0$ for all $\boldsymbol{w} \in \mathcal{W}_k$. The orthogonality follows by taking $\boldsymbol{w} = \boldsymbol{r}_j \in \mathcal{W}_{j+1}$ for $j = 0, 1, \ldots, k-1$.

d) Show that if $k \le m+1$ then $\mathcal{W}_k = \operatorname{span}(\boldsymbol{r}_0, \boldsymbol{r}_1, \ldots, \boldsymbol{r}_{k-1})$ and $\dim \mathcal{W}_k = k$.

Solution. Let $\mathcal{V}_k := \operatorname{span}(\boldsymbol{r}_0, \boldsymbol{r}_1, \ldots, \boldsymbol{r}_{k-1})$. Since $\boldsymbol{r}_0, \boldsymbol{r}_1, \ldots, \boldsymbol{r}_{k-1}$ are orthogonal and nonzero they are linearly independent. Indeed, If $\sum_{i=0}^{k-1} c_i \boldsymbol{r}_i = \boldsymbol{0}$ for some $c_0, \ldots, c_{k-1}$, then

$$\left\langle \sum_{i=0}^{k-1} c_i \boldsymbol{r}_i, \boldsymbol{r}_j \right\rangle = c_j \langle \boldsymbol{r}_j, \boldsymbol{r}_j \rangle = 0, \qquad j = 0, \ldots, k-1,$$

so that $c_j = 0$. It follows that $\dim \mathcal{V}_k = k$. Also $\boldsymbol{r}_j \in \mathcal{W}_{j+1} \subset \mathcal{W}_k$, $j = 0, \ldots, k-1$ implies that $\mathcal{V}_k \subset \mathcal{W}_k$. Since $\dim \mathcal{W}_k \le k$ (since $\mathcal{W}_k$ is spanned by k vectors), it follows that actually $\dim \mathcal{W}_k = k$. It follows that $\{\boldsymbol{r}_0, \boldsymbol{r}_1, \ldots, \boldsymbol{r}_{k-1}\}$ also must be a basis for $\mathcal{W}_k$, so that $\mathcal{V}_k = \mathcal{W}_k$.

e) Show that if $1 \le k \le m-1$ then

$$\boldsymbol{B}\boldsymbol{r}_k = \alpha_k \boldsymbol{r}_{k+1} + \beta_k \boldsymbol{r}_{k-1}, \tag{13.71}$$

where $\alpha_k := \langle \boldsymbol{B}\boldsymbol{r}_k, \boldsymbol{r}_{k+1} \rangle / \rho_{k+1}$ and $\beta_k := \langle \boldsymbol{B}\boldsymbol{r}_k, \boldsymbol{r}_{k-1} \rangle / \rho_{k-1}$.

Solution. Since $\boldsymbol{B}\boldsymbol{r}_k \in \mathcal{W}_{k+2} = \operatorname{span}(\boldsymbol{r}_0, \boldsymbol{r}_1, \ldots, \boldsymbol{r}_{k+1})$ and the $\boldsymbol{r}_j$ are nonzero and orthogonal we have

$$\boldsymbol{B}\boldsymbol{r}_k = \sum_{j=0}^{k+1} \frac{\langle \boldsymbol{B}\boldsymbol{r}_k, \boldsymbol{r}_j \rangle}{\rho_j} \boldsymbol{r}_j.$$

By **c)**

$$\langle \boldsymbol{B}\boldsymbol{r}_k, \boldsymbol{r}_j \rangle = -\langle \boldsymbol{r}_k, \boldsymbol{B}\boldsymbol{r}_j \rangle = 0, \qquad j = 0, \ldots, k-2$$

and $\langle \boldsymbol{B}\boldsymbol{r}_k, \boldsymbol{r}_k \rangle = 0$ by Exercise 13.13**a)**. Thus we obtain (13.71), since only the indices $k-1$ and $k+1$ contribute.

f) Define $\alpha_0 := \langle \boldsymbol{B}\boldsymbol{r}_0, \boldsymbol{r}_1 \rangle / \rho_1$ and show that $\alpha_0 = 1$.

Solution. Since $\boldsymbol{x}_1 \in \mathcal{W}_1$ we have $\boldsymbol{x}_1 = \gamma \boldsymbol{b}$ for some $\gamma \in \mathbb{R}$. By (13.64) and definition of $\boldsymbol{x}_1$,

$$\gamma\langle \boldsymbol{b}, \boldsymbol{b}\rangle = \gamma\langle \boldsymbol{A}\boldsymbol{b}, \boldsymbol{b}\rangle = \langle \boldsymbol{A}\boldsymbol{x}_1, \boldsymbol{b}\rangle = \langle \boldsymbol{b}, \boldsymbol{b}\rangle \qquad \Longrightarrow \qquad \gamma = 1,$$

showing that $\boldsymbol{x}_1 = \boldsymbol{b}$. Since $\boldsymbol{r}_0 = \boldsymbol{b}$ we find

$$\alpha_0 = \frac{\langle \boldsymbol{B}\boldsymbol{b}, \boldsymbol{r}_1\rangle}{\rho_1} = \frac{\langle \boldsymbol{b} - \boldsymbol{A}\boldsymbol{b}, \boldsymbol{r}_1\rangle}{\rho_1} = \frac{\langle \boldsymbol{r}_1, \boldsymbol{r}_1\rangle}{\rho_1} = 1.$$

g) Show that if $1 \le k \le m-1$ then $\beta_k = -\alpha_{k-1}\rho_k/\rho_{k-1}$.

Solution. Since $\boldsymbol{B}^{\mathrm{T}} = -\boldsymbol{B}$ and $\boldsymbol{x}^{\mathrm{T}}\boldsymbol{y} = \boldsymbol{y}^{\mathrm{T}}\boldsymbol{x}$ for any $\boldsymbol{x}, \boldsymbol{y} \in \mathbb{R}^n$

$$\begin{aligned}\beta_k\rho_{k-1} &= \langle \boldsymbol{B}\boldsymbol{r}_k, \boldsymbol{r}_{k-1}\rangle = \langle \boldsymbol{r}_k, \boldsymbol{B}^{\mathrm{T}}\boldsymbol{r}_{k-1}\rangle = -\langle \boldsymbol{r}_k, \boldsymbol{B}\boldsymbol{r}_{k-1}\rangle = \\ &= -\langle \boldsymbol{B}\boldsymbol{r}_{k-1}, \boldsymbol{r}_k\rangle = -\alpha_{k-1}\rho_k.\end{aligned}$$

The result follows after dividing by ρ_{k-1}.

h) Show that

$$\langle \boldsymbol{r}_{k+1}, \boldsymbol{A}^{-1}\boldsymbol{r}_{k+1}\rangle = \langle \boldsymbol{r}_{k+1}, \boldsymbol{A}^{-1}\boldsymbol{r}_j\rangle, \qquad j = 0, 1, \ldots, k. \tag{13.72}$$

Hint.
Show that $\boldsymbol{A}^{-1}(\boldsymbol{r}_{k+1} - \boldsymbol{r}_j) \in \mathcal{W}_{k+1}$.

Solution. We have

$$\boldsymbol{A}^{-1}(\boldsymbol{r}_{k+1} - \boldsymbol{r}_j) = \boldsymbol{A}^{-1}(\boldsymbol{b} - \boldsymbol{A}\boldsymbol{x}_{k+1} - \boldsymbol{b} + \boldsymbol{A}\boldsymbol{x}_j) = \boldsymbol{x}_j - \boldsymbol{x}_{k+1} \in \mathcal{W}_{k+1}$$

But since $\langle \boldsymbol{r}_{k+1}, \boldsymbol{w}\rangle = 0$ for all $\boldsymbol{w} \in \mathcal{W}_{k+1}$

$$\begin{aligned}\langle \boldsymbol{r}_{k+1}, \boldsymbol{A}^{-1}\boldsymbol{r}_{k+1}\rangle &= \langle \boldsymbol{r}_{k+1}, \boldsymbol{A}^{-1}(\boldsymbol{r}_{k+1} - \boldsymbol{r}_j)\rangle + \langle \boldsymbol{r}_{k+1}, \boldsymbol{A}^{-1}\boldsymbol{r}_j\rangle \\ &= \langle \boldsymbol{r}_{k+1}, \boldsymbol{A}^{-1}\boldsymbol{r}_j\rangle,\end{aligned}$$

where the first inner product is zero since $\boldsymbol{r}_{k+1} \in \mathcal{W}_{k+2}$, $\boldsymbol{A}^{-1}(\boldsymbol{r}_{k+1} - \boldsymbol{r}_j) \in \mathcal{W}_{k+1}$.

i) Use (13.71) and (13.72) to show that $\alpha_k + \beta_k = 1$ for $k = 1, 2, \ldots, m-1$.

Solution. By (13.71)

$$\boldsymbol{A}\boldsymbol{r}_k = \boldsymbol{r}_k - \boldsymbol{B}\boldsymbol{r}_k = \boldsymbol{r}_k - \alpha_k\boldsymbol{r}_{k+1} - \beta_k\boldsymbol{r}_{k-1},$$

so that

$$\boldsymbol{r}_k = \boldsymbol{A}^{-1}\boldsymbol{r}_k - \alpha_k\boldsymbol{A}^{-1}\boldsymbol{r}_{k+1} - \beta_k\boldsymbol{A}^{-1}\boldsymbol{r}_{k-1}$$

By (13.72) we have $\langle \boldsymbol{r}_{k+1}, \boldsymbol{A}^{-1}\boldsymbol{r}_{k-1}\rangle = \langle \boldsymbol{r}_{k+1}, \boldsymbol{A}^{-1}\boldsymbol{r}_{k+1}\rangle$. By orthogonality of the residuals

$$\begin{aligned}0 = \langle \boldsymbol{r}_{k+1}, \boldsymbol{r}_k\rangle &= \langle \boldsymbol{r}_{k+1}, \boldsymbol{A}^{-1}\boldsymbol{r}_k\rangle - (\alpha_k + \beta_k)\langle \boldsymbol{r}_{k+1}, \boldsymbol{A}^{-1}\boldsymbol{r}_{k+1}\rangle \\ &= \langle \boldsymbol{r}_{k+1}, \boldsymbol{x}^* - \boldsymbol{x}_k\rangle - (\alpha_k + \beta_k)\langle \boldsymbol{r}_{k+1}, \boldsymbol{x}^* - \boldsymbol{x}_{k+1}\rangle \\ &= \langle \boldsymbol{r}_{k+1}, \boldsymbol{x}^*\rangle - (\alpha_k + \beta_k)\langle \boldsymbol{r}_{k+1}, \boldsymbol{x}^*\rangle.\end{aligned}$$

Since $\langle \boldsymbol{r}_{k+1}, \boldsymbol{x}^*\rangle \neq 0$ we obtain $\alpha_k + \beta_k = 1$.

j) Show that $\alpha_k \geq 1$ for $k = 1, 2, \ldots, m-1$.

Solution. We have $\alpha_k = 1 - \beta_k = 1 + \alpha_{k-1}\rho_k/\rho_{k-1}$. Since $\alpha_0 = 1$ and $\rho_j > 0$ for $j = 1, \ldots, m-1$, the result follows by induction.

k) Show that $\boldsymbol{x}_k$, $\boldsymbol{r}_k$ and $\boldsymbol{\omega}_k$ satisfy the recurrence relations (13.68), (13.69) and (13.70).

Solution. *Proof of* (13.70)*:* By (13.71) we have

$$\boldsymbol{r}_{k+1} = -\frac{\beta_k}{\alpha_k}\boldsymbol{r}_{k-1} + \frac{1}{\alpha_k}\boldsymbol{B}\boldsymbol{r}_k = \left(1 - \frac{1}{\alpha_k}\right)\boldsymbol{r}_{k-1} + \frac{1}{\alpha_k}\boldsymbol{B}\boldsymbol{r}_k.$$

With $\omega_k := 1/\alpha_k$ we obtain (13.70) for $k \geq 1$. For $k = 0$ we obtain from (13.71) that $\boldsymbol{r}_1 = \boldsymbol{B}\boldsymbol{r}_0 = \boldsymbol{B}\boldsymbol{b} = \boldsymbol{b} - \boldsymbol{A}\boldsymbol{b}$, and this is consistent since $\boldsymbol{x}_1 = \boldsymbol{b}$.

Proof of (13.68)*:* We have $\omega_0 = 1/\alpha_0 = 1$. Moreover, the proof of **j)** implies (13.68) directly.

Proof of (13.69)*:* Using (13.70) we find

$$\begin{aligned}\boldsymbol{b} - \boldsymbol{A}\boldsymbol{x}_{k+1} &= (1 - \omega_k)(\boldsymbol{b} - \boldsymbol{A}\boldsymbol{x}_{k-1}) + \omega_k(\boldsymbol{b} - \boldsymbol{A}\boldsymbol{x}_k) - \omega_k\boldsymbol{A}\boldsymbol{r}_k \\ &= \boldsymbol{b} - (1 - \omega_k)\boldsymbol{A}\boldsymbol{x}_{k-1} - \omega_k\boldsymbol{A}\boldsymbol{x}_k - \omega_k\boldsymbol{A}\boldsymbol{r}_k,\end{aligned}$$

from which (13.69) follows after canceling $\boldsymbol{b}$ and multiplying with $\boldsymbol{A}^{-1}$ on both sides. This holds for $k = 0$ since (13.70) holds for $k = 0$. Moreover, since $\boldsymbol{r}_0 = \boldsymbol{r}_{-1} = \boldsymbol{b}$ we have $\boldsymbol{x}_0 = \boldsymbol{x}_{-1} = \boldsymbol{0}$.

Exercises section 13.4

Exercise 13.15: Another explicit formula for Chebyshev polynomials ?

Show that

$$T_n(t) = \cosh(n \operatorname{arccosh} t) \text{ for } t \geq 1,$$

where arccosh is the inverse function of $\cosh x := (e^x + e^{-x})/2$.

Solution. For any integer $n \geq 0$, write

$$P_n(t) := \cosh\big(n\phi(t)\big), \qquad \phi(t) := \operatorname{arccosh}(t), \qquad t \in [1, \infty).$$

It is well known, and easily verified, that

$$\cosh(x+y) = \cosh(x)\cosh(y) + \sinh(x)\sinh(y).$$

Using this and that cosh is even and sinh is odd, one finds

$$\begin{aligned} P_{n+1}(t) + P_{n-1}(t) &= \cosh\big((n+1)\phi\big) + \cosh\big((n-1)\phi\big) \\ &= \cosh(n\phi)\cosh(\phi) + \sinh(n\phi)\sinh(\phi) + \\ &\quad \cosh(n\phi)\cosh(\phi) - \sinh(n\phi)\sinh(\phi) \\ &= 2\cosh(\phi)\cosh(n\phi) \\ &= 2tP_n(t). \end{aligned}$$

It follows that P_n and T_n satisfy the same recurrence relation. Since they also share initial terms $P_0(t) = 1 = T_0(t)$ and $P_1(t) = t = T_1(t)$, necessarily $P_n = T_n$ for any $n \geq 0$.

Exercise 13.16: Maximum of a convex function

Show that if $f : [a,b] \to \mathbb{R}$ is convex then $\max_{a \leq x \leq b} f(x) \leq \max\{f(a), f(b)\}$.

Solution. This is a special case of the *maximum principle* in convex analysis, which states that a convex function, defined on a compact convex set Ω, attains its maximum on the boundary of Ω.

Let $f : [a,b] \to \mathbb{R}$ be a convex function. Consider an arbitrary point $x = (1-\lambda)a + \lambda b \in [a,b]$, with $0 \leq \lambda \leq 1$. Since f is convex,

$$\begin{aligned} f(x) = f\big((1-\lambda)a + \lambda b\big) &\leq (1-\lambda)f(a) + \lambda f(b) \\ &\leq (1-\lambda)\max\{f(a), f(b)\} + \lambda \max\{f(a), f(b)\} = \max\{f(a), f(b)\}, \end{aligned}$$

see Figure 13.2. It follows that $f(x) \leq \max\{f(a), f(b)\}$ and that f attains its maximum on the boundary of its domain of definition.

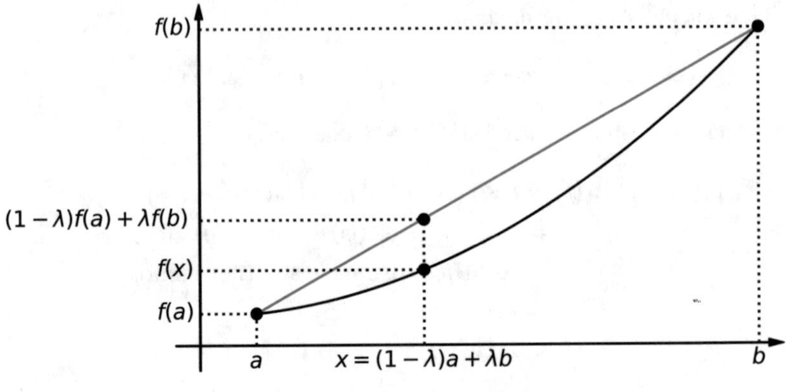

Figure 13.2: A convex function f is bounded from above by convex combinations.

Exercises section 13.5

Exercise 13.17: Variable coefficient

For $m = 2$, show that (13.57) takes the form

$$\boldsymbol{A}\boldsymbol{x} = \begin{bmatrix} a_{1,1} & -c_{3/2,1} & -c_{1,3/2} & 0 \\ -c_{3/2,1} & a_{2,2} & 0 & -c_{2,3/2} \\ -c_{1,3/2} & 0 & a_{3,3} & -c_{3/2,2} \\ 0 & -c_{2,3/2} & -c_{3/2,2} & a_{4,4} \end{bmatrix} \begin{bmatrix} v_{1,1} \\ v_{2,1} \\ v_{1,2} \\ v_{2,2} \end{bmatrix} = \begin{bmatrix} (dv)_{1,1} \\ (dv)_{2,1} \\ (dv)_{1,2} \\ (dv)_{2,2} \end{bmatrix},$$

where

$$\begin{bmatrix} a_{1,1} \\ a_{2,2} \\ a_{3,3} \\ a_{4,4} \end{bmatrix} = \begin{bmatrix} c_{1/2,1} + c_{1,1/2} + c_{1,3/2} + c_{3/2,1} \\ c_{3/2,1} + c_{2,1/2} + c_{2,3/2} + c_{5/2,1} \\ c_{1/2,2} + c_{1,3/2} + c_{1,5/2} + c_{3/2,2} \\ c_{3/2,2} + c_{2,3/2} + c_{2,5/2} + c_{5/2,2} \end{bmatrix}.$$

Show that the matrix $\boldsymbol{A}$ is symmetric, and if $c(x, y) > 0$ for all $(x, y) \in \Omega$ then it is strictly diagonally dominant.

Solution. In the equation for entry (j, k), $v_{j,k}$ contributes with $c_{j-1/2,k} + c_{j,k-1/2} - c_{j+1/2,k} - c_{j,k+1/2}$. This means that the contributions in entries $(1, 1)$, $(2, 1)$, $(1, 2)$, and $(2, 2)$ (i.e., in the order obtained by stacking the columns) are

$$\begin{bmatrix} c_{1/2,1} + c_{1,1/2} + c_{3/2,1} + c_{1,3/2} \\ c_{3/2,1} + c_{2,1/2} + c_{5/2,1} + c_{2,3/2} \\ c_{1/2,2} + c_{1,3/2} + c_{3/2,2} + c_{1,5/2} \\ c_{3/2,2} + c_{2,3/2} + c_{5/2,2} + c_{2,5/2} \end{bmatrix}.$$

These are the expressions for $a_{i,i}$, and they are placed on the diagonal. Also, in the equation for entry (j,k),

- $v_{j-1,k}$ contributes with $-c_{j-1/2,k}$. This means that $-c_{3/2,1}$ and $-c_{3/2,2}$ contribute in the first subdiagonal.
- $v_{j+1,k}$ contributes with $-c_{j+1/2,k}$. This means that $-c_{3/2,1}$ and $-c_{3/2,2}$ contribute in the first superdiagonal.
- $v_{j,k-1}$ contributes with $-c_{j,k-1/2}$. This means that $-c_{1,3/2}$ and $-c_{2,3/2}$ contribute in the second subdiagonal.
- $v_{j,k+1}$ contributes with $-c_{j,k+1/2}$. This means that $-c_{1,3/2}$ and $-c_{2,3/2}$ contribute in the second superdiagonal.

This accounts for all entries in the matrix.

The matrix is clearly symmetric. To check diagonal dominance one goes through all four rows. For the first row, $|a_{1,1}| = c_{1/2,1} + c_{1,1/2} + c_{3/2,1} + c_{1,3/2}$, and the off-diagonal contribution is $c_{3/2,1} + c_{1,3/2}$. The difference is $c_{1/2,1} + c_{1,1/2} > 0$. For all other rows one sees in the same way that the two nonzero off-diagonal entries appear in $a_{i,i}$, so that terms cancel in the same way. It follows that the matrix is strictly diagonally dominant.

Chapter 14
Numerical Eigenvalue Problems

Exercises section 14.1

Exercise 14.1: Yes or No (Exam exercise 2006-1)

Answer simply yes or no to the following questions:

a) Every matrix $\boldsymbol{A} \in \mathbb{C}^{m\times n}$ has a singular value decomposition?

Solution. Yes, we have seen that this is the case.

b) The algebraic multiplicity of an eigenvalue is always less than or equal to the geometric multiplicity?

Solution. No, we have seen that it is the other way around.

c) The QR factorization of a matrix $\boldsymbol{A} \in \mathbb{R}^{n\times n}$ can be determined by Householder transformations in $\mathcal{O}(n^2)$ arithmetic operations?

Solution. No, we have seen that $4n^3/3$ operations are required in order to bring a matrix to upper triangular form using Householder transformations.

d) Let $\rho(\boldsymbol{A})$ be the spectral radius of $\boldsymbol{A} \in \mathbb{C}^{n\times n}$. Then $\lim_{k\to\infty} \boldsymbol{A}^k = \boldsymbol{0}$ if and only if $\rho(\boldsymbol{A}) < 1$?

Solution. Yes, we have seen in Theorem 12.10 that this is the case.

T. Lyche et al., *Exercises in Numerical Linear Algebra and Matrix Factorizations*, Texts in Computational Science and Engineering 23, https://doi.org/10.1007/978-3-030-59789-4_14

Exercises section 14.2

Exercise 14.2: Nonsingularity using Gershgorin

Consider the matrix

$$\boldsymbol{A} = \begin{bmatrix} 4 & 1 & 0 & 0 \\ 1 & 4 & 1 & 0 \\ 0 & 1 & 4 & 1 \\ 0 & 0 & 1 & 4 \end{bmatrix}.$$

Show using the Gershgorin circle theorem (Theorem 14.1) that $\boldsymbol{A}$ is nonsingular.

Solution. We compute the Gershgorin disks

$$R_1 = R_4 = C_1 = C_4 = \{z \in \mathbb{C} : |z - 4| \le 1\},$$

$$R_2 = R_3 = C_2 = C_3 = \{z \in \mathbb{C} : |z - 4| \le 2\},$$

see Figure 14.1. Then, by the Gershgorin circle theorem, each eigenvalue of $\boldsymbol{A}$ lies in

$$(R_1 \cup \cdots \cup R_4) \cap (C_1 \cup \cdots \cup C_4) = \{z \in \mathbb{C} : |z - 4| \le 2\}.$$

In particular $\boldsymbol{A}$ has only nonzero eigenvalues, implying that $\boldsymbol{A}$ must be nonsingular.

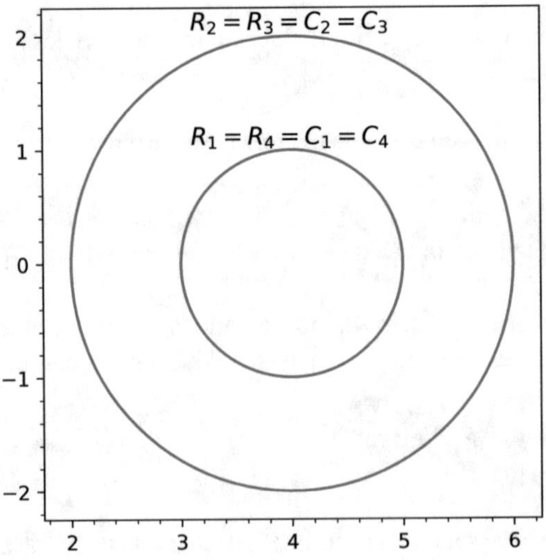

Figure 14.1: Gershgorin disks in Exercise 14.2.

Exercise 14.3: Gershgorin, strictly diagonally dominant matrix

Show using the Gershgorin circle theorem (Theorem 14.1) that a strictly diagonally dominant matrix $\boldsymbol{A}$ ($|a_{i,i}| > \sum_{j\neq i} |a_{i,j}|$ for all i) is nonsingular.

Solution. Suppose $\boldsymbol{A}$ is a strictly diagonally dominant matrix. For such a matrix, one finds Gershgorin disks

$$R_i = \left\{ z \in \mathbb{C} : |z - a_{ii}| \leq \sum_{j\neq i} |a_{ij}| \right\}.$$

Since $|a_{ii}| > \sum_{j\neq i} |a_{ij}|$ for all i, the origin is not an element of any of the disks R_i, and therefore neither of the union $\bigcup R_i$, nor of the intersection $(\bigcup R_i) \cap (\bigcup C_i)$. Then, by the Gershgorin circle theorem, $\boldsymbol{A}$ only has nonzero eigenvalues, implying that $\det(\boldsymbol{A}) = \det(\boldsymbol{A} - 0 \cdot \boldsymbol{I}) \neq 0$ and $\boldsymbol{A}$ is nonsingular.

Exercise 14.4: Gershgorin disks (Exam exercise 2009-2)

The eigenvalues of $\boldsymbol{A} \in \mathbb{R}^{n,n}$ lie inside $R \cap C$, where $R := R_1 \cup \cdots \cup R_n$ is the union of the row disks R_i of $\boldsymbol{A}$, and $C := C_1 \cup \cdots \cup C_n$ is the union of the column disks C_j. You do not need to prove this. Write a MATLAB function `[s,r,c]=gershgorin(A)` that computes the centers $\boldsymbol{s} = [s_1, \ldots, s_n] \in \mathbb{R}^n$ of the row and column disks, and their radii $\boldsymbol{r} = [r_1, \ldots, r_n] \in \mathbb{R}^n$ and $\boldsymbol{c} = [c_1, \ldots, c_n] \in \mathbb{R}^n$, respectively.

Solution.

code/gershgorin.m

```
function [s,r,c] = gershgorin(A)
n=length(A);
s=diag(A); r=zeros(n,1); c=r;
for i=1:n
   for j=1:n
      r(i)=r(i)+abs(A(i,j));
      c(i)=c(i)+abs(A(j,i));
   end;
   r(i)=r(i)-abs(s(i)); c(i)=c(i)-abs(s(i));
end
```

Listing 14.1: Given a matrix $\boldsymbol{A}$, compute the centers $\boldsymbol{s}$ and radii $\boldsymbol{r}$ and $\boldsymbol{c}$ of the row and column Gershgorin disks — for-loops

code/gershgorinv.m

```
function [s,r,c] = gershgorinv(A)
n=length(A);
s=diag(A); e=ones(n,1);
r=abs(A)*e-abs(s);
c=(abs(A))'*e-abs(s);
```

Listing 14.2: Given a matrix $\boldsymbol{A}$, compute the centers $\boldsymbol{s}$ and radii $\boldsymbol{r}$ and $\boldsymbol{c}$ of the row and column Gershgorin disks — vectorized implementation

Exercises section 14.3

Exercise 14.5: Continuity of eigenvalues

Suppose

$$\boldsymbol{A}(t) := \boldsymbol{D} + t(\boldsymbol{A} - \boldsymbol{D}), \qquad \boldsymbol{D} := \operatorname{diag}(a_{11}, \ldots, a_{nn}), \qquad t \in \mathbb{R},$$

$0 \leq t_1 < t_2 \leq 1$ and that μ is an eigenvalue of $\boldsymbol{A}(t_2)$. Show, using Theorem 14.2 with $\boldsymbol{A} = \boldsymbol{A}(t_1)$ and $\boldsymbol{E} = \boldsymbol{A}(t_2) - \boldsymbol{A}(t_1)$, that $\boldsymbol{A}(t_1)$ has an eigenvalue λ such that

$$|\lambda - \mu| \leq C(t_2 - t_1)^{1/n}, \qquad C \leq 2\big(\|\boldsymbol{D}\|_2 + \|\boldsymbol{A} - \boldsymbol{D}\|_2\big). \qquad (\star)$$

Thus, as a function of t, every eigenvalue of $\boldsymbol{A}(t)$ is a continuous function of t.

Solution. Applying Theorem 14.2[1] to the matrix $\boldsymbol{A}(t_1)$ with perturbation $\boldsymbol{E} := \boldsymbol{A}(t_2) - \boldsymbol{A}(t_1)$, one finds that $\boldsymbol{A}(t_1)$ has an eigenvalue λ such that

$$|\lambda - \mu| \leq \big(\|\boldsymbol{A}(t_1)\|_2 + \|\boldsymbol{A}(t_2)\|_2\big)^{1-1/n} \|\boldsymbol{A}(t_2) - \boldsymbol{A}(t_1)\|_2^{1/n}.$$

Applying the triangle inequality to the definition of $\boldsymbol{A}(t_1)$ and $\boldsymbol{A}(t_2)$, and using that the function $x \longmapsto x^{1-1/n}$ is monotone increasing,

$$|\lambda - \mu| \leq \Big(2\|\boldsymbol{D}\|_2 + (t_1 + t_2)\|\boldsymbol{A} - \boldsymbol{D}\|_2\Big)^{1-1/n} \|(\boldsymbol{A} - \boldsymbol{D})\|_2^{1/n} (t_2 - t_1)^{1/n}.$$

Finally, using that $t_1 + t_2 \leq 2$, that the function $x \longmapsto x^{1/n}$ is monotone increasing, and that $\|(\boldsymbol{A} - \boldsymbol{D})\|_2 \leq 2\|\boldsymbol{D}\|_2 + 2\|(\boldsymbol{A} - \boldsymbol{D})\|_2$, one obtains $(\star)$.

Exercise 14.6: ∞-norm of a diagonal matrix

Give a direct proof that $\|\boldsymbol{A}\|_\infty = \rho(\boldsymbol{A})$ if $\boldsymbol{A}$ is diagonal.

[1] L. Elsner, "An optimal bound for the spectral variation of two matrices", Linear Algebra and its Applications **71** (1985), 77–80.

Solution. Let $\boldsymbol{A} = \operatorname{diag}(\lambda_1, \ldots, \lambda_n)$ be a diagonal matrix. The spectral radius $\rho(\boldsymbol{A})$ is the absolute value of the biggest eigenvalue, say λ_i, of $\boldsymbol{A}$. One has

$$\|\boldsymbol{A}\|_\infty = \max_{\|\boldsymbol{x}\|_\infty=1} \|\boldsymbol{A}\boldsymbol{x}\|_\infty = \max_{\|\boldsymbol{x}\|_\infty=1} \max\{|\lambda_1 x_1|, \ldots, |\lambda_n x_n|\} \leq \rho(\boldsymbol{A}),$$

as $|\lambda_1|, \ldots, |\lambda_n| \leq |\lambda_i| = \rho(\boldsymbol{A})$ and since the components of any vector $\boldsymbol{x}$ satisfy $x_1, \ldots, x_n \leq \|\boldsymbol{x}\|_\infty$. Moreover, this bound is attained for the standard basis vector $\boldsymbol{x} = \boldsymbol{e}_i$, since $\|\boldsymbol{A}\boldsymbol{e}_i\|_\infty = |\lambda_i| = \rho(\boldsymbol{A})$.

Exercise 14.7: Eigenvalue perturbations (Exam exercise 2010-2)

Let $\boldsymbol{A} = [a_{kj}]$, $\boldsymbol{E} = [e_{kj}]$, and $\boldsymbol{B} = [b_{kj}]$ be matrices in $\mathbb{R}^{n,n}$ with

$$a_{kj} = \begin{cases} 1, & j = k+1, \\ 0, & \text{otherwise}, \end{cases} \qquad e_{kj} = \begin{cases} \epsilon, & k = n, j = 1, \\ 0, & \text{otherwise}, \end{cases} \tag{14.10}$$

and $\boldsymbol{B} = \boldsymbol{A} + \boldsymbol{E}$, where $0 < \epsilon < 1$. Thus for $n = 4$,

$$\boldsymbol{A} := \begin{bmatrix} 0\,1\,0\,0 \\ 0\,0\,1\,0 \\ 0\,0\,0\,1 \\ 0\,0\,0\,0 \end{bmatrix}, \quad \boldsymbol{E} := \begin{bmatrix} 0\,0\,0\,0 \\ 0\,0\,0\,0 \\ 0\,0\,0\,0 \\ \epsilon\,0\,0\,0 \end{bmatrix}, \quad \boldsymbol{B} := \begin{bmatrix} 0\,1\,0\,0 \\ 0\,0\,1\,0 \\ 0\,0\,0\,1 \\ \epsilon\,0\,0\,0 \end{bmatrix}.$$

a) Find the eigenvalues of $\boldsymbol{A}$ and $\boldsymbol{B}$.

Solution. Since $\boldsymbol{A}$ is triangular its eigenvalues λ_j are its diagonal elements. Thus the eigenvalues of $\boldsymbol{A}$ are $\lambda_j = 0, j = 1, 2, \ldots, n$.

For $\boldsymbol{B}$ the equation $\boldsymbol{B}\boldsymbol{x} = \lambda\boldsymbol{x}$ gives $\lambda x_k = x_{k+1}$, $k = 1, \ldots, n-1$ and $\lambda x_n = \epsilon x_1$. Thus

$$\lambda^n x_n = \epsilon\lambda^{n-1} x_1 = \epsilon\lambda^{n-2} x_2 = \epsilon\lambda^{n-3} x_3 = \cdots = \epsilon x_n.$$

We must have $x_n \neq 0$ since otherwise $\boldsymbol{x} = \boldsymbol{0}$. Canceling x_n we find $\lambda^n = \epsilon$ and the eigenvalues of $\boldsymbol{B}$ are

$$\mu_j = \mu e^{2\pi i j/n}, \qquad j = 1, \ldots, n,$$

where $\mu = \epsilon^{1/n}$ and $i = \sqrt{-1}$. Alternatively, expanding the determinant of $\boldsymbol{B}$ by its first column yields its characteristic polynomial, $\pi_{\boldsymbol{B}}(\lambda) = (-1)^n(\lambda^n - \epsilon)$, leading to the same eigenvalues.

b) Show that $\|\boldsymbol{A}\|_2 = \|\boldsymbol{B}\|_2 = 1$ for arbitrary $n \in \mathbb{N}$.

Solution. The spectral norm of $\boldsymbol{A}$ is the largest singular value, which is the square root of the largest eigenvalue of $\boldsymbol{A}^{\mathrm{T}}\boldsymbol{A}$. Now $\boldsymbol{A}^{\mathrm{T}}\boldsymbol{A} = \operatorname{diag}(0, 1, \ldots, 1)$ is diagonal with eigenvalues 0 and 1. Thus $\|\boldsymbol{A}\|_2 = 1$. For $\boldsymbol{B}$ we find $\boldsymbol{B}^{\mathrm{T}}\boldsymbol{B} = \mathrm{diag}(\epsilon^2, 1, \ldots, 1)$ and $\|\boldsymbol{B}\|_2 = 1$.

c) Recall Elsner's Theorem (Theorem 14.2). Let $\boldsymbol{A}, \boldsymbol{E}, \boldsymbol{B}$ be given by (14.10). What upper bound does (14.1) in Elsner's theorem give for the eigenvalue $\mu = \epsilon^{1/n}$ of $\boldsymbol{B}$? How sharp is this upper bound?

Solution. We have $\|\boldsymbol{A}\|_2 = \|\boldsymbol{B}\|_2 = 1$. Now $\boldsymbol{E}^{\mathrm{T}}\boldsymbol{E} = \mathrm{diag}(\epsilon^2, 0, \ldots, 0)$ so $\|\boldsymbol{E}\|_2 = \epsilon$. Elsner's theorem says that there is an eigenvalue λ of $\boldsymbol{A}$ such that

$$|\mu - \lambda| \leq \left(\|\boldsymbol{A}\|_2 + \|\boldsymbol{B}\|_2\right)^{1-1/n} \|\boldsymbol{E}\|_2^{1/n} = 2^{1-1/n}\epsilon^{1/n} \leq 2\epsilon^{1/n}.$$

Since $\lambda = 0$ the exact difference is

$$|\mu - \lambda| = \mu = \epsilon^{1/n}.$$

The upper bound differs from the exact value by a constant less than 2, so it is quite sharp. It captures the $\epsilon^{1/n}$ behavior.

Exercises section 14.4

Exercise 14.8 : Number of arithmetic operations, Hessenberg reduction

?

Show that the number of arithmetic operations for Algorithm 14.1 in the book is of the order $\frac{10}{3}n^3 = 5G_n$.

Solution. An arithmetic operation is a floating point operation, so we need not bother with any integer operations, like the computation of $k+1$ in the indices. As we are only interested in the overall complexity, we count only terms that can contribute to this.

For the first line involving C, the multiplication `v'*C` involves $(n-k)^2$ floating point multiplications and about $(n-k)^2$ floating point sums. Next, computing the outer product `v*(v'*C)` involves $(n-k)^2$ floating point multiplications, and subtracting `C - v*(v'*C)` needs $(n-k)^2$ subtractions. This line therefore involves (almost) $4(n-k)^2$ arithmetic operations. Similarly we find $4n(n-k)$ arithmetic operations for the line after that.

These $4(n-k)^2 + 4n(n-k)$ arithmetic operations need to be carried out for $k = 1, \ldots, n-2$, meaning that the algorithm requires of the order

$$N := \sum_{k=1}^{n-2} \left(4(n-k)^2 + 4n(n-k)\right)$$

arithmetic operations. This sum can be computed by either using the formulas for $\sum_{k=1}^{n-2} k$ and $\sum_{k=1}^{n-2} k^2$, or using that the highest order term can be found by evaluating an associated integral. One finds that the algorithm requires of the order

$$N \sim \int_0^n \left(4(n-k)^2 + 4n(n-k)\right) dk = \frac{10}{3}n^3$$

arithmetic operations.

Exercise 14.9: Assemble Householder transformations

Show that the number of arithmetic operations required by Algorithm 14.2 in the book is of the order $\frac{4}{3}n^3 = 2G_n$.

Solution. The multiplication `v'*C` involves $(n-k)^2$ floating point multiplications and about $(n-k)^2$ floating point sums. Next, computing the outer product `v*(v'*C)` involves $(n-k)^2$ floating point multiplications, and subtracting `C - v*(v'*C)` needs $(n-k)^2$ subtractions. In total we find (almost) $4(n-k)^2$ arithmetic operations, which have to be carried out for $k = 1, \ldots, n-2$, meaning that the algorithm requires of the order

$$N := \sum_{k=1}^{n-2} 4(n-k)^2$$

arithmetic operations. This sum can be computed by either using the formulae for $\sum_{k=1}^{n-2} k$ and $\sum_{k=1}^{n-2} k^2$, or using that the highest order term can be found by evaluating an associated integral. One finds that the algorithm requires of the order

$$N \sim \int_0^n 4(n-k)^2 dk = \frac{4}{3}n^3$$

arithmetic operations.

Exercise 14.10: Tridiagonalize a symmetric matrix

If $\boldsymbol{A}$ is real and symmetric we can modify Algorithm 14.1 as follows. To find $\boldsymbol{A}_{k+1}$ from $\boldsymbol{A}_k$ we have to compute $\boldsymbol{V}_k\boldsymbol{E}_k\boldsymbol{V}_k$ where $\boldsymbol{E}_k$ is symmetric. Dropping subscripts we have to compute a product of the form $\boldsymbol{G} = (\boldsymbol{I} - \boldsymbol{v}\boldsymbol{v}^{\mathrm{T}})\boldsymbol{E}(\boldsymbol{I} - \boldsymbol{v}\boldsymbol{v}^{\mathrm{T}})$. Let $\boldsymbol{w} := \boldsymbol{E}\boldsymbol{v}$, $\beta := \frac{1}{2}\boldsymbol{v}^{\mathrm{T}}\boldsymbol{w}$ and $\boldsymbol{z} := \boldsymbol{w} - \beta\boldsymbol{v}$. Show that $\boldsymbol{G} = \boldsymbol{E} - \boldsymbol{v}\boldsymbol{z}^{\mathrm{T}} - \boldsymbol{z}\boldsymbol{v}^{\mathrm{T}}$. Since $\boldsymbol{G}$ is symmetric, only the sub- or superdiagonal elements of $\boldsymbol{G}$ need to be computed. Computing $\boldsymbol{G}$ in this way, it can be shown that we need $\mathcal{O}(4n^3/3)$ operations to tridiagonalize a symmetric matrix by orthonormal similarity transformations. This is less than half the work to reduce a nonsymmetric matrix to upper Hessenberg form[a].

[a] We refer to G. W. Stewart, "Matrix Algorithms Volume II: Eigensystems", SIAM, Philadelphia, 2001, for a detailed algorithm.

Solution. We get $\boldsymbol{z} = \boldsymbol{w} - \beta\boldsymbol{v} = \boldsymbol{E}\boldsymbol{v} - \frac{1}{2}\boldsymbol{v}\boldsymbol{v}^{\mathrm{T}}\boldsymbol{E}\boldsymbol{v}$ and $\boldsymbol{z}^{\mathrm{T}} = \boldsymbol{v}^{\mathrm{T}}\boldsymbol{E} - \frac{1}{2}\boldsymbol{v}^{\mathrm{T}}\boldsymbol{E}\boldsymbol{v}\boldsymbol{v}^{\mathrm{T}}$, which yields

$$\begin{aligned} G &= (\boldsymbol{I} - \boldsymbol{v}\boldsymbol{v}^{\mathrm{T}})\boldsymbol{E}(\boldsymbol{I} - \boldsymbol{v}\boldsymbol{v}^{\mathrm{T}}) = \boldsymbol{E} - \boldsymbol{v}\boldsymbol{v}^{\mathrm{T}}\boldsymbol{E} - \boldsymbol{E}\boldsymbol{v}\boldsymbol{v}^{\mathrm{T}} + \boldsymbol{v}\boldsymbol{v}^{\mathrm{T}}\boldsymbol{E}\boldsymbol{v}\boldsymbol{v}^{\mathrm{T}} \\ &= \boldsymbol{E} - \boldsymbol{v}(\boldsymbol{v}^{\mathrm{T}}\boldsymbol{E} - \frac{1}{2}\boldsymbol{v}^{\mathrm{T}}\boldsymbol{E}\boldsymbol{v}\boldsymbol{v}^{\mathrm{T}}) - (\boldsymbol{E}\boldsymbol{v} - \frac{1}{2}\boldsymbol{v}\boldsymbol{v}^{\mathrm{T}}\boldsymbol{E}\boldsymbol{v})\boldsymbol{v}^{\mathrm{T}} \\ &= \boldsymbol{E} - \boldsymbol{v}\boldsymbol{z}^{\mathrm{T}} - \boldsymbol{z}\boldsymbol{v}^{\mathrm{T}}. \end{aligned}$$

Exercises section 14.5

Exercise 14.11: Counting eigenvalues

Consider the matrix $\boldsymbol{A}$ in Exercise 14.2. Determine the number of eigenvalues greater than 4.5.

Solution. Let

$$\boldsymbol{A} = \begin{bmatrix} 4 & 1 & 0 & 0 \\ 1 & 4 & 1 & 0 \\ 0 & 1 & 4 & 1 \\ 0 & 0 & 1 & 4 \end{bmatrix}, \qquad \alpha = 4.5.$$

Applying the recursive procedure described in Corollary 14.2, we find the diagonal elements $d_1(\alpha), d_2(\alpha), d_3(\alpha), d_4(\alpha)$ of the matrix $\boldsymbol{D}$ in the factorization $\boldsymbol{A} - \alpha\boldsymbol{I} = \boldsymbol{L}\boldsymbol{D}\boldsymbol{L}^{\mathrm{T}}$,

$$\begin{aligned} d_1(\alpha) &= 4 - 9/2 = -1/2, \\ d_2(\alpha) &= 4 - 9/2 - 1^2/(-1/2) = +3/2, \\ d_3(\alpha) &= 4 - 9/2 - 1^2/(+3/2) = -7/6, \\ d_4(\alpha) &= 4 - 9/2 - 1^2/(-7/6) = +5/14. \end{aligned}$$

As precisely two of these are negative, Corollary 14.2 implies that there are precisely two eigenvalues of $\boldsymbol{A}$ strictly smaller than $\alpha = 4.5$. As

$$\det(\boldsymbol{A} - 4.5\boldsymbol{I}) = \det(\boldsymbol{L}\boldsymbol{D}\boldsymbol{L}^{\mathrm{T}}) = d_1(\alpha)d_2(\alpha)d_3(\alpha)d_4(\alpha) \neq 0,$$

the matrix $\boldsymbol{A}$ does not have an eigenvalue equal to 4.5. We conclude that the remaining two eigenvalues must be bigger than 4.5.

Exercise 14.12: Overflow in LDL* factorization

Let for $n \in \mathbb{N}$

$$A_n = \begin{bmatrix} 10 & 1 & 0 & \cdots & 0 \\ 1 & 10 & 1 & \ddots & \vdots \\ 0 & \ddots & \ddots & \ddots & 0 \\ \vdots & \ddots & 1 & 10 & 1 \\ 0 & \cdots & 0 & 1 & 10 \end{bmatrix} \in \mathbb{R}^{n\times n}.$$

a) Let d_k be the diagonal elements of D in an LDL* factorization of A_n. Show that $5+\sqrt{24} < d_k \le 10$, $k = 1, 2, \ldots, n$.

Solution. Since A_n is tridiagonal and strictly diagonally dominant, it has a unique LU factorization by Theorem 2.3. From Equations (2.16), one can determine the corresponding LDL* factorization. For $n = 1, 2, \ldots$, let $d_{n,k}$, with $k = 1, \ldots, n$, be the diagonal elements of the diagonal matrix D_n in a symmetric factorization of A_n.

We proceed by induction. Let $n \ge 1$ be any positive integer. For the first diagonal element, corresponding to $k = 1$, Equations (2.16) immediately yield $5 + \sqrt{24} < d_{n,1} = 10 \le 10$. Next, assume that $5 + \sqrt{24} < d_{n,k} \le 10$ for some $1 \le k < n$. We show that this implies that $5 + \sqrt{24} < d_{n,k+1} \le 10$. First observe that $\left(5+\sqrt{24}\right)^2 = 25 + 10\sqrt{24} + 24 = 49 + 10\sqrt{24}$. From Equations (2.16) we know that $d_{n,k+1} = 10 - 1/d_{n,k}$, which yields $d_{n,k+1} < 10$ since $d_{n,k} > 0$. Moreover, $5 + \sqrt{24} < d_{n,k}$ implies

$$d_{n,k+1} = 10 - \frac{1}{d_{n,k}} > 10 - \frac{1}{5+\sqrt{24}} = \frac{50 + 10\sqrt{24} - 1}{5+\sqrt{24}} = 5 + \sqrt{24}.$$

Hence $5 + \sqrt{24} < d_{n,k+1} \le 10$, and we conclude that $5 + \sqrt{24} < d_{n,k} \le 10$ for any $n \ge 1$ and $1 \le k \le n$.

b) Show that $D_n := \det(A_n) > (5 + \sqrt{24})^n$. Give $n_0 \in \mathbb{N}$ such that your computer gives an overflow when D_{n_0} is computed in floating point arithmetic.

Solution. We have $A = LDL^{\mathrm{T}}$ with L triangular and with ones on the diagonal. As a consequence,

$$\det(A) = \det(L)\det(D)\det(L) = \det(D) = \prod_{i=1}^{n} d_i > \left(5+\sqrt{24}\right)^n.$$

An overflow in MATLAB is indicated by a return of `Inf`. This will occur for some large n, depending on your platform.

Exercise 14.13: Simultaneous diagonalization

(Simultaneous diagonalization of two symmetric matrices by a congruence transformation). Let $\boldsymbol{A}, \boldsymbol{B} \in \mathbb{R}^{n\times n}$ where $\boldsymbol{A}^{\mathrm{T}} = \boldsymbol{A}$ and $\boldsymbol{B}$ is symmetric positive definite. Then $\boldsymbol{B} = \boldsymbol{U}^{\mathrm{T}}\boldsymbol{D}\boldsymbol{U}$ for some orthogonal matrix $\boldsymbol{U}$ and a diagonal matrix $\boldsymbol{D} = \mathrm{diag}(d_1, \ldots, d_n)$ with positive diagonal elements. Let $\hat{\boldsymbol{A}} = \boldsymbol{D}^{-1/2}\boldsymbol{U}\boldsymbol{A}\boldsymbol{U}^{\mathrm{T}}\boldsymbol{D}^{-1/2}$ where

$$\boldsymbol{D}^{-1/2} := \mathrm{diag}\left(d_1^{-1/2}, \ldots, d_n^{-1/2}\right).$$

a) Show that $\hat{\boldsymbol{A}}$ is symmetric.

Solution. Since $\boldsymbol{D}^{-\frac{1}{2}}$, like any diagonal matrix, and $\boldsymbol{A}$ are symmetric, one has

$$\hat{\boldsymbol{A}}^{\mathrm{T}} = {\boldsymbol{D}^{-\frac{1}{2}}}^{\mathrm{T}}\boldsymbol{U}\boldsymbol{A}^{\mathrm{T}}\boldsymbol{U}^{\mathrm{T}}{\boldsymbol{D}^{-\frac{1}{2}}}^{\mathrm{T}} = \boldsymbol{D}^{-\frac{1}{2}}\boldsymbol{U}\boldsymbol{A}\boldsymbol{U}^{\mathrm{T}}\boldsymbol{D}^{-\frac{1}{2}} = \hat{\boldsymbol{A}}.$$

b) Write $\hat{\boldsymbol{A}} = \hat{\boldsymbol{U}}^{\mathrm{T}}\hat{\boldsymbol{D}}\hat{\boldsymbol{U}}$ where $\hat{\boldsymbol{U}}$ is orthogonal and $\hat{\boldsymbol{D}}$ is diagonal. Set $\boldsymbol{E} := \boldsymbol{U}^{\mathrm{T}}\boldsymbol{D}^{-1/2}\hat{\boldsymbol{U}}^{\mathrm{T}}$. Show that $\boldsymbol{E}$ is nonsingular and that

$$\boldsymbol{E}^{\mathrm{T}}\boldsymbol{A}\boldsymbol{E} = \hat{\boldsymbol{D}}, \qquad \boldsymbol{E}^{\mathrm{T}}\boldsymbol{B}\boldsymbol{E} = \boldsymbol{I}.$$

Solution. Since $\hat{\boldsymbol{A}}$ is symmetric, it admits an orthogonal diagonalization $\hat{\boldsymbol{A}} = \hat{\boldsymbol{U}}^{\mathrm{T}}\hat{\boldsymbol{D}}\hat{\boldsymbol{U}}$. Let $\boldsymbol{E} := \boldsymbol{U}^{\mathrm{T}}\boldsymbol{D}^{-\frac{1}{2}}\hat{\boldsymbol{U}}^{\mathrm{T}}$. Then $\boldsymbol{E}$, as the product of three nonsingular matrices, is nonsingular. Its inverse is given by $\boldsymbol{F} := \hat{\boldsymbol{U}}\boldsymbol{D}^{\frac{1}{2}}\boldsymbol{U}$, since

$$\boldsymbol{F}\boldsymbol{E} = \hat{\boldsymbol{U}}\boldsymbol{D}^{\frac{1}{2}}\boldsymbol{U}\boldsymbol{U}^{\mathrm{T}}\boldsymbol{D}^{-\frac{1}{2}}\hat{\boldsymbol{U}}^{\mathrm{T}} = \hat{\boldsymbol{U}}\boldsymbol{D}^{\frac{1}{2}}\boldsymbol{D}^{-\frac{1}{2}}\hat{\boldsymbol{U}}^{\mathrm{T}} = \hat{\boldsymbol{U}}\hat{\boldsymbol{U}}^{\mathrm{T}} = \boldsymbol{I}$$

and similarly $\boldsymbol{E}\boldsymbol{F} = \boldsymbol{I}$. Moreover, from $\hat{\boldsymbol{A}} = \hat{\boldsymbol{U}}^{\mathrm{T}}\hat{\boldsymbol{D}}\hat{\boldsymbol{U}}$ it follows that $\hat{\boldsymbol{U}}\hat{\boldsymbol{A}}\hat{\boldsymbol{U}}^{\mathrm{T}} = \hat{\boldsymbol{D}}$, which gives

$$\boldsymbol{E}^{\mathrm{T}}\boldsymbol{A}\boldsymbol{E} = \hat{\boldsymbol{U}}\boldsymbol{D}^{-\frac{1}{2}}\boldsymbol{U}\boldsymbol{A}\boldsymbol{U}^{\mathrm{T}}\boldsymbol{D}^{-\frac{1}{2}}\hat{\boldsymbol{U}}^{\mathrm{T}} = \hat{\boldsymbol{U}}\hat{\boldsymbol{A}}\hat{\boldsymbol{U}}^{\mathrm{T}} = \hat{\boldsymbol{D}}.$$

Similarly $\boldsymbol{B} = \boldsymbol{U}^{\mathrm{T}}\boldsymbol{D}\boldsymbol{U}$ implies $\boldsymbol{U}\boldsymbol{B}\boldsymbol{U}^{\mathrm{T}} = \boldsymbol{D}$, which yields

$$\boldsymbol{E}^{\mathrm{T}}\boldsymbol{B}\boldsymbol{E} = \hat{\boldsymbol{U}}\boldsymbol{D}^{-\frac{1}{2}}\boldsymbol{U}\boldsymbol{B}\boldsymbol{U}^{\mathrm{T}}\boldsymbol{D}^{-\frac{1}{2}}\hat{\boldsymbol{U}}^{\mathrm{T}} = \hat{\boldsymbol{U}}\boldsymbol{D}^{-\frac{1}{2}}\boldsymbol{D}^{\frac{1}{2}}\boldsymbol{D}^{\frac{1}{2}}\boldsymbol{D}^{-\frac{1}{2}}\hat{\boldsymbol{U}}^{\mathrm{T}} = \boldsymbol{I}.$$

We conclude that for a symmetric matrix $\boldsymbol{A}$ and symmetric positive definite matrix $\boldsymbol{B}$, the congruence transformation $\boldsymbol{X} \longmapsto \boldsymbol{E}^{\mathrm{T}}\boldsymbol{X}\boldsymbol{E}$ simultaneously diagonalizes the matrices $\boldsymbol{A}$ and $\boldsymbol{B}$, and even maps $\boldsymbol{B}$ to the identity matrix[2]

[2] For a more general result see Theorem 10.1 in P. Lancaster and L.Rodman, "Canonical forms for Hermitian matrix pairs under strict equivalence and congruence", SIAM Review **47** (2005), 407–443.

Exercise 14.14: Program code for one eigenvalue

Suppose $\boldsymbol{A} = \text{tridiag}(\boldsymbol{c}, \boldsymbol{d}, \boldsymbol{c})$ is symmetric and tridiagonal with elements $d_1, \ldots, d_n$ on the diagonal and $c_1, \ldots, c_{n-1}$ on the neighboring subdiagonals. Let $\lambda_1 \geq \lambda_2 \geq \cdots \geq \lambda_n$ be the eigenvalues of $\boldsymbol{A}$. We shall write a program to compute one eigenvalue λ_m for a given m using bisection and the method outlined in (14.9).

a) Write a `function k=counting(c,d,x)` which for given x counts the number of eigenvalues of $\boldsymbol{A}$ strictly greater than x. Use the replacement described above Exercise 14.14 in the book if one of the $d_j(x)$ is close to zero.

Solution. Let $\boldsymbol{A} = \text{tridiag}(\boldsymbol{c}, \boldsymbol{d}, \boldsymbol{c})$ and x be as in the exercise. If $d_j(x)$ is close to zero, e.g. smaller (in absolute value) than $\delta_j := c_j\epsilon_M$ with ϵ_M the Machine epsilon in MATLAB, it is suggested to replace $d_j(x)$ by δ_j. The following MATLAB program counts the number k of eigenvalues of $\boldsymbol{A}$ strictly less than x.

code/count.m

```
function k=count(c,d,x)
n = length(d);
k = 0; u = d(1)-x;
if u < 0
    k = k+1;
end
for i = 2:n
    umin = abs(c(i-1))*eps;
    if abs(u) < umin
        if u < 0
        u = -umin;
    else
        u = umin;
        end
    end
    u = d(i)-x-c(i-1)^2/u;
    if u < 0
        k = k+1;
    end
end
```

Listing 14.3: Count the number k of eigenvalues strictly less than x of a tridiagonal matrix $\boldsymbol{A} = \text{tridiag}(\boldsymbol{c}, \boldsymbol{d}, \boldsymbol{c})$.

b) Write a `function lambda=findeigv(c,d,m)` which first estimates an interval $(a, b]$ containing all eigenvalues of $\boldsymbol{A}$ and then generates a sequence $\{(a_j, b_j]\}$ of intervals each containing λ_m. Iterate until $b_j - a_j \leq (b-a)\epsilon_M$, where ϵ_M is MATLAB's machine epsilon `eps`. Typically $\epsilon_M \approx 2.22 \times 10^{-16}$.

Solution. Let $\boldsymbol{A} = \text{tridiag}(\boldsymbol{c}, \boldsymbol{d}, \boldsymbol{c})$ and m be as in the exercise. The following MATLAB program computes a small interval $[a, b]$ around the mth eigenvalue λ_m of $\boldsymbol{A}$ and returns the point λ in the middle of this interval.

code/findeigv.m

```
function lambda = findeigv(c,d,m)
n = length(d);
a = d(1)-abs(c(1)); b = d(1)+abs(c(1));
for i = 2:n-1
   a = min(a, d(i)-abs(c(i-1))-abs(c(i)));
   b = max(b, d(i)+abs(c(i-1))+abs(c(i)));
end
a = min(a, d(n)-abs(c(n-1)));
b = max(b, d(n)+abs(c(n-1)));
h = b-a;
while abs(b-a) > eps*h
    c0 = (a+b)/2;
    k = count(c,d,c0);
    if k < m
        a = c0;
    else
        b = c0;
    end
end
lambda = (a+b)/2;
```

Listing 14.4: Compute a small interval around the mth eigenvalue λ_m of a tridiagonal matrix $\boldsymbol{A} = \text{tridiag}(\boldsymbol{c}, \boldsymbol{d}, \boldsymbol{c})$ and return the point λ in the middle of this interval.

c) Test the program on $\boldsymbol{T} := \text{tridiag}(-1, 2, -1)$ of size 100. Compare the exact value of λ_5 with your result and the result obtained by using MATLAB's built-in function `eig`.

Solution. A comparison between the values and errors obtained by the different methods is shown in Table 14.1.

Exercise 14.15: Determinant of upper Hessenberg matrix

Suppose $\boldsymbol{A} \in \mathbb{C}^{n \times n}$ is upper Hessenberg and $x \in \mathbb{C}$. We will study two algorithms to compute $f(x) = \det(\boldsymbol{A} - x\boldsymbol{I})$.

a) Show that Gaussian elimination without pivoting requires $\mathcal{O}(n^2)$ arithmetic operations.

Solution. Scaling row k requires $n - k$ multiplications. Zeroing the $(k+1, k)$-entry (and correspondingly scaling the remaining entries in the row) requires $2(n-k)$ additions/multiplications. The total number of operations in bringing the matrix to

method	value	error
exact	0.02413912051848666	0
`findeigv`	0.02413912051848621	$4.44 \cdot 10^{-16}$
MATLAB `eig`	0.02413912051848647	$1.84 \cdot 10^{-16}$

Table 14.1: A comparison between the exact value of λ_5 and the values returned by the functions `findeigv` and `eig`, as well as the errors.

upper triangular form is thus

$$\sum_{k=1}^{n-1} 3(n-k) = \frac{3}{2}n(n-1),$$

which is $\mathcal{O}(n^2)$.

b) Show that the number of arithmetic operations is the same if partial pivoting is used.

Solution. There is only one possible row interchange at each step in Gaussian elimination of an upper Hessenberg matrix, and this does not affect the upper Hessenberg structure of the lower right submatrix. The number of operations is therefore the same.

c) Estimate the number of arithmetic operations if Givens rotations are used.

Solution. According to Algorithm 5.3 and the solution of Exercise 5.18, the number of operations to bring the matrix to upper triangular form is $\mathcal{O}(n^2)$.

d) Compare the two methods discussing advantages and disadvantages.

Solution. We have seen that the two methods are comparable in complexity. Givens rotations, however, give better guarantees when it comes to stability, as pivoting can be unstable.

Chapter 15
The QR Algorithm

Exercise 15.1: Orthogonal vectors

Show that $\boldsymbol{u}$ and $\boldsymbol{A}\boldsymbol{u} - \lambda\boldsymbol{u}$ are orthogonal when $\lambda = \frac{\boldsymbol{u}^*\boldsymbol{A}\boldsymbol{u}}{\boldsymbol{u}^*\boldsymbol{u}}$.

Solution. In the exercise it is implicitly assumed that $\boldsymbol{u}^*\boldsymbol{u} \neq 0$ and therefore $\boldsymbol{u} \neq 0$. The vectors $\boldsymbol{u}$ and $\boldsymbol{A}\boldsymbol{u} - \lambda\boldsymbol{u}$ are orthogonal precisely when

$$0 = \langle \boldsymbol{u}, \boldsymbol{A}\boldsymbol{u} - \lambda\boldsymbol{u} \rangle = \boldsymbol{u}^*(\boldsymbol{A}\boldsymbol{u} - \lambda\boldsymbol{u}) = \boldsymbol{u}^*\boldsymbol{A}\boldsymbol{u} - \lambda\boldsymbol{u}^*\boldsymbol{u}.$$

Dividing by $\boldsymbol{u}^*\boldsymbol{u}$ yields

$$\lambda = \frac{\boldsymbol{u}^*\boldsymbol{A}\boldsymbol{u}}{\boldsymbol{u}^*\boldsymbol{u}}.$$

T. Lyche et al., *Exercises in Numerical Linear Algebra and Matrix Factorizations*, Texts in Computational Science and Engineering 23, https://doi.org/10.1007/978-3-030-59789-4_15

Editorial Policy

1. Textbooks on topics in the field of computational science and engineering will be considered. They should be written for courses in CSE education. Both graduate and undergraduate textbooks will be published in TCSE. Multidisciplinary topics and multidisciplinary teams of authors are especially welcome.

2. Format: Only works in English will be considered. For evaluation purposes, manuscripts may be submitted in print or electronic form, in the latter case, preferably as pdf- or zipped ps-files. Authors are requested to use the LaTeX style files available from Springer at: http://www.springer.com/gp/authors-editors/book-authors-editors/resources-guidelines/rights-permissions-licensing/manuscript-preparation/5636#c3324 (Layout & templates – LaTeX template – contributed books).
   Electronic material can be included if appropriate. Please contact the publisher.

3. Those considering a book which might be suitable for the series are strongly advised to contact the publisher or the series editors at an early stage.

General Remarks

Careful preparation of manuscripts will help keep production time short and ensure a satisfactory appearance of the finished book.

The following terms and conditions hold:

Regarding free copies and royalties, the standard terms for Springer mathematics textbooks hold. Please write to martin.peters@springer.com for details.

Authors are entitled to purchase further copies of their book and other Springer books for their personal use, at a discount of 33.3% directly from Springer-Verlag.

Series Editors

Timothy J. Barth
NASA Ames Research Center
NAS Division
Moffett Field, CA 94035, USA
barth@nas.nasa.gov

Michael Griebel
Institut für Numerische Simulation
der Universität Bonn
Wegelerstr. 6
53115 Bonn, Germany
griebel@ins.uni-bonn.de

David E. Keyes
Mathematical and Computer Sciences
and Engineering
King Abdullah University of Science
and Technology
P.O. Box 55455
Jeddah 21534, Saudi Arabia
david.keyes@kaust.edu.sa

and

Department of Applied Physics
and Applied Mathematics
Columbia University
500 W. 120 th Street
New York, NY 10027, USA
kd2112@columbia.edu

Risto M. Nieminen
Department of Applied Physics
Aalto University School of Science
and Technology
00076 Aalto, Finland
risto.nieminen@tkk.fi

Dirk Roose
Department of Computer Science
Katholieke Universiteit Leuven
Celestijnenlaan 200A
3001 Leuven-Heverlee, Belgium
dirk.roose@cs.kuleuven.be

Tamar Schlick
Department of Chemistry
and Courant Institute
of Mathematical Sciences
New York University
251 Mercer Street
New York, NY 10012, USA
schlick@nyu.edu

Editor for Computational Science
and Engineering at Springer:
Martin Peters
Springer-Verlag
Mathematics Editorial
Tiergartenstrasse 17
69121 Heidelberg, Germany
martin.peters@springer.com

Texts in Computational Science and Engineering

1. H. P. Langtangen, *Computational Partial Differential Equations.* Numerical Methods and Diffpack Programming. 2nd Edition
2. A. Quarteroni, F. Saleri, P. Gervasio, *Scientific Computing with MATLAB and Octave.* 4th Edition
3. H. P. Langtangen, *Python Scripting for Computational Science.* 3rd Edition
4. H. Gardner, G. Manduchi, *Design Patterns for e-Science.*
5. M. Griebel, S. Knapek, G. Zumbusch, *Numerical Simulation in Molecular Dynamics.*
6. H. P. Langtangen, *A Primer on Scientific Programming with Python.* 5th Edition
7. A. Tveito, H. P. Langtangen, B. F. Nielsen, X. Cai, *Elements of Scientific Computing.*
8. B. Gustafsson, *Fundamentals of Scientific Computing.*
9. M. Bader, *Space-Filling Curves.*
10. M. Larson, F. Bengzon, *The Finite Element Method: Theory, Implementation and Applications.*
11. W. Gander, M. Gander, F. Kwok, *Scientific Computing: An Introduction using Maple and MATLAB.*
12. P. Deuflhard, S. Röblitz, *A Guide to Numerical Modelling in Systems Biology.*
13. M. H. Holmes, *Introduction to Scientific Computing and Data Analysis.*
14. S. Linge, H. P. Langtangen, *Programming for Computations* - A Gentle Introduction to Numerical Simulations with MATLAB/Octave.
15. S. Linge, H. P. Langtangen, *Programming for Computations* - A Gentle Introduction to Numerical Simulations with Python.
16. H.P. Langtangen, S. Linge, *Finite Difference Computing with PDEs* - A Modern Software Approach.
17. B. Gustafsson, *Scientific Computing from a Historical Perspective*
18. J.A. Trangenstein, *Scientific Computing* - Vol. I. – Linear and Nonlinear Equations
19. J.A. Trangenstein, *Scientific Computing* - Vol. II. – Eigenvalues and Optimization
20. J.A. Trangenstein, *Scientific Computing* - Vol. III. – Approximation and Integration
21. H.P. Langtangen, K.-A. Mardal, *Introduction to Numerical Methods for Variational Problems.*
22. T. Lyche, *Numerical Linear Algebra and Matrix Factorizations.*

For further information on these books please have a look at our mathematics catalogue at the following URL: `www.springer.com/series/5151`

Monographs in Computational Science and Engineering

1. J. Sundnes, G.T. Lines, X. Cai, B.F. Nielsen, K.-A. Mardal, A. Tveito, *Computing the Electrical Activity in the Heart.*

For further information on this book, please have a look at our mathematics catalogue at the following URL: www.springer.com/series/7417

Lecture Notes in Computational Science and Engineering

1. D. Funaro, *Spectral Elements for Transport-Dominated Equations.*
2. H.P. Langtangen, *Computational Partial Differential Equations.* Numerical Methods and Diffpack Programming.
3. W. Hackbusch, G. Wittum (eds.), *Multigrid Methods V.*
4. P. Deuflhard, J. Hermans, B. Leimkuhler, A.E. Mark, S. Reich, R.D. Skeel (eds.), *Computational Molecular Dynamics: Challenges, Methods, Ideas.*
5. D. Kröner, M. Ohlberger, C. Rohde (eds.), *An Introduction to Recent Developments in Theory and Numerics for Conservation Laws.*
6. S. Turek, *Efficient Solvers for Incompressible Flow Problems.* An Algorithmic and Computational Approach.
7. R. von Schwerin, ***M****ulti* ***B****ody* ***S****ystem* ***SIM****ulation.* Numerical Methods, Algorithms, and Software.
8. H.-J. Bungartz, F. Durst, C. Zenger (eds.), *High Performance Scientific and Engineering Computing.*
9. T.J. Barth, H. Deconinck (eds.), *High-Order Methods for Computational Physics.*
10. H.P. Langtangen, A.M. Bruaset, E. Quak (eds.), *Advances in Software Tools for Scientific Computing.*
11. B. Cockburn, G.E. Karniadakis, C.-W. Shu (eds.), *Discontinuous Galerkin Methods.* Theory, Computation and Applications.
12. U. van Rienen, *Numerical Methods in Computational Electrodynamics.* Linear Systems in Practical Applications.
13. B. Engquist, L. Johnsson, M. Hammill, F. Short (eds.), *Simulation and Visualization on the Grid.*
14. E. Dick, K. Riemslagh, J. Vierendeels (eds.), *Multigrid Methods VI.*
15. A. Frommer, T. Lippert, B. Medeke, K. Schilling (eds.), *Numerical Challenges in Lattice Quantum Chromodynamics.*
16. J. Lang, *Adaptive Multilevel Solution of Nonlinear Parabolic PDE Systems.* Theory, Algorithm, and Applications.

17. B.I. Wohlmuth, *Discretization Methods and Iterative Solvers Based on Domain Decomposition.*

18. U. van Rienen, M. Günther, D. Hecht (eds.), *Scientific Computing in Electrical Engineering.*

19. I. Babuška, P.G. Ciarlet, T. Miyoshi (eds.), *Mathematical Modeling and Numerical Simulation in Continuum Mechanics.*

20. T.J. Barth, T. Chan, R. Haimes (eds.), *Multiscale and Multiresolution Methods.* Theory and Applications.

21. M. Breuer, F. Durst, C. Zenger (eds.), *High Performance Scientific and Engineering Computing.*

22. K. Urban, *Wavelets in Numerical Simulation.* Problem Adapted Construction and Applications.

23. L.F. Pavarino, A. Toselli (eds.), *Recent Developments in Domain Decomposition Methods.*

24. T. Schlick, H.H. Gan (eds.), *Computational Methods for Macromolecules: Challenges and Applications.*

25. T.J. Barth, H. Deconinck (eds.), *Error Estimation and Adaptive Discretization Methods in Computational Fluid Dynamics.*

26. M. Griebel, M.A. Schweitzer (eds.), *Meshfree Methods for Partial Differential Equations.*

27. S. Müller, *Adaptive Multiscale Schemes for Conservation Laws.*

28. C. Carstensen, S. Funken, W. Hackbusch, R.H.W. Hoppe, P. Monk (eds.), *Computational Electromagnetics.*

29. M.A. Schweitzer, *A Parallel Multilevel Partition of Unity Method for Elliptic Partial Differential Equations.*

30. T. Biegler, O. Ghattas, M. Heinkenschloss, B. van Bloemen Waanders (eds.), *Large-Scale PDE-Constrained Optimization.*

31. M. Ainsworth, P. Davies, D. Duncan, P. Martin, B. Rynne (eds.), *Topics in Computational Wave Propagation.* Direct and Inverse Problems.

32. H. Emmerich, B. Nestler, M. Schreckenberg (eds.), *Interface and Transport Dynamics.* Computational Modelling.

33. H.P. Langtangen, A. Tveito (eds.), *Advanced Topics in Computational Partial Differential Equations.* Numerical Methods and Diffpack Programming.

34. V. John, *Large Eddy Simulation of Turbulent Incompressible Flows.* Analytical and Numerical Results for a Class of LES Models.

35. E. Bänsch (ed.), *Challenges in Scientific Computing - CISC 2002.*

36. B.N. Khoromskij, G. Wittum, *Numerical Solution of Elliptic Differential Equations by Reduction to the Interface.*

37. A. Iske, *Multiresolution Methods in Scattered Data Modelling.*

38. S.-I. Niculescu, K. Gu (eds.), *Advances in Time-Delay Systems.*

39. S. Attinger, P. Koumoutsakos (eds.), *Multiscale Modelling and Simulation.*

40. R. Kornhuber, R. Hoppe, J. Périaux, O. Pironneau, O. Wildlund, J. Xu (eds.), *Domain Decomposition Methods in Science and Engineering.*

41. T. Plewa, T. Linde, V.G. Weirs (eds.), *Adaptive Mesh Refinement – Theory and Applications.*

42. A. Schmidt, K.G. Siebert, *Design of Adaptive Finite Element Software.* The Finite Element Toolbox ALBERTA.

43. M. Griebel, M.A. Schweitzer (eds.), *Meshfree Methods for Partial Differential Equations II.*

44. B. Engquist, P. Lötstedt, O. Runborg (eds.), *Multiscale Methods in Science and Engineering.*

45. P. Benner, V. Mehrmann, D.C. Sorensen (eds.), *Dimension Reduction of Large-Scale Systems.*

46. D. Kressner, *Numerical Methods for General and Structured Eigenvalue Problems.*

47. A. Boriçi, A. Frommer, B. Joó, A. Kennedy, B. Pendleton (eds.), *QCD and Numerical Analysis III.*

48. F. Graziani (ed.), *Computational Methods in Transport.*

49. B. Leimkuhler, C. Chipot, R. Elber, A. Laaksonen, A. Mark, T. Schlick, C. Schütte, R. Skeel (eds.), *New Algorithms for Macromolecular Simulation.*

50. M. Bücker, G. Corliss, P. Hovland, U. Naumann, B. Norris (eds.), *Automatic Differentiation: Applications, Theory, and Implementations.*

51. A.M. Bruaset, A. Tveito (eds.), *Numerical Solution of Partial Differential Equations on Parallel Computers.*

52. K.H. Hoffmann, A. Meyer (eds.), *Parallel Algorithms and Cluster Computing.*

53. H.-J. Bungartz, M. Schäfer (eds.), *Fluid-Structure Interaction.*

54. J. Behrens, *Adaptive Atmospheric Modeling.*

55. O. Widlund, D. Keyes (eds.), *Domain Decomposition Methods in Science and Engineering XVI.*

56. S. Kassinos, C. Langer, G. Iaccarino, P. Moin (eds.), *Complex Effects in Large Eddy Simulations.*

57. M. Griebel, M.A Schweitzer (eds.), *Meshfree Methods for Partial Differential Equations III.*

58. A.N. Gorban, B. Kégl, D.C. Wunsch, A. Zinovyev (eds.), *Principal Manifolds for Data Visualization and Dimension Reduction.*

59. H. Ammari (ed.), *Modeling and Computations in Electromagnetics: A Volume Dedicated to Jean-Claude Nédélec.*

60. U. Langer, M. Discacciati, D. Keyes, O. Widlund, W. Zulehner (eds.), *Domain Decomposition Methods in Science and Engineering XVII.*

61. T. Mathew, *Domain Decomposition Methods for the Numerical Solution of Partial Differential Equations.*

62. F. Graziani (ed.), *Computational Methods in Transport: Verification and Validation.*

63. M. Bebendorf, *Hierarchical Matrices.* A Means to Efficiently Solve Elliptic Boundary Value Problems.

64. C.H. Bischof, H.M. Bücker, P. Hovland, U. Naumann, J. Utke (eds.), *Advances in Automatic Differentiation.*

65. M. Griebel, M.A. Schweitzer (eds.), *Meshfree Methods for Partial Differential Equations IV.*

66. B. Engquist, P. Lötstedt, O. Runborg (eds.), *Multiscale Modeling and Simulation in Science.*

67. I.H. Tuncer, Ü. Gülcat, D.R. Emerson, K. Matsuno (eds.), *Parallel Computational Fluid Dynamics 2007.*

68. S. Yip, T. Diaz de la Rubia (eds.), *Scientific Modeling and Simulations.*

69. A. Hegarty, N. Kopteva, E. O'Riordan, M. Stynes (eds.), *BAIL* 2008 – *Boundary and Interior Layers.*

70. M. Bercovier, M.J. Gander, R. Kornhuber, O. Widlund (eds.), *Domain Decomposition Methods in Science and Engineering XVIII.*

71. B. Koren, C. Vuik (eds.), *Advanced Computational Methods in Science and Engineering.*

72. M. Peters (ed.), *Computational Fluid Dynamics for Sport Simulation.*

73. H.-J. Bungartz, M. Mehl, M. Schäfer (eds.), *Fluid Structure Interaction II - Modelling, Simulation, Optimization.*

74. D. Tromeur-Dervout, G. Brenner, D.R. Emerson, J. Erhel (eds.), *Parallel Computational Fluid Dynamics 2008.*

75. A.N. Gorban, D. Roose (eds.), *Coping with Complexity: Model Reduction and Data Analysis.*

76. J.S. Hesthaven, E.M. Rønquist (eds.), *Spectral and High Order Methods for Partial Differential Equations.*

77. M. Holtz, *Sparse Grid Quadrature in High Dimensions with Applications in Finance and Insurance.*

78. Y. Huang, R. Kornhuber, O.Widlund, J. Xu (eds.), *Domain Decomposition Methods in Science and Engineering XIX.*

79. M. Griebel, M.A. Schweitzer (eds.), *Meshfree Methods for Partial Differential Equations V.*

80. P.H. Lauritzen, C. Jablonowski, M.A. Taylor, R.D. Nair (eds.), *Numerical Techniques for Global Atmospheric Models.*

81. C. Clavero, J.L. Gracia, F.J. Lisbona (eds.), *BAIL 2010 – Boundary and Interior Layers, Computational and Asymptotic Methods.*

82. B. Engquist, O. Runborg, Y.R. Tsai (eds.), *Numerical Analysis and Multiscale Computations.*

83. I.G. Graham, T.Y. Hou, O. Lakkis, R. Scheichl (eds.), *Numerical Analysis of Multiscale Problems.*

84. A. Logg, K.-A. Mardal, G. Wells (eds.), *Automated Solution of Differential Equations by the Finite Element Method.*

85. J. Blowey, M. Jensen (eds.), *Frontiers in Numerical Analysis - Durham 2010.*

86. O. Kolditz, U.-J. Gorke, H. Shao, W. Wang (eds.), *Thermo-Hydro-Mechanical-Chemical Processes in Fractured Porous Media - Benchmarks and Examples.*

87. S. Forth, P. Hovland, E. Phipps, J. Utke, A. Walther (eds.), *Recent Advances in Algorithmic Differentiation.*

88. J. Garcke, M. Griebel (eds.), *Sparse Grids and Applications.*

89. M. Griebel, M.A. Schweitzer (eds.), *Meshfree Methods for Partial Differential Equations VI.*

90. C. Pechstein, *Finite and Boundary Element Tearing and Interconnecting Solvers for Multiscale Problems.*

91. R. Bank, M. Holst, O. Widlund, J. Xu (eds.), *Domain Decomposition Methods in Science and Engineering XX.*

92. H. Bijl, D. Lucor, S. Mishra, C. Schwab (eds.), *Uncertainty Quantification in Computational Fluid Dynamics.*

93. M. Bader, H.-J. Bungartz, T. Weinzierl (eds.), *Advanced Computing.*

94. M. Ehrhardt, T. Koprucki (eds.), *Advanced Mathematical Models and Numerical Techniques for Multi-Band Effective Mass Approximations.*

95. M. Azaïez, H. El Fekih, J.S. Hesthaven (eds.), *Spectral and High Order Methods for Partial Differential Equations ICOSAHOM 2012.*

96. F. Graziani, M.P. Desjarlais, R. Redmer, S.B. Trickey (eds.), *Frontiers and Challenges in Warm Dense Matter.*

97. J. Garcke, D. Pflüger (eds.), *Sparse Grids and Applications – Munich 2012.*

98. J. Erhel, M. Gander, L. Halpern, G. Pichot, T. Sassi, O. Widlund (eds.), *Domain Decomposition Methods in Science and Engineering XXI.*

99. R. Abgrall, H. Beaugendre, P.M. Congedo, C. Dobrzynski, V. Perrier, M. Ricchiuto (eds.), *High Order Nonlinear Numerical Methods for Evolutionary PDEs - HONOM 2013.*

100. M. Griebel, M.A. Schweitzer (eds.), *Meshfree Methods for Partial Differential Equations VII.*

101. R. Hoppe (ed.), *Optimization with PDE Constraints - OPTPDE 2014.*

102. S. Dahlke, W. Dahmen, M. Griebel, W. Hackbusch, K. Ritter, R. Schneider, C. Schwab, H. Yserentant (eds.), *Extraction of Quantifiable Information from Complex Systems.*

103. A. Abdulle, S. Deparis, D. Kressner, F. Nobile, M. Picasso (eds.), *Numerical Mathematics and Advanced Applications - ENUMATH 2013.*

104. T. Dickopf, M.J. Gander, L. Halpern, R. Krause, L.F. Pavarino (eds.), *Domain Decomposition Methods in Science and Engineering XXII.*

105. M. Mehl, M. Bischoff, M. Schäfer (eds.), *Recent Trends in Computational Engineering - CE2014.* Optimization, Uncertainty, Parallel Algorithms, Coupled and Complex Problems.

106. R.M. Kirby, M. Berzins, J.S. Hesthaven (eds.), *Spectral and High Order Methods for Partial Differential Equations - ICOSAHOM'14.*

107. B. Jüttler, B. Simeon (eds.), *Isogeometric Analysis and Applications 2014.*

108. P. Knobloch (ed.), *Boundary and Interior Layers, Computational and Asymptotic Methods – BAIL 2014.*

109. J. Garcke, D. Pflüger (eds.), *Sparse Grids and Applications – Stuttgart 2014.*

110. H. P. Langtangen, *Finite Difference Computing with Exponential Decay Models.*

111. A. Tveito, G.T. Lines, *Computing Characterizations of Drugs for Ion Channels and Receptors Using Markov Models.*

112. B. Karazösen, M. Manguoğlu, M. Tezer-Sezgin, S. Göktepe, Ö. Uğur (eds.), *Numerical Mathematics and Advanced Applications - ENUMATH 2015.*

113. H.-J. Bungartz, P. Neumann, W.E. Nagel (eds.), *Software for Exascale Computing - SPPEXA 2013-2015.*

114. G.R. Barrenechea, F. Brezzi, A. Cangiani, E.H. Georgoulis (eds.), *Building Bridges: Connections and Challenges in Modern Approaches to Numerical Partial Differential Equations.*

115. M. Griebel, M.A. Schweitzer (eds.), *Meshfree Methods for Partial Differential Equations VIII.*

116. C.-O. Lee, X.-C. Cai, D.E. Keyes, H.H. Kim, A. Klawonn, E.-J. Park, O.B. Widlund (eds.), *Domain Decomposition Methods in Science and Engineering XXIII.*

117. T. Sakurai, S. Zhang, T. Imamura, Y. Yusaku, K. Yoshinobu, H. Takeo (eds.), *Eigenvalue Problems: Algorithms, Software and Applications, in Petascale Computing.* EPASA 2015, Tsukuba, Japan, September 2015.

118. T. Richter (ed.), *Fluid-structure Interactions.* Models, Analysis and Finite Elements.

119. M.L. Bittencourt, N.A. Dumont, J.S. Hesthaven (eds.), *Spectral and High Order Methods for Partial Differential Equations ICOSAHOM 2016.*

120. Z. Huang, M. Stynes, Z. Zhang (eds.), *Boundary and Interior Layers, Computational and Asymptotic Methods BAIL 2016.*

121. S.P.A. Bordas, E.N. Burman, M.G. Larson, M.A. Olshanskii (eds.), *Geometrically Unfitted Finite Element Methods and Applications*. Proceedings of the UCL Workshop 2016.

122. A. Gerisch, R. Penta, J. Lang (eds.), *Multiscale Models in Mechano and Tumor Biology*. Modeling, Homogenization, and Applications.

123. J. Garcke, D. Pflüger, C.G. Webster, G. Zhang (eds.), *Sparse Grids and Applications - Miami 2016.*

124. M. Schäfer, M. Behr, M. Mehl, B. Wohlmuth (eds.), *Recent Advances in Computational Engineering*. Proceedings of the 4th International Conference on Computational Engineering (ICCE 2017) in Darmstadt.

125. P.E. Bjørstad, S.C. Brenner, L. Halpern, R. Kornhuber, H.H. Kim, T. Rahman, O.B. Widlund (eds.), *Domain Decomposition Methods in Science and Engineering XXIV*. 24th International Conference on Domain Decomposition Methods, Svalbard, Norway, February 6–10, 2017.

126. F.A. Radu, K. Kumar, I. Berre, J.M. Nordbotten, I.S. Pop (eds.), *Numerical Mathematics and Advanced Applications – ENUMATH 2017.*

127. X. Roca, A. Loseille (eds.), *27th International Meshing Roundtable.*

128. Th. Apel, U. Langer, A. Meyer, O. Steinbach (eds.), *Advanced Finite Element Methods with Applications*. Selected Papers from the 30th Chemnitz Finite Element Symposium 2017.

129. M. Griebel, M. A. Schweitzer (eds.), *Meshfree Methods for Partial Differencial Equations IX.*

130. S. Weißer, BEM-based Finite Element *Approaches on Polytopal Meshes.*

131. V. A. Garanzha, L. Kamenski, H. Si (eds.), *Numerical Geometry, Grid Generation and Scientific Computing*. Proceedings of the 9th International Conference, NUMGRID 2018/Voronoi 150, Celebrating the 150th Anniversary of G. F. Voronoi, Moscow, Russia, December 2018.

132. E. H. van Brummelen, A. Corsini, S. Perotto, G. Rozza (eds.), *Numerical Methods for Flows.*

For further information on these books please have a look at our mathematics catalogue at the following URL: www.springer.com/series/3527

Printed in the United States
by Baker & Taylor Publisher Services